中兽医药国际培训教材

Traditional Chinese Veterinary Medicine

郑继方　杨志强　王学智　主编

中国农业科学技术出版社

图书在版编目（CIP）数据

中兽医药国际培训教材 / 郑继方，杨志强，王学智主编. —北京：中国农业科学技术出版社，2014.12

ISBN 978-7-5116-1927-3

Ⅰ. ①中… Ⅱ. ①郑…②杨…③王… Ⅲ. ①中兽医学-技术培训-教材-汉、英 Ⅳ. ①S853

中国版本图书馆 CIP 数据核字（2014）第 283026 号

国家科技基础性工作专项“传统中兽医药资源抢救和整理”（编号：2013FY110600）

责任编辑 闫庆健 胡晓蕾
责任校对 贾晓红

出 版 者 中国农业科学技术出版社
北京市中关村南大街 12 号 邮编：100081
电　　话 (010)82106632(编辑室) (010)82109704(发行部)
(010)82109703(读者服务部)
传　　真 (010) 82106625
网　　址 http://www.castp.cn
经 销 者 各地新华书店
印 刷 者 北京富泰印刷有限责任公司
开　　本 880 mm×1 230 mm 1/16
印　　张 15.875 彩插 12 面
字　　数 480 千字
版　　次 2014 年 12 月第 1 版 2014 年 12 月第 1 次印刷
定　　价 60.00 元

《中兽医药国际培训教材》
编 委 会

主　编　郑继方　杨志强　王学智

副主编　罗超应　张继瑜　杨锐乐　李建喜
　　　　李滋睿　周　磊

编　委　罗永江　谢家声　董鹏程　师　音
　　　　李锦宇　辛蕊华　王贵波　张景艳
　　　　曾玉峰　秦　哲　王　磊　孔晓军
　　　　刘丽娟

前　言

随着世界中兽医药热潮的兴起，前来我国学习、研究中兽医药学的各国学者日益增多。已经形成了一支学习、继承、传播、发展中兽医药事业不可忽视的力量。中兽医药学走向世界，尤其是中兽医药学传统理论走向世界，面临从实践技能的介绍到思维模式的沟通等不同层面上的“跨文化传播”问题。为适应中兽医药学国际交流形势的迫切需要，在我国科技部的支持下，本所编写了这套《中兽医中英文培训教材》，以期为促进中兽医药体系的弘扬传播，为进—步拓展中兽医药教育的国际交流与合作做出应有的贡献。

本套教材的编写，既坚持中兽医药学的系统性、科学性、独特性和实践性，又注重协调学术发展中的继承与创新、理论与实践的关系，融合了本所多年的科技成果，并结合留学生对教材讲授内容的需求，同时还注意到阅读对象在学习时间、语言及民族文化、心理等方面的特殊性。我们在2005年至2011年教学的基础上，对教材进行了修改。修改后的教材，力求文字简洁、通俗易懂，篇幅精练、阐述明确、利于实用。

本套教材，充分注意到留学生语言及文化差异等特点，对名词术语的英译，反复斟酌修改，尽可能统一规范。限于条件及水平，不足之处敬待读者指正。

中国农业科学院兰州畜牧与兽药研究所

2014年10月

Preface

Alongside the worldwide enormous rise of interest in traditional Chinese veterinary medicine (TCVM) more and more overseas (advanced) scholars come to China to pursue it, and they have become a force not to be ignored in study, inheritance, dissemination and development of TCVM. To meet the urgent high-leveled international exchange in this area, supported by the technological ministry of PRC, we have compiled this English textbook of TCVM for overseas students It is expected the textbook will push dissemination of the scientific system of TCVM and make due contribution to international exchange and cooperation in traditional Chinese medical education.

The textbook adheres to the systematic scientific and practical nature and uniqueness of TCVM theory, and the theory and practice in academic progress, and the scientific research fruits of Lanzhou Institute of Animal and Pharmaceutics Science, CAAS. The textbook has gone through repeated revision and final revision and the characteristics of the foreign students' language and cultural difference are constantly stressed.

This textbook is revised and supplemented based on the feedback of students and teachers on the six times International Training Workshop on Traditional Chinese Veterinary Medicine and Techniques from 2005 to 2011 in Lanzhou. Because of the present condition and level, the editors hope that readers will comment on the shortcomings of the textbook to help promote the advance of TCVM education for foreign students.

Lanzhou Institute of Animal and Pharmaceutics Science, CAAS

October, 2014

目　　录

Lecture One

The Basic Theory of the Traditional Chinese Veterinary Medicine

Traditional Chinese Veterinary Medicine (TCVM), a great treasure house of culture, is an indispensable part of the splendid classic Chinese culture. In its long course of development, it has absorbed the quintessence of classical Chinese philosophy, culture and science, and summarized the experience of the Chinese people in fighting against animal disease. It is rich in theory and practical in treatment. Today modern veterinary medicine is quite advanced, but TCVM is still widely used because of its significant clinical curative effect. In recent decades, TCVM is understood and adopted in more and more countries and regions in the world.

1 The Basic Characteristics of TCVM

Concept of holism, and syndrome differentiation and treatment are the two basic characteristics of TCVM in understanding animal physiology and pathology as well as diagnosis, treatment and prevention of disease.

1.1 Concept of holism

The concept of holism means that the animal body is an organic whole and that animals are interrelated with nature.

1.1.1 Organic wholeness of the body

TCVM believes that the animal body is composed of various tissues and organs, including the viscera, the meridians, the five sensory organs, the nine orifices, the four limbs and all the skeletal parts. These different tissues and organs are united into an organic whole because they are closely related to each other in structure, physiology and pathology. So in clinical diagnosis and treatment of disease, thorough examination of the five sensory organs, physical condition, mouth, tongue and pulse must be made to analyze the pathological changes of the viscera so as to decide correct therapeutic principles and methods.

1.1.2 Correlation between animal and nature

Animals live in nature and nature provides them with various necessities, such as sunlight, air and water. On the other hand, various changes taking place in nature may directly or indirectly affect the animal body and bring on corresponding physiological or pathological responses. Geographical conditions, similar to the seasonal variations, also affect the physiological activity and pathological state of the animals.

1.2 Bianzheng Lunzhi (Syndrome differentiation and treatment)

1.2.1 Implication of syndrome differentiation and treatment

Syndrome differentiation and treatment means to analyze, induce, synthesize, judge and summarize the clinical data of symptoms and signs collected with the four diagnostic methods (namely inspection, listening and smelling, inquiry, taking pulse and palpation) into certain syndrome. Then the therapeutic methods are decided according to the result of syndrome differentiation. Syndrome differentiation and treatment is a basic principle in TCVM to understand and treat disease. "Zheng" (syndrome) is a summarization of the pathological changes of a disease at a certain stage in its course of development, including the location, cause and nature of the disease as well as the state of Xie (pathogenic factors) and Zheng (the healthy Qi). Compared with single symptom, syndrome can more extensively, completely and correctly demonstrate the nature of a disease. For example, if a animal has clinical manifestations of serious aversion to cold, light fever, headache, body pain, anhidrosis and floating-tense pulse, it can be analyzed that the disease is mused by wind-cold, the location is superficial, the nature of the disease is cold, and the relationship between the pathogenic factors and healthy Qi is excess. According to such a differentiation it can be induced that the animal is suffering from external excess syndrome due to wind-cold, which can be treated by dispersing wind and dissipating cold with drugs acrid in flavor and warm in property for relieving superficial pathogenic factors. It is obvious that syndrome differentiation is prerequisite to treatment, while treatment is the aim of syndrome differentiation and the method to test whether syndrome differentiation is correct. In fact, syndrome differentiation and treatment are two interrelated and indispensable aspects in the diagnosis and treatment of disease.

1.2.2 Differentiation of syndrome and differentiation of disease

Clinically differentiation of syndrome and differentiation of disease are intrinsically interrelated on the one hand, and different on the other. It is generally thought that disease includes the whole pathological course while syndrome is just the summarization of disease at a certain stage of its development. For this reason one disease may display different syndromes while different diseases may demonstrate the same syndrome in their course of development. In TCVM, the understanding and treatment of disease mainly focuses on differentiating syndrome and analyzing the common nature or difference of syndrome in the course of differentiating disease. For example, common cold is clinically divided into wind cold syndrome and wind-heat syndrome due to difference in pathogenic factors. The former is treated by relieving superficial pathogenic factors with drugs acrid in taste and warm in property while the latter with drugs acrid in taste and cool in property, which is known as "treating the same disease with different methods" in TCVM. However, same therapeutic method can be used to treat different diseases with the emergence of the same syndrome in their course of development. Take dysentery and jaundice for example. They are two different diseases. But if they all demonstrate damp-heat syndrome, both of them can be treated by the therapeutic method for clearing away damp heat, which is known as "treating different diseases with the same method" in TCVM. It is clear that syndrome differentiation and treatment emphasizes the role of syndrome in treatment and advocates the idea of "treating the same syndrome with the same method" and "treating different syndromes with different methods."

2 Yin-yang and the Five Elements Doctrines

Yin-yang and the five elements can be traced to ancient Chinese philosophy. In ancient China, TCVM doctors applied yin-yang and the five elements doctrines to physiological functions, pathological changes of the animal body, diagnosis and treatment.

2.1 The Concepts of Yin-yang Theory

Yin and yang symbolize two opposite and related sides. Generally speaking, anything that is active, ascending, bright, progressive, hyper functioning or functional aspect is yang; while anything that is static, descending, dim, degenerate, hypo-functioning, or organic aspect is yin. For example, in viewing the attributes of things, heaven being in the upper is attributed to yang, and the earth being in the lower is attributed to yin; water being characterized by cold and downward-flawing is attributed to yin, and fire being characterized by hotness and flaring-up is attributed to yang. When speaking of changes and motion of things, motion lessness belongs to yin and motion to yang, as a thing is in a static state, it is attributed to yin, and if a thing is in excited state, it is attributed to yang. If there is transformation or qi, it is attributed to yang, and as it becomes visible or tangible, it is attributed to yin.

Yin or yang attribute of a thing is by no means absolute, but relative. The relativity of yin and yang embodies inter transformation of yin and yang under a given condition. Also, yin may transform into yang and yang into yin. On the other forelimb, within yin or yang there still might be further subdivided yin and yang. For example, during the day belongs to yang and night to yin, and morning is yang within yang and afternoon is yin within yang.

2.2 Concepts of the Five Elements

Over thousands of years people in ancient China recognized that wood, fire, earth, metal and water are the five indispensable materials in people life. Later, people looked to the properties of these five materials to explain the whole physical world. They also believed that the five materials not only inter-promotion and inter-restriction, but are also in a constant state of motion and change. This soon be came what is now known as "the five-element doctrine" or theory.

2.2.1 Characteristics of the Five Elements

Ancient TCVM practitioners used the five elements to study and classify tissue structure, physiology and pathology of the animal body, and nature. They also categorized things into wood, fire, earth, metal, water and five elements according to their different natures, actions, shapes or forms. This is known as the method of "categorization according to picture." Studies and explanation of the complex relations among the viscera and tissues of the animal body physiologically and pathologically, and the relation between the animal body and its external environment were also developed.

2.2.2 Promotion, Restriction, Subjugation and Violation of the Five Elements

Promotion and Restriction Promotion indicates promoting and improving the orderly function among the five elements. The order of promotion is as follows: wood promotes fire, fire promotes earth, earth promotes metal, metal promotes water, and water in turn promotes wood. In the promotion of the five elements, each element has two aspects of "being promoted" and "promoting." The promoting element is the "mother," and the promoted element is the "child." The promotion relationship is thus called the "mother-child relationship." As an example, because wood promotes fire, wood is the mother of fire; and wood is promoted by water, thus it is the child of water.

Restriction implies checking and bringing under control among the five elements. The order of restriction is: wood restricts earth, earth restricts water, water restricts fire, fire restricts metal, and metal restricts wood. In the restriction of the five elements, each element possesses two aspects of "being restricted" and "restricting." The restricting element is the "dominator," and the promoted element is the "subordinate." With wood as an example, the element that restricts wood is metal, so metal is the dominator of wood. The element that is restricted by wood is earth, thus earth is the subordinate of wood.

Invasion by Another Element Invasion indicates, "encroaching on that which is weak." It is an abnormal manifestation in the normal co-coordinative relationship of elements. For example, when wood is excessive, and metal fails to check wood, then the excessive wood will invade earth and make the latter even more deficient.

Violation in relation to the elements is "using one's strength to bully another one." That is, one of the five elements becomes too excessive, thus the element that originally restricts it cannot restrict it any more, and is restricted by it. This is called "counter-restriction." For example, in normal restriction relations, metal restricts wood, if wood becomes excessive or metal becomes deficient, then wood will reverse and violate metal. The order of violation is just opposite to the order of restriction in the five elements.

3 Visceral Theories

TCVM classifies the internal organs of the body into two kinds: The five zang-viscera, or the heart, liver, spleen, lung and kidney, all of which have the common physiological function of producing and storing essential qi; The six fu-viscera, or the gallbladder, stomach, small intestine, large intestine, urinary bladder and sanjiao, which have the common physiological functions of receiving, transforming, and transporting fodder and water.

3.1 The Five Zang-viscera

3.1.1 Heart

The heart is located in the thorax and guarded externally by the pericardium. Its physiological function is governing blood and the vessels, and giving motive power for blood circulation. The heart opens to the tongue.

3.1.2 Lungs

The lungs are situated in the thorax. They communicate with the heart, and open to the nose. The major functions of the lungs include governing qi, controlling respiration, dispersing and descending, and smoothing water passages.

3.1.3 Spleen

The spleen is located in the middle-jiao. Its major physiological functions include governing transformation and transportation, controlling blood, and ascending essence. It dominates the muscles and limbs, opens to the mouth, and manifests on the lips.

3.1.4 Liver

The liver is located in the right hypochondrium. It is responsible for regulating the smooth flow of qi, storage of blood, and dominates the tendons. The liver opens to the eyes, and manifests on the nails.

3.1.5 Kidneys

The kidneys are located in the lumbar region. The kidneys store total essence, manufacture marrow, dominate bones, and are the source for growth, development, and reproduction. They also govern water metabolism and the reception of qi. The kidneys open to the ears, and manifest in the hair.

3.2 Six Fu-viscera

3.2.1 Gallbladder

The gallbladder is attached to the liver. It stores an "essential fluid" (bile), which originates from the liver. Bile is excreted into the intestine and has the action of promoting digestion of fodder and fluids; one of the most important conditions for the normal spleen and stomach transformation and transportation functions.

Since in physiology the gallbladder mainly stores and discharges bile, which helps digestion of fodder and fluids, the gallbladder is considered as one of the six fu-viscera. However, the gallbladder itself has no physiological function of transmitting fodder and drink, and in storing the essential fluid it is different from the stomach and intestines. It is also referred to as one of the extraordinary fu viscera.

3.2.2 Stomach

The stomach is located under the diaphragm and in the epigastrium. Its upper opening is the cardia, by which it connects to the esophagus and its lower opening is in the pylorus through which it opens into the small intestine.

The major physiological functions of the stomach include receiving and digesting fodder and fluids. The fodder and fluids, after being taken in through the mouth and esophagus, are then held in the stomach. Therefore, the stomach is referred to as "the sea of fodder and drink." The fodder and drink held in the stomach, through the digestion of the stomach, are transformed into chyme, which then goes down into the small intestine. If this function of the stomach becomes weakened, the result will be poor appetite, indigestion, epigastric and a distending pain. The function of the stomach to receive and digest fodder and fluids depends upon the transformation and transportation functions of the spleen.

3.2.3 Small Intestine

The small intestine is located in the abdomen. Its upper end connects with the pylorus and communicates with the stomach, and its lower end connects with the ileocecal junction thus opening to the large intestine. The major functions of the small intestine are transforming chyme and separating the clear from the turbid. The small intestine receives chyme from the stomach and digests and further transforms it. The small intestine then separates the digested fodder into two parts. The essential substance (the clear) is transported by spleen to entire body and the waste (the turbid) is sent down through the ileocecal junction to the large intestine. The useless water is poured into the urinary bladder. Thus, when both water and the waste go smoothly to their own passages urine and bowel movements are normal.

3.2.4 Large Intestine

The large intestine connects to the small intestine through the ileocecal junction, and its lower outlet is the anus. The large intestine receives waste material sent down from the small intestine, and after the remained water it still contains absorbed stool is formed and finally it discharges the stool through the anus. The large intestine is the passageway of transmitting waste. If the large intestine is dysfunctioned in transmitting waste, diarrhea or constipation may occur.

3.2.5 Urinary Bladder

The major functions of the urinary bladder include storing and discharging urine. In the process of water metabolism in the body, water nourishes and moistens the entire body through the functions of the lung, spleen and kidneys. The used water by the body is sent down to the kidneys where, through qi transformation, the clear (useful) part flows hack into the body, and the turbid (waste) is changed into urine and sent down to the urinary bladder. When the urine reaches a certain volume, it is discharged out of the body through the qi transformation of both the kidneys and urinary bladder. If qi transformation of the urinary bladder is weakened or inhibited, dysuria or anuria occur. If the bladder loses its control due to kidney qi deficiency, frequent urination or incontinence may result. If a damp-heat evil invades the urinary bladder, frequent and urgent urination or urethralgia may occur.

3.2.6 San-jiao

The san-jiao is a collective term including the upper-jiao, middle-jiao and lower-jiao. Regarding the viscera, the upper-jiao includes the heart and lungs, the middle-jiao includes the spleen and stomach, and the lower-jiao includes the liver, kidneys, intestines and urinary bladder.

The physiological functions of the san-jiao include controlling the qi transformation of the entire body, transforming essential substances from fodder and drink, and water metabolism. The reception, digestion and absorption of fodder and fluids, distribution of essential qi, and discharge of metabolic materials are related to the san-jiao.

3.3 Relations Among Zang-viscera and Fu-viscera

Each zang-viscus and fu-viscus is externally-internally related. The zang viscus belongs to yin and the fu-viscus to yang. The heart and small intestine, lungs and large intestine, spleen and stomach, liver and gallbladder, kidneys and urinary bladder are respective pairs of external-internal organs. Through affiliation and connection of their meridians, one zang-viscus and one fu-viscus, one yin and one yang, one interior and one exterior, all of them mutually cooperate. This kind of relationship not only expresses that they are related physiologically, but also determines their mutual affection pathologically.

3.3.1 Heart and Small Intestine

The meridian of the heart belongs to the heart and connects to the small intestine, the meridian of small intestine belongs to the small intestine and connects to the heart, through which the two viscera form an external-internal relationship. The internal relation of the two viscera is remarkably reflected in pathology. For example, if heart-fire spreads down to the small intestine, it will often cause a syndrome of excess-heat in the small intestine manifesting oliguria, and deep red and hot urine. Conversely, if the heat in the small intestine steams the heart through the meridians, aphthous, and stomtitis might occur.

3.3.2 Lungs and Large Intestine

The meridians of the lung and the large intestine are connected to each other, by which the lung and the large intestine form an external-internal relationship. So, the two viscera influence each other both physiologically and pathologically. When the lung qi descends, the large intestine will function normally and defecation is smooth. If the large intestine is blocked by stasis, it can also affect lung qi in its descent. Clinically, if the lung qi fails to descend and the body fluids cannot go downward, it may lead to dyschesia. If there is excess-heat in the large intestine, it may also cause failure of lung qi to descend and thus asthma and chest fullness will occur.

3.3.3 Spleen and Stomach

The spleen and stomach are both in the middle-jiao. There are meridians mutually belonging to and connecting them, making up an external-internal relationship. The spleen governs transformation and transportation, and the stomach governs reception, the spleen transports the body fluids to the stomach. While spleen qi ascends, stomach qi descends. The spleen likes dryness and dislikes dampness, while the stomach likes moisture and dislikes dryness. The two viscera share in the work of digesting fodder and fluids, along with absorption and transportation of essential substances. Pathologically, the two viscera also affect each other. For example, a case of deficient cold of the spleen is often tied to a deficiency of stomach yang, known as "deficiency cold in the middle-jiao." Clinically the spleen and stomach are frequently treated simultaneously, for example, by warming the middle-jiao to eliminate cold.

3.3.4 Liver and Gallbladder

The gallbladder is attached to the liver, and the meridians are connected through an external-internal relationship. It is to aid the digestion of the stomach and intestines that gallbladder stores and discharges bile. Bile originates in the liver and if liver qi does not flow smoothly, the normal excretion of bile will be affected. Conversely, any disorder involving bile may also affect the liver. Therefore, the disorders of the liver and gallbladder often occur simultaneously.

3.3.5 Kidneys and Urinary Bladder

The meridians of the kidney and urinary bladder are connected to each other externally and internally. The function of urinary bladder qi transformation depends upon kidney qi, which aids qi transformation of body fluids by the urinary bladder, and controls the urinary bladder's closing and opening functions in passing urine. When kidney qi is abundant it has control of urine and the urinary bladder will open or close appropriately to maintain normal water metabolism. If kidney qi is deficient and fails to control the closing and opening functions of the urinary bladder, it will cause urinary incontinence, and polyuria. Thus, abnormalities in storing and discharging urine are disorders of not only the urinary bladder but are closely related to the kidneys.

4 Qi, Blood and Body Fluids

Qi, blood and body fluids are the basic substances constituting the body and are the material base for the organs and meridians in carrying out their respective physiological activities. The energy for the physiological activities of the viscera and meridians is derived from qi, blood and body fluids. The production and metabolism of such substances are dependent upon the normal physiological activities of tissues, organs and the meridians. So, there are close relations both physiologically and pathologically among qi, blood and body fluids and the organs, tissues and meridians. Qi, blood and body fluids, either from the congenital or essential substances of fodder and drink, are the material base for maintaining the vital activities of the body. By the nourishment and the replenishment of these substances the viscera and meridians perform their normal physiological activities while at the same time these substances are constantly consumed and produced.

In TCVM, "essence" is also one of the basic substances constituting the body. It has two meanings. In a narrow sense, "essence" means genital essence that is stored in the kidneys. "In a broad sense," essence " includes all essential substances-qi, blood, body fluids and the essential substances of fodder and drink. These are collectively called "essential qi."

4.1 Qi

4.1.1 The Concept of Qi

The concept of qi in TCVM is derived from qi in ancient Chinese philosophy. In ancient times philosophers believed that qi was the most essential substance making up the world and that everything in the universe was generated by the motion and change of qi, the concept of qi as the root of a myriad of things was introduced into TCVM as a theory and concept. Qi was then looked upon as the essential substance constituting the body and maintaining the life activities of the body.

4.1.2 Functions of Qi

4.1.2.1 Driving Action

Qi is an essential substance with a strong activity. It plays a major role in activating and driving physiological activities of various viscera and meridians, the generation and circulation of blood, generation, and the distribution and excretion of body fluid. If this activating and driving action of qi gets weak, it will influence growth and development of the animals, functional activities of the viscera and meridians, and generation and circulation of blood and body fluids as well.

4.1.2.2 Warming Action

Qi is the source of thermal energy of the body. The constancy of temperature of the body also depends upon qi level of warmth. The warming action of qi is also closely related to the physiological activities of such organs and tissues as the viscera and meridians, and normal circulation of blood and qi as well.

4.1.2.3 Defending Action

Defending from attack by pathogenic evils is mainly dependent upon "anti-pathogenic qi" of the body. The anti-pathogenic-qi includes functional activities of qi, blood and body fluids, as well as the viscera and meridians. Of them, qi is the most important one. The defending action of qi mainly embodies protecting the body surface and defending from attack of exopathic evils.

4.1.2.4 Containing Action

The containing action of qi mainly embodies preventing losses of blood and body fluids. For example, containing blood from extravasation out of the vessels resulting in bleeding is mainly the function of spleen qi. Containing sweat, urine, saliva, sperm and blood are the respective functions of defensive qi, kidney qi, spleen qi and they control the excretion and prevent undue loss of such substances.

The containing action and driving action of qi are cooperative, interdependent and complementary. Driving action provides motive force for blood circulation and body fluid metabolism, while the containing action of qi guarantees normal circulation and metabolism.

4.1.2.5 Qi Transformation

Qi transformation means changes as a consequence of the movements of qi. It implies the metabolism of qi, blood, body fluids, essence and inter-transformation resulting from the action of qi. For example, production of qi, blood and body fluids depend upon the transformation of fodder and drink into essential substances, which then transform into qi, blood and body fluids. Body fluids, after metabolism, transform into sweat and urine; fodder, after digestion and absorption, has its waste transformed into feces. All of the previous are examples of qi transformation. The process of qi transformation, therefore, is actually a process of substance metabolism and a process of substance and energy conversion.

4.2 Blood

4.2.1 The Concept of Blood

Blood is a red liquid circulating within the vessels, being one of the basic substances constituting the body and maintaining its vital activities.

4.2.2 Functions of Blood

Blood has the function of nourishing and moistening the whole body. When healthy it is mainly reflected as a ruddy and lustrous mouth and tongue, well-developed and strong muscles, lustrous skin and hair, and well coordinated motor skills.

4.3 Body Fluids

4.3.1 The Concept of Body Fluids

Body fluid is a general term for all normal liquids in the body, including interstitial fluid in the viscera and other tissues and organs, and normal secretions such as gastric juice, intestinal juice, tears, etc.. Body fluid, like qi and blood, is a basic substance constituting the body and plays an important in maintaining its normal life activities.

Because of differences in quality, function and distribution, body fluid is again divided into two types: generally, what is clear, thin and distributing to the skin, muscles, orifices, fluid infiltrating into the blood vessels, and moistening is called "jin"; while that which is nourishing, turbid, thick and distributed to the joints, viscera, and brain marrow is called "ye." Jin and ye are both liquids, come from drinks and fodder and are dependent upon the transformation and transportation of the spleen and stomach. As they can transform each other, they are commonly collectively included as body fluids.

4.3.2 Functions of Body Fluids

Body fluids have the physiological functions of moistening and nourishing. For example, the fluid distributing onto the body surface moistens the skin and hairs; the fluid flowing into the orifices moisten and safeguards the eyes, mouth, and nose.

5 The Meridians

Meridian theories in TCVM include the histological structure, physiological function and pathological changes of the animals meridian system, relationship with the viscera and qi, blood, and body fluids. Meridian theory is an important component of the theoretical system of TCVM and has a guiding significance in a number of modalities, especially acupuncture and moxibustion.

5.1 Concepts of the Meridians

The meridians are passages through which the qi and blood of the body circulate, and by which the viscera and the other various tissues and organs are interrelated. The upper, lower, interior and exterior parts of the body are related. The meridian is a general term used to include both meridians and their collaterals. Meridians are tracks, virtual vertical trunk lines that generally run for the most part through the deeper parts of the body. The term collaterals imply a network, running mainly through the more superficial parts of the body. Meridians have specific routes, while collaterals cross to reach and connect every part of the body. Hence the meridian system makes the body an organic whole.

5.2 Physiological Functions of the Meridians

5.2.1 Communication in the Body Superficially, Inferiorly, Superficially and Internally

The animal body is composed of five zang-viscera, six fu-viscera, four extremities, hundreds of bones, the five sense organs and nine orifices, and skin and muscles. While they have different respective physiological functions, they jointly carry out the organic activities of the entire body, coordinating the interior, exterior, upper and lower parts to form an organic whole. This organic cooperation is accomplished by the communication and linkage of the meridian system. The twelve meridians and their branches belong to and connect with the viscera in a cross manner, the eight extra-meridians communicates with the twelve regular meridians, the muscles and skin areas of the twelve regular meridians communicate with skin, flesh, and tendons. As a result, various viscera, tissues and organs are organically related, forming a unity in which all parts of the body are superficially, internally, superiorly and inferiorly related to, cooperative with and supplemented.

5.2.2 The Circulation of Qi and Blood in Nourishing the Body

Various tissues and organs of the body are dependent on the moistening and nourishing qualities of qi and blood in sustaining their own physiological activities, viscera and the other tissues receive their nourishment, and they resist exogenous evils by defending the body. For example, nutritive qi spreads over the viscera, and provides a material basis not only for the five zang-viscera to store essences, but also for the six fu-viscera to transform and transport fodder. The meridians can circulate qi and blood, regulate yin and yang, and nourish the body. So, the meridians have the action of fighting against exopathic evils and guarding the body.

5.2.3 Reaction and Conduction

The meridians possess the function to sense and respond to various internal and external stimulations, to transmit and conduct them to the relevant viscera, and make the viscera change correspondingly in physiological and pathological terms. The "arrival of qi" (needling sensation) and "induction of qi" in acupuncture therapy are two examples of the phenomena of reaction and conduction by the meridians. Conversely, the changes of visceral functions caused by some factors may also be reflected in the body surface through the transmission of the meridians.

5.2.4 Regulation of Functional Balance in the Body

Under normal conditions, the meridians can circulate qi and blood, and regulate yin and yang. Pathologically, for the syndrome of disharmony of qi and blood as well as an excess or deficiency of either yin or yang, we can use acupuncture and moxibustion to trigger the regulating action of the meridians to reduce the excess and reinforce the deficient, and thus create a state of balance in the body. Practice and clinic experience have proven that acupuncture at the points of relevant meridians may indeed regulate visceral functions, and inhibit that which is hyperactive, and excite that which is inhibited.

6 Etiologies

There are many factors that can cause diseases, including the six exopathic factors, improper diet, overstrain and over ease.

TCVM holds that there is no syndrome without cause. Any syndrome is a disharmonious reflection of the body under the influence and action of some factors. TCVM studies the cause of a disease, through analysis of symptoms and signs of the disease to infer its cause. This process or method gives a basis for treatment and prescription and is called "seeking the cause by syndrome differentiation." Therefore, etiology in TCVM not only studies the nature and pathogenic character of pathogenic factors, but also probes into clinical manifestations of syndromes they cause. Thus it can play a role in clinical diagnosis and treatment.

The Six Exopathic Factors.

The six-exopathic factors are a collective term used for six kinds of exogenous pathogenic factors including wind, cold, summer-heat, dampness, dryness and fire. Under normal conditions, wind, cold, summer-heat, dampness, dryness and fire are called "six climatic factors," being normal climatic changes in nature. As animals have developed certain adaptive powers in normal conditions the six climatic factors do not normally cause disease. Only when the climatic change is sharp or the resistance of the body becomes weak, the six climatic factors will become pathogenic factors, thus causing the body to fall ill. The six climatic factors are then called the six-exopathic factors. Since the six-exopathic factors are abnormal, they are also referred to as the "six evils", and are causes of exogenous diseases.

6.1 Wind

Wind prevails in spring, but it may occur in any season. While invasion by wind evil may often occur in spring, it is not limited to spring. Wind evil mostly invades the body from the surface to cause exogenous diseases.

6.2 Cold

Cold is prevalent in winter. There is a difference between exogenous cold and endogenous cold in disease. Exogenous cold means that cold evil attacks the body from outside. In diseases caused by exogenous cold there is a difference between cold-attack and cold-stroke. A case in which cold evil attacks the body surface and depresses the defensive-yang is known as a "cold-attack", while a case in which cold evil directly invades the interior and damages the visceral yang-qi is called "cold-stroke". Endogenous cold is the pathological state resulting from the failure of yang-qi to warm the body due to its deficiency. While exogenous cold and endogenous cold are different, they are mutually related and influence each other. A animal with endogenous cold due to deficiency of yang is predisposed to an invasion of an exogenous cold evil; whereas exogenous cold evil invading and staying in the body for a long period of time is likely to damage yang-qi, and can lead to endogenous cold.

6.3 Summer-heat

Summer-heat prevails in summer, being derived from fire and heat. It is remarkable in that it only appears in summer. Summer-heat is a pure exogenous evil, and there is no such thing as endogenous summer-heat.

6.4 Dampness

Dampness is prevalent in late summer. The period when summer is changing into autumn is the time of the year with the most humidity. There is a difference between exogenous clamp and endogenous damp in disease. Exogenous damp is a pathogenic factor invading the body from outside when the climate is damp, for example being caught in rain, or living in a damp condition. Endogenous damp is a pathological state when water-damp accumulates internally, which is usually caused by failure of the spleen in transportation. Exogenous damp and endogenous damp are different, but they often influence each other in the onset of disease. Exogenous damp that invades the body from outside usually affects the spleen, making the spleen fail in transportation, and thus causes formation of damp internally. However, a animal with retention of water-damp internally due to deficiency of spleen-yang is predisposed to invasion of exogenous damp.

6.5 Dryness

Dryness is prevalent in autumn. The climate in autumn is dry, and the atmosphere is lacking in moisture. As a result, dry-diseases often occur. Dryness invades the body mostly via the mouth and nose to attack the lung-defense. There are differences between warm dryness and cool-dryness diseases. In early autumn summer heat lingers, and dryness combines with warm-heat to invade the body, leading to a warm-dryness disease. In late autumn, early winter cold appears, dryness and cold usually associate with each other to attack the body, and thus cool-dryness disease.

6.6 Fire

Fire and heat are both yang in nature and may be collectively looked upon as fire-heat. Though they are similar, yet here are still differences. Heat is the lesser stage of fire, and strong heat results in fire. Heat is mostly from the outside, such as wind-heat, summer-heat, and damp-heat, while fire is usually generated from the inside, such as flaring up of the heat-fire, hyperactivity of liver-fire, and gallbladder-fire. Both exogenous and endogenous diseases are caused by fire and heat.

7 Principles for Disease Prevention and Treatment

7.1 Prevention

Prevention includes taking certain steps in advance to stop the occurrence and development of diseases. TCVM has always attached great importance to the prevention of disease. As early as two thousand years ago, the Huangdi Neijing (Huangdi's Internal Classic) put forward the concept of "treating the undiseased", which includes treatment to prevent disease and before disease develops.

7.1.1 Prevention Before Disease Occurs

This implies that before a disease occurs, various measures should be taken to prevent its occurrence. The onset of a disease involves two aspects: first, a deficiency or functional disturbance of anti-pathogenic-qi; second, an invasion of the body by pathogenic factors. The pathogenic factor is an important condition while the anti-pathogenic-qi is seen as the internal cause, with the former becoming operative through the latter. Therefore, for prevention of occurrence of a disease both aspects should be taken into consideration.

7.1.2 Treatment Before Disease Develops

Prevention before a disease occurs is the ideal preventative measure. However, once a disease already occurs, one should strive for an early diagnosis and treatment so to stop the disease from further the development and progress.

7.2 Treatment Principles

The treatment principle is the general method in treating disease. It is formulated under the guidance of the concept of wholism and treatment determination based on syndrome differentiation. It is a universal guiding principle for the determination of methods, formula and dosage. Under the guidance of the general rule, the methods of replenishing qi, nourishing blood, supplementing yin and strengthening yang are methods to strengthen the vital, while the methods of diaphoresis, emesis and purgation are methods to eliminate the evil.

Disease syndromes are varied with complicated pathological changes. Practitioners must be adept at understanding the essential aspects of a disease in with the complicated and changeable manifestations and give treatment aimed at its root aspect; strengthen the vital and dispel the evil based on changes of deficiency or excess, regulate yin and yang according to changes of imbalance of yin and yang, and prescribe the treatment to varied according to different season, locality and individual. In paying attention to such factors a satisfactory effect will be achieved.

7.2.1 Treatment Aimed at the Root of Disease

Treatment aimed at the root aspect of a disease means a cardinal principle of treatment based on syndrome differentiation. The root is judged by comparing it with the branch. The root and branch are relative concepts with multiple meanings. They may be used to explain the primary and the secondary in relation to various contradictions in the course of a disease. For instance, viewing the two sides of the vital and the evil, vital-qi is the root and evil-qi, the branch. In terms of cause and symptom, the cause is the root and the symptom is the branch. Considering the order of diseases, the old or primary disease is the root while the new or secondary disease is the branch.

7.2.2 Strengthening the Vital and Dispelling the Evil

The course of a disease, in a certain sense, is the process of the struggle between the two contradictory aspects, the vital and the evil. The outcome of the struggle between the vital and the evil decides the progress of a disease. When the evil gains the upper hand, the disease will progress, when the vital gains the upper hand, the disease will weaken. Thus in treatment strengthening the vital and dispelling the evil should be carried out to change the ratio in strength of the two sides of the vital and evil in order to gain recovery. Therefore, strengthening the vital and dispelling the evil are important principles in clinical treatment.

7.2.3 Regulating Yin and Yang

The occurrence of a disease is essentially the outcome of the superiority or inferiority of yin or yang and destruction of the relative balance they share. Therefore, regulating yin and yang to remedy and restore their relative balance is a cardinal principle in treatment.

7.2.4 Treatment and the Season, Locality and Individual

These concepts involve determining a suitable treatment for disease according to differences in season, local and the animal's constitution, sex, and age.

Lecture Two

Science of the Traditional Chinese Veterinary Materia Medica

The properties and actions of Chinese medicinal herbs, which are the essential basis of the analyses and clinical usage of Chinese medicinal herbs, are summarized in medical practice and on the basis of the theories of yin yang, zangfu, meridians, and therapeutic principles of traditional Chinese Veterinary medicine, etc.. The theories of their properties are mainly summarized as the four natures and five flavors, floating and sinking, meridian tropism, and toxicity, etc..

1 Four Natures

Four properties of Chinese medicinal herbs, cold, hot, warm and cool, are also called the four natures or four xing in TCVM. Cold-cool and warm-hot are two completely different categories of natures, cold-cool belonging to yin, and warm-hot to yang, whereas cold and cool or hot and warm are only different, to some degrees, in their variance. Chinese medicinal herbs with cold-cool nature can clear away heat, purge fire and eliminate toxic materials, which are mainly used for heat-syndrome; Chinese medicinal herbs with warm-hot nature have the actions of expelling cold and restoring yang, which are mainly used for cold-syndrome. In addition, there are also some Chinese medicinal herbs known as neutral ones whose cold or hot nature is not so remarkable and whose action is relatively mild. But actually they still have differences in the tendency to cool or warm so that they are still in the range of four natures.

2 Five Flavors

The five flavors of Chinese medicinal herbs refer to the five different tastes, pungent, sweet, sour, bitter and salty, which can be tasted by the tongue. With the development of the theory dealing with the medicinal properties, some flavors are summarized out of clinical actions of Chinese medicinal herbs, therefore, there is a little difference between the flavors of medicinal herbs and the tastes got by tongue. The Chinese medicinal herbs with same flavor mostly possess similar actions while the medicinal herbs with different flavors show different actions in the treatment.

3 Lifting, Lowering, Floating and Sinking

By "lifting" of Chinese medicinal herbs we mean that the direction of Chinese herbal action is toward the upper parts. Those that possess the lifting action are indicated for a disease in lower and deeper parts. For example, Huangqi (Radix Astragali) and Shengma (Rhizoma Cimicifugae) can raise splenic qi and are indicated for syndrome of visceroptosis with hyposplenic qi such as chronic diarrhea and lingering dysentery, prolapse of the rectum, prolapse of uterus and gastroptosis. By "lowering" of Chinese medicinal herbs we mean that they function toward the lower parts and possess the action of descending adverse qi and are indicated for the disease due to adverse ascending of pathogenic factors. For example, Diathesis (Haematitum), Chenxiang (Lignum Aquilariae Resinatum) and Shijueming (Concha Haliotidis) can descend adverse flow of qi and fire, subdue exuberant yang of the liver and descend adverse qi of the lung and stomach and are indicated for bleeding, painful swollen gum and aphtbae due to ascendancy of rebellious qi and fire; cough and dyspnea due to abnormal rising of lung-qi, nausea and vomiting and eructation due to abnormal rising of stomach qi. By "floating" of Chinese medicinal herbs we mean that they function toward the upper and outward parts, generally exert the effects of sweating and dispersing and are indicated for the disease in the upper and superficial parts. For instance, Mahuang (Herba Ephedrae), Zisu (Folium Perillae), Fangfeng (Radix Saposhnikoviae) and Duhuo (Radix Angelicae Pubescentis) can dispel wind-cold and dampness from the exterior and are indicated for wind and cold exterior-syndrome, wind-damp type of Bi-syndrome, etc.. By "sinking" of Chinese medicinal herbs we mean that they function toward the lower and inward parts, have the effects of lowering the adverse flow of qi and relaxing bowels and are indicated for the disease in the lower and interior. For instance, Dahuang (Radix et Rhizoma Rhei) and Mutong (Caulis Akebiae) separately have the effects of relaxing the bowels and promoting diuresis and are used to treat constipation, abdominal distention and pain, and dysuria, etc..

4 Meridian Tropism of Chinese Medicinal Herbs

Meridian tropism refers to that medicinal herbs may often produce their therapeutic effects on some portion of a animal body in preference, in other words, their therapeutic action is mainly related to some viscous or channel, or some channels in predominance but it may seem to produce fewer effects on or seem not related to the other viscera and channels. Meridian tropism takes the theory of viscera and meridians, and the indication of syndromes as a basis. For instance, Mahuang (Herba Ephedrae) and Xingren (Semen Armeniacae Amarae) effective to syndromes of the disorder of the lung meridian marked by cough and dyspnea are attributed to the lung-meridian; Qingpi (Pericarpium Citri Reticulatae Viride) and Xiangfu (Rhizoma Cyperi) indicated for syndromes of the disorder of the liver-meridian marked by distending pain of breast and hypochondrium and hernia pain are attributed to the liver-meridian. So generally speaking, what meridian or meridians a medicinal herb is attributed to is just related to the certain meridian or meridians on which the herb may work. If certain medicinal herb can work on several meridians, which means the medicinal herb can be used widely to treat the disorders of these meridians. From the above, we can see that meridian tropism of Chinese medicinal herbs is summarized from the therapeutic effects through a long time of clinical observation, and being practiced repeatedly, gradually develops into a theory.

5 Toxicity

Toxicity refers to a harmful effect of a medicinal herb to the animal body, a poisonous medicinal material being known as a toxin. The medicinal herbs drastic or poisonous in nature, if used improperly, can do harm to the body for the light and can cause death for the severe. In order to ensure safety in the use of medicinal herbs, their toxicity must be thoroughly understood.

Dosage of poisonous medicinal herbs in the treatment is close or same to poisoning dosage, so the safety margin is small and poisoning is easily resulted. Whereas the dosage of non-poisonous medicinal herbs in treatment is much farther from the poisoning dosage, and the safety margin is also larger. But it is not absolute whether they can result in poisonous reaction or not. In order to ensure the safety in administration of medicinal herbs and bring therapeutic effects into play and avoid poisonous reaction, you should pay attention to the follows as you use the poisonous ones: Strictly processing, Control of dosage, Notes of application.

6 Processing of Chinese Medicinal Herbs

By processing of medicinal herbs, termed as Paozhi in TCVM, we refer to various processes of preparing crude medicinal materials according to the theories of TCVM and requirements of therapy, prescription, preparation of forms, and storage of Chinese medicinal herbs. It includes common or special treatment of crude or part of crude medicinal materials. Since most Chinese medicinal herbs come from the crude plants, of which some are mixed with impurity in the collecting process, some may easily change their properties and are not to be stored for a long time, some are violently toxic and are not to be taken directly and may need to be treated in special way so as to tally with the needs of treatment. Therefore, it is necessary for them to be processed so as to produce satisfactory medicinal effects and ensure safety in clinical practice before their use and preparation.

6.1 Processing with water

Processing herbs with water is a kind of method of treating pharmaceutical materials for the purposes of cleaning, softening, or making them easy to cut or regulate medicinal properties, reducing their toxicity, or making mineral drugs pure, fine and smooth, etc.. The commonly-used methods are listed as follows: wash, cleaning, steep, moisten, ulcer, pickle, lift, spray, etc..

6.2 Processing with fire

Processing with fire is a method, which is used in treating crude medicinal materials by heating with fire. This method includes the following commonly used ones, such as parching, stir-baking with adjuvants, calcining, and roasting in ashes, etc..

7 Compatibility of Chinese Medicinal Herbs

Compatibility of Chinese medicinal herbs refers to the combination of more than two herbs with purpose in the light of the clinical requirement and medicinal properties and actions. It is the main method of medicinal application in clinic and also the basis of making up formulae of Chinese medicinal herbs.

Chinese medicinal herbs may have complicated changes by combination. Some may reinforce or decrease their effects, moderate or eliminate their original toxic side effects, whereas others may produce toxicity and poor reactions. Ancient physicians as seven aspects, namely, singular application, mutual reinforcement, mutual assistance, mutual restraint, mutual detoxication, mutual inhibition and incompatibility, generalized the relationship between a single and the compatible ingredients previously.

8 Prescription Incompatibility

Prescription incompatibility refers to that some medicinal herbs cannot be used together in a prescription, otherwise the toxic effect will be produced harming the animal health, and even his life. Incompatibility also denotes incompatible medicinal herbs, especially denotes "the 18 incompatible medicaments", and "19 medicaments of mutual antagonisms".

In the 18 incompatible medicaments the following herbs are believed to be incompatible in their actions if given in combination: Wutou (Radix Aconiti) being incompatible with Banxia (Rhizoma Pinelliae), Gualou (Fructus Trichosanthis), Beimu (Bulbus Fritillariae), Bailian (Radix Ampelopsis) and Baiji (Rhizoma Bletinae); Gancao (Radix Glycyrrhizae) incompatible with Haizao (Sargassum), Daji (Radix Eauphorbiae Pekinensis), Yuanhua (Flos Genkwa) and Gansui (Radix Euphorbiae Kansui); Lilu (Rhizoma et Radix Veratri) incompatible with Dangshen (Radix Codonopsitis Pilosulae), Shashen (Radix Adenophorae Strictae), Danshen (Radix Salviae Miltiorrhizae), Xuanshen (Radix Scrophulariae), Kushen (Radix Sophorae Flavescentis), Xixin (Herba Asari Cum Radice) and Shaoyao (Radix Paeoniae). Nineteen medic aments of mutual antagonisms include Liuhuang (Sulfur) being antagonistic Qianniuzi (Semen Pharbitidis), Dingxiang (Flos Caryorphylli) antagonistic to Yujin (Radix Curcumae), Chuanwu (Radix Aconiti) and Caowu (Radix Aconiti Ferus) to Xijiao (Cornu Rhinoceri Asiatici), Yaxiao (Crystallized Mirabilite) to Sanleng (Rhizoma Sparganii) and Guangui (Cortex Cinnamomi) to Wulingzhi (Faeces Trogopterori). It must be denoted that "being antagonistic" here means "being loath" or "being dislike", which is different from "Mutual Restraint" in the compatibility of Chinese medicinal herbs.

As to the 18 incompatible medicaments and 19 medicaments of mutual antagonisms, they are regarded as ingredients, which are incompatible, but some of them were still used in combination by some physician in various dynasties. The conclusion of the 18 incompatible medicaments and 19 medicaments of antagonisms got in modern experiments and research is not completely similar. Therefore, the conclusion has not been confirmed and further research will be made. So we should use them cautiously and generally we should avoid using them in combination.

9 Contraindication of Chinese Medicinal Herbs in Pregnancy

Some medicinal herbs should be regarded as those that are contraindicated or used with cautions in pregnant animal, otherwise, side effects of damaging fetus, inducing abortion or miscarriage may be brought about. They may be divided into two kinds, one being contraindicated in pregnancy, which involve poisonous medicinal herbs and drastic purgatives; the other being used with caution, which mainly includes those for eliminating blood stasis and promoting circulation of qi, the purgatives and part of herbs warming the interior. If this kind of medicinal herbs is used improperly, the fetus may be damaged or abortion (miscarriage) may be induced.

Chinese medicinal herbs contraindicated are Shoo-in (Hydmrgyrum), Pishuang (Arsenicum), Xionghuang (Realgar), Qingfen (Calomelas), Banmao (Radix Sacchari Arundinacei), Wugong (Scolopendra), Maqianzi (Semen Strychni), Chansu (Venenum Bufonis), Chuanwu (Radix Aconiti), Caowu (Radix Aconiti Ferus), Lilu (Radix et Rhizoma Veratri), Danfan (Chalcanthium), Guadi (Pedicculus Melo Fructus), Badou (Fructus Crotonis), Gansui (Radix Euphorbiae Kansui), Daji (Radix Euphorbiae Pekinensis), Yuanhua (Flos Genkwa), Qianniuzi (Semen Pharbitidis), Shanglu (Radix Phytolaccae), Shexiang (Moschus), Ganqi (Resina Rhois Praeparata), Shuizhi (Hirudo), Mengchong (Tabanus), Sanleng (Rhizoma Sparganii), and Ezhu (Rhizoma Curcumae Zedoariae), etc..

Chinese medicinal herbs used with caution are Niuxi (Radix Achyranthis Bidentatae), Chuanxiong (Rhizoma Chuanxiong), Honghua (Flos Carthami), Taoren (Semen Petsicae), Jianghuang (Rhizoma Curcumae Longae), Mudanpi (Cortex Moutan Radicis), Zhishi (Fructus Aurantii Immaturus), Zhike (Fructus Aurantii), Dahuang (Radix et Rhizoma Rhei), Fanxieye (Folium Cassiae), Luhui (Aloe), Tiannanxing (Tuber Arisaematis), Mangxiao (Natrii Sulfas), Fuzi (Radix Aconiti Lateralis Praeparata), and

Rougui (Cortex Cinnamomi), etc..

10 Dosage

The amount of medicinal herbs to be taken is called dosage, which signifies the daily amount of one particular herb by adults, and secondly shows the comparative measurements of medicinal herbs in the same prescription and is also known as relative dosage. Generally speaking, the amount following each herb refers to the daily amount of one dried mw herb by adults in making decoction or the daily amount of powder by adults, which is ground from dry raw herbs. As to the amount of Chinese medicinal herbs to be taken, we mostly take weight as the unit of measurement, and take quantity and capacity as the unit for the special ones. Now we take the metric system (g) as the unit of measurement, which is now stipulated in the mainland of China.

11 Methods of Decocting Chinese Medicinal Herbs

Stewing utensils available are a clay pot or earthen jar, or a piece of enamelware. The water available must be clean and without peculiar smell. First put Chinese medicinal herbs into the enamelware and add water to it, the water being usually over the surface of the herbs. Before being decocted, the Chinese medicinal hebs need immersing in water for half an hour so as to make their medicinal components easily dissolve in the solution. Fire used in decocting the herbs should be controlled in the light of medicinal properties and qualities. The medicinal herbs with aromatic smell should be decocted with strong fire until the solution is boiled for several minutes, then a small fire is followed until the decoction is done, otherwise the medicinal effects will reduce, nourishing medicinal herbs, since their qualities are greasy, should be decocted with a small fire for a long time or the effective factors are not easily decocted out. A dose of Chinese medicinal herbs is taken daily, which is usually decocted twice while nourishing ones may be decocted three times.

Since their qualities and properties are usually obviously different, different medicinal herbs should be given different treatment in decocting method and time. When a prescription is made out, the methods should be noted, so as it be followed by drug store or sick animals when the solution is decocted. The chief methods are shown as follows.

12 Methods of Taking Chinese Medicinal Herbs

Generally speaking, decoction must be taken warm. Chinese medicinal herbs in a prescription or a dose may be decocted twice, and the decocted juice is mixed together, being divided into two parts for daily use. An acute case must take two doses a day or even three doses, that is, once for every four hours. A chronic sick animal may take a dose a day or two days. Those used for stopping vomiting should be taken frequently in small amount; diaphoretics should be taken warmly so as to promote the medicinal actions until sweating; purging Chinese medicinal herbs are taken until reducing diarrhea or vomiting. Pill or powder may be taken with warm water. As far as treatments are concerned, Chinese medicinal herbs warm in nature should be taken in cold or those cold in nature should be taken in warm.

13 Diaphoretics or Exterior Syndrome Relieving Chinese Medicinal Herbs

Any Chinese medicinal herb that has the actions of dispersing or expelling pathogens from the superficies and relieving the exterior syndrome by means of perspiration is considered to be diaphoretic medicinal herbs. When diaphoretics are used, their dosage should be controlled and the application should be stopped as soon as the syndrome disappears, otherwise, profuse-sweating will consume yang and the body fluids. Therefore, they should be contraindicated or used with great care for spontaneous perspiration due to superficial asthenia, night sweat due to yin-deficiency, and prolonged pyocutaneous disease, stranguria or blood loss with exterior syndrome. Diaphoretics mostly with active volatile oil, if added to decoction, should not be decocted or boiled for a long time to prevent effective constituents from volatilizing so as to decrease the therapeutic results.

13.1 Mahuang (Herba Ephedrae)

The source is from the herbaceous stem of *Ephedra sinica* Stapf, *E. intermedia* Schrenk etc. *A. Mey*, and *E. equisetina* Bunge, family Ephedraceae. The producing areas are mainly in Hebei, Shanxi, Gansu provinces and Inner Mongolia Autonomous Region, etc.. The medicinal material is collected in autumn, dried in shade and cut into pieces. Its commonly used form is crude one or the one prepared with honey.

Medicinal Properties They are pungent and slightly bitter in flavor, warm in nature, and attributive to the lung and bladder meridians.

Actions Induce sweating to relieve superficies disperse the lung to relieve asthma and promote diuresis to subside edema.

Application

a. It is used for superficial syndrome due to wind and cold. Since the medicinal herb can strongly promote sweating, it is often used for superficial sthenia-syndrome due to wind and cold, which is manifested as aversion to cold, fever, anhidrosis, floating and tight pulse. It is often used in combination with Guizhi (Ramulus Cinnamomi), such as Mahuang Tang (Decoction).

b. It is used for sthenia-syndromes with cough and dyspnea. It is indicated for all syndromes such as cough and dyspnea due to stagnation of lung-qi, or cold, heat, phlegm or retention of fluids whether they have superficial syndrome or not. It is combined with Xingren (Semen Pruni Armeniacae) and Gancao (Radix Glycyrrhizae), such as San'ao Tang (Decoction) for cough and dyspnea due to accumulation of wind and cold in the lung. It is combined with Ganjiang (Rhizoma Zingiberis), Xixin (Herba Asari), Banxia (Rhizoma Pinelliae), etc., like Xiaoqinglong Tang (Decoction) for retention of cold-fluid in the lung manifested as profuse watery sputum. It may also be combined with Shigao (Gypsum Fibrosum), Xingren (Semen Pruni Armeniacae) and Gancao (Radix Glycyrrhizae) to treat syndrome with asthma and yellowishthick sputum due to stagnancy of lung-heat, such as Mahuang Xingren Gancao Shigao Tang (Decoction) to clear away lung heat and relieve dyspnea and cough.

Usage and Dosage It is used in decoction for oral use. The dosage for horse and cattle is 15 ~ 30g, for pig, sheep and goat is 3 ~ 10g. The crude form is suitable for inducing sweating to relieve superficies and the prepared form with honey for relieving cough and dyspnea.

Notes It is not used excessively and contraindicated in conditions of spontaneous perspiration due to exterior deficiency, night sweating due to yin deficiency, asthenia-dyspnea.

13.2 Guizhi (Ramulus Cinnamomi)

The source is from the tender branch of *Cinnamomum cassia* Presl, family Lauraceae. Its producing areas are mainly in the provinces of Guangdong, Guangxi and Yunnan. The medicinal material is collected in spring and summer, and dried either in shade or in the sun, then is cut into pieces or segments for use.

Medicinal Properties Pungent and sweet in flavor, warm in nature, and attributive to the heart, lung and bladder meridians.

Actions Induce sweating to relieve superficies and activate yang and circulate qi by warming meridian.

Application

a. It is used for the superficial syndrome due to wind and cold whether the syndrome is asthenic or sthenic. For the wind-cold syndrome with spontaneous perspiration due to superficial deficiency, it is often combined with Baishaoyao (Radix Paeoniae Alba), etc., such as Guizhi Tang (Decoction) to keep the actions of yingqi and weiqi in balance. It is combined with Mahuang (Herba Ephedrae) to treat sthenia-syndrome of the superficies with anhidrosis.

b. It is used for pains due to blood stagnancy due to stagnation of cold. It is often combined with Zhishi (Fructus Aurantii Immaturus), Xiebai (*Alliium Macrostemon*), etc., such as Gualou Xiebai Guizhi Tang (Decoction) for the treatment of the weakness of thoracic yang, the obstruction in heart channel and thoracic pain due to thoracic obstruction; combined with Fuzi (Radix Aconiti Carmichaeli Praeparata), such as Guizhi Fuzi Tang (Decoction) for the treatment of arthralgia due to stagnation of wind-cold-damp.

c. It is combined with Zhuling (*Polyporus*), Zexie (Rhizoma Alismatis), etc., such as Wuling San (Powder) for dysuria and edema.

Usage and Dosage It is used in decoction for oral use. The dosage for horse and cattle is 15 ~ 45g, for pig, sheep and goat is 3 ~ 10g.

Notes Since it is pungent and warm in nature and easily damages yin and blood, it is contraindicated for the syndromes such as exuberant heat impairing yin in seasonal febrile disease, hyperactivity of yang due to yin-deficiency in miscellaneous diseases and bleeding due to bloodheat, and must be used with caution in pregnant animal.

Explanation Both Guizhi (Ramulus Cinnamomi) and Mahuang (Herba Ephedrae) are diaphoretics for dispersing wind-cold, whereas Guizhi (Ramulus Cinnamorai) is moderate in action and weaker than Mahuang (Herba Ephedrae) in inducing sweating. It can be used for spontaneous perspiration due to superficial deficiency and can be combined with Baishaoyao (Radix Paeoniae Alba) to warm the meridians, activate yang, resolve fluid retention and induce diuresis, etc. Since Mahuang (Herba Ephedrae) is a medicinal herb mainly attributive to the lung meridian and has the strong action of perspiration, it is often used for exterior sthenia syndrome with anhidrosis due to wind and cold and can disperse the lung and relieve asthma, promote diuresis to relieve edema.

13.3 Chaihu (Radix Bupleuri)

Its source is from the dried root or the herb of the Perennial herbaceous plant, *Bupleurum chinense Dc.* or *B. scorzonerifolium* Willd, family Umbelliferae. The former is mainly produced in the provinces of Liaoning, Gansu, Hebei, and Henan, etc., the latter in Hubei, Jiangsu, and Sichuan, etc.. The medicinal material is collected in spring and autumn, dried in the sun and cut into pieces. The raw or the one fried with vinegar is used for medication.

Medicinal Properties They are bitter and pungent in flavor, slightly cold in nature and attributive to the liver and gallbladder meridians.

Actions Regulate the functional relation of internal organs to relieve fever, disperse the stagnated liver-qi and uplift yang-qi to rise sinking.

Application

a. It is used for fever due to exogenous pathogenic factors with alternating episodes of chills and fever. It is effective in eliminating the pathogenic factors located in the half-superficial and half-interior, hence it is an indispensable medicine for treating shaoyang disease. It is usually combined with Huangqin (Radix Scutellariae), Banxia (Rhizoma Pinelliae), etc., such as Xiao Chaihu Tang (Decoction). For fever due to exogenous pathogenic factors, it can be combined with Gegen (Radix Puerariae), Huangqin (Radix Scutellariae), and Shigao (Gypsum Fibrosum), etc., such as Chai Ge Jieji Tang (Decoction).

b. It is used for the treatment of prolapses due

to deficiency of qi, such as prolapse of rectum, or uterus, and shortness of breath and weakness, etc. since it is effective in uplifting qi of the stomach and spleen. It can be combined with Shengma (Rhizoma cimicifugae), Huangqi (Radix Astragali) and other herbs that can uplift the qi of middle energizer, such as Buzhong Yiqi Tang (Decoction).

Usage and Dosage It is used in decoction for oral use. The dosage for horse and cattle is 15 ~ 45g, for pig, sheep and goat is 6 ~ 15g.

Notes Contraindicated for the syndromes of liver-wind stirring inside, hyperactivity of fire due to yin-deficiency or adverse flow of qi.

14 Heat Clearing Chinese Medicinal Herbs

Any Chinese medicinal herb that has the main action of clearing away interior heat is considered to be an antipyretic herb.

Antipyretic herbs are mostly cold or cool in nature and have the actions to clear away heat, purge fire, dry dampness, cool the blood, relieve toxic materials, clear away deficiency-heat, etc.. They are chiefly used for various internal heat syndromes without exterior pathogenic factors and stagnation marked by febrile disease due to exogenous pathogenic factors with high fever, excessive thirst and dysentery due to damp and heat; maculae due to warm-toxin, carbuncle, ulcer, skin disease, and fever due to yin-deficiency.

Heat clearing medicinal herbs are both cold and cool in nature, easily injuring the spleen and stomach, so they are used with caution for deficiency of spleen-qi and stomach-qi, which is manifested as poor appetite and loose stool; furthermore, heat clearing and dampness drying medicinal herbs should be also used with caution for yin deficiency and consumption of the body fluids when they are used to treat febrile diseases since they are bitter in flavor and cold in nature and easily injure yin and consume body fluids as well as remove dryness, and febrile disease easily consumes body fluids, too.

14.1 Shigao (Gypsum Fibrosum)

It is the monoclinic system of gypsum ore, containing hydrous calcium sulfate, its producing areas are in the provinces of Hubei, Gansu and Sichuan and can be mined all year round. The crude one may be pounded or calcined for being used as a medication.

Medicinal Properties Bitter and sweet in flavor, extreme cold in nature and attributive to the lung and stomach meridians.

Actions Clear away heat and purge fire, relieve restlessness and thirst, and induce astringent and promote tissue regeneration.

Application

a. It is used for high fever and excessive thirst. With strong actions of clearing away heat and purging fire, and relieving restlessness and thirst, it is the essential medicine to clear away sthenic heat in qi-fen of lung and stomach meridians. For sthenic-heat syndrome manifested as seasonal febrile disease in qi-fen, high fever, excessive thirst, and perspiration, it is often used together with Zhimu (Rhizoma Anemarrhenae), such as Baihu Tang (Decoction).

b. For stagnation of heat in the lung manifested as cough, thick sputum, fever and thirst, it is often combined with Mahuang (Herba Ephedrae), Xingren (Semen Armeniacae Amarum) to strengthen its action of relieving cough and asthma, such as Ma Xing Shi Gan Tang (Decoction).

c. For up-rising of stomach-fire with swelling, it is often combined with Huanglian (Rhizoma Coptidis), Shengma (Rhizoma Cimicifugae) and others that can clear away stomach-heat, such as Qingwei San (Powder).

d. Externally used for inducing astringent and prorooting tissue regeneration, pyocutaneous disease, eczema, burned or boiled wounds and bleeding due to trauma. It can be used in single or together with Qingdai (Indigo Pulverata Levis), Huangbai (Cortex Phellodendri), etc. to clear away heat and remove toxic materials, eliminate dampness and treat ulcer.

Usage and Dosage It is used in decoction for oral use. The dosage for horse and cattle is 30 ~ 120g, for pig, sheep and goat is 15 ~ 30g.

Notes It is contraindicated in the sick animal with deficiency and cold of the spleen and stomach, or interior heat due to yin deficiency.

14. 2 Huangqin (Radix Scutellariae)

The source is from the root of *Scutellaria baicalensis* Georgi, family Labiatae. The medicinal material is mainly produced in the provinces of Hebei, Shanxi, and inner Mongolia autonomous region, etc. , digged up and collected in spring and autumn, usually cut into pieces, the crude one and the one stir-baked with wine or the carbonized one may be used.

Medicinal Properties Bitter in flavor, cold in nature and attributive to the lung, stomach, gallbladder and large intestine meridians.

Actions Clear away heat and remove dampness, purge the sthenic fire and remove toxic materials, cool blood and stop bleeding to prevent miscarriage.

Application

a. It is used for seasonal febrile diseases of dampness type, jaundice, dysentery and stranguria of dampheat type, especially effective in clearing away damp-heat of both the middle and upper energizers. For the treatment of seasonal febrile disease of damp type with fever, perspiration, greasy fur, it is used together with Huashi (Talcum), Tongcao (Medulla Tetrapanacis), and Baidoukou (Fructus Amomi Rotundus), etc. such as Huangqin Huashi Tang (Decoction). For jaundice of damp-heat type, it is often combined with Yinchen (Herba Artemisiae Scopariae), Zhizi (Fructus Gardeniae), and Dahuang (Radix et Rhizoma Rhei), etc. so as to strengthen the effects of draining the gallbladder to relieve jaundice. For the treatment of dysentery of damp-heat type, it is combined with Jungian (Rhizoma Cupids) and Gagmen (Radix Pereira), such as Gagmen In Lain Tang (Decoction), for the treatment of damp-heat of the bladder with dribbling and painful urination, used together with Muting (Radix Acedias), Hash (Talcum), Cheqianzi (Semen Plantaginis) and others that can eliminate heat and dampness.

b. It is used for heat-heat syndrome of qi-fen and cough due to lung-heat and good at clearing away lung-fire and sthenic-heat of the upper energizer. It can be used alone, namely Lignin Wan (Pill), or combined with Sangbaipi (Cortex Mori Radicis), Zhimu (Rhizoma Anemarrhenae), and Maimendong (Radix Ophiopogonis) for stagnation of lung-heat and failure of the lung to depurate and descend marked with cough and thick sputum. For sthenic-heat of qi-fen with high fever, excessive thirst, urination with redness and constipation, it is often combined with Dahuang (Radix et Rhizoma Rhei), Zhizi (Fructus Gardeniae) and others that can purge fire and relax the bowels, such as Liangge San (Powder).

c. For exterior sores, abscess of internal organs and other heat-toxin syndromes of internal medicine, surgery and five sensory organs, it is used together with Huanglian (Rhizoma Coptidis), Lianqiao (Fructus Forsythiae), and Pugongying (Herba Taraxaci), etc. .

d. For hemopyretic bleeding manifested as hematemesis, hemoptysis, hematochezia, and hemafecia, it can be used alone, or combined with Sanqi (Radix Notoginseng), Huaihua (Flos Sophorae), and Baimaogen (Rhizoma Imperatae), etc. .

e. For excessive fetal movement due to heat-syndrome in pregnancy, it can be used together with Baizhu (Rhizoma Atractylodis Albalae), Danggui (Radix Angelicae Sinensis), and others, such as Danggui San (Powder).

Usage and Dosage It is used in decoction for oral use. The dosage for horse and cattle is 15 ~ 90g, for pig, sheep and goat is 9 ~ 15g. The crude one is used for clearing away heat, the stir-baked one for preventing miscarriage, the carbonized one for relieving bleeding, the stir-baked with wine for clearing away heat of the upper energizer.

Notes It is bitter and cold and easy to injure the stomach, so it is contraindicated for deficiency and cold of the spleen and stomach.

14.3 Kushen (Radix Sophorae Flavescentis)

The source is from the root of the Sophora flarescerts Ait., family Leguminosae. The medicinal material is produced in all parts of China, collected in spring or autumn, cut into pieces, dried in the sun and the raw is used.

Medicinal Properties Bitter in flavor, cold in nature and attributive to the liver, stomach, large intestine and bladder meridians.

Actions Clear away heat and dry dampness, promote diuresis and kill worms

Application

a. It is suitable for dysentery of dampness-heat type and jaundice with brownish urine. It can be used alone or combined with Muxiang (Radix Aucklandiae) and Gancao (Radix Glycyrrhizae), such as Xiang Shen Wan Pill) to treat the former syndrome; it is usually combined with Zhizi (Fructus Gardeniae), Yinchen (Herba Artemisiae Scopariae) and Longdan (Radix Gentianae), etc. so as to remove dampness to cure jaundice in the treatment of jaundice; combined with Cheqianzi (Semen Plantaginis), Zexie (Rhizoma Alismatis) and others that promote diuresis in the treatment of stranguria of dampness-heat type and dysuria.

b. It is used in the treatment of pruritus vuluae, eczema, and scabies. It can kill worms and arrest itching as well as clear away heat and eliminate dampness, so it is a common medicine to treat the above skin diseases, and can be used orally or externally. It is usually combined with Huangbai (Cortex Phellodendri), Shechuangzi (Fructus Cnidii), etc..

Usage and Dosage It is used in decoction for oral use. The dosage for horse and cattle is 15 ~ 60g, for pig, sheep and goat is 6 ~ 15g.

Notes Contraindicated for deficiency and cold of the spleen and stomach since it is more bitter and colder in property, and it's incompatible with Lilu (Rhizoma et Radix Veratri).

14.4 Shengdihuang (Radix Rehmanniae)

The source is from the root of Rehmannia glutinosa Libosch, family Scrophulariaceae. The medicinal material is mainly produced in the provinces of Henan, Hebei, Inner Mongolia Autonomous Region and the Northeast China, collected in autumn, fresh or dry crude one cut into pieces can be used.

Medicinal Properties Sweet and bitter in flavor, cold in nature and attributive to the heart, liver and kidney meridians.

Actions Clear away heat and cool the blood, nourish yin and promote the production of the body fluids.

Application

a. It is indicated for seasonal febrile disease involving ying-fen and xue-fen. It is effective in clearing away heat in xue-fen and can nourish yin and promote the production of the body fluids as well. It is usually combined with Xuanshen (Radix Scrophulariae), Maimendong (Radix Ophiopogonis) and others that can clear away heat, promote the production of the body fluids and cool the blood, such as Qingying Tang (Decoction) for seasonal febrile diseases involving ying-fen and xue-fen, which are manifested as fever being more severe at night, restlessness, macules and crimson tongue.

b. For bleeding due to blood-heat manifested as hematemesis, epistaxis, hemafecia, and hematuria, etc., it is usually used together with Cebaiye (Cacumen Biotae), Shenheye (Folium Nelumbinis, unprepared), Sbeng'aiye (Folium *Artemisiae argyi*), etc., such as Sisheng Wan (Pill); for macules due to blood-heat, it is used together with Mudanpi (Cortex Moutan Radicis), Chishaoyao (Radix Paeoniae Rubra).

c. It is used for syndrome of consumption of the body fluids due to yin-deficiency. For febrile disease consuming yin manifested as red tongue and oral dryness, it is combined with Shashen (Radix Adenophorae), Maimendong (Radix Ophiopogonis), and Yuzhu (Rhizoma Polygonati Odorati), etc.; for febrile diseases with consumption of the body fluids manifested as constipation due to dryness of intestine, it is usually used together with Xuanshen (Radix Scrophulariae) and Maimendong (Radix Ophiopogonis), such as Zengye Tang (Decoction).

Usage and Dosage It is used in decoction for oral use. The dosage for horse and cattle is 30 ~ 60g, for pig, sheep and goat is 10 ~ 15g.

Notes Contraindicated for spleen deficiency with dampness with diarrhea since the medicinal herb is cold, cool and moist in nature.

14.5 Jinyinhua (Flos Lonicerae)

The source is from the flower bud of *Lonicera japonica* Thunb., *L. hypogtauca* Miq. and *L. confusa* DC., family Caprifoliaceae. The medicinal material is produced in all parts of China and collected in summer and dried in shade. The crude, the carbonized one or distillation form can be used.

Medicinal Properties They are sweet in flavor, cold in nature and attributive to the lung, heart and stomach meridians.

Actions Clear away heat and relieve toxin, disperse wind and heat, eliminate summer-heat by cooling.

Application

a. It can be used for seasonal febrile diseases or exterior syndrome of exogenous wind-heat type, no matter whether the heat of seasonal febile diseases is in weifen, qi-fen, or involving ying-fen and xue-fen. For early stage of seasonal febrile diseases, pathogenic factors being in weifen, or exterior syndrome of exogenous wind-heat type with fever and light aversion to cold, it is usually combined with Jingjie (Herba Schizonepetae), Bohe (Herba Menthae), Niubangzi (Fructus Arctii), etc., such as Yin Qiao San (Powder); for heat in qi-fen with high fever and excessive thirst, combined with Shigao (Gypsum Fibrosum), Zhimu (Rhizoma Anemarrhenae), etc.; for heat in ying-fen and xue-fen with eruptions, crimson and dry tongue, combined with Shengdihuang (Radix Rehmanniae) and Mudanpi (Cortex Moutan Radicis), etc..

b. It is used for carbuncle and pyocutaneous disease, and most suitable for yang syndrome of heat-toxin type, whether the carbuncle is ripe or not or the beginning of rupture. It can be used orally or the fresh is pounded for external application, or combined with Pugongying (Herba Taxaxaci), Yejuhua (Flos Chrysanthemi Indici), and Zihuadiding (Herba Violae), etc.; for intestinal abscess, combined with Yiyiren (Semen Coicis), Huangqin (Radix Scutellariae), Danggui (Radix Angelicae Sinensis), etc.; for lung abscess, combined with Yuxingcao (Herba Houttuyniae), Lugen (Rhizoma Phragmitis), and Taoren (Semen Persicae), etc. to clear away toxic heat and remove pus.

c. For blood dysentery of heat type or purulent hematochezia, it can be decocted alone into thick decoction or used together with Huangqin (Radix Scutellariae), Huanglian (Rhizoma Coptidis), Baitouweng (Radix Pulsatillae), etc. to strengthen the action of relieving dysentery.

d. For summer-heat syndrome with excessive thirst, sore throat, summer carbuncle, and heat rash, etc., it can be steamed with water into Jinyinhua Lu (Fluid) for oral or external use.

Usage and Dosage It is used in decoction for

oral use. The dosage for horse and cattle is 15 ~ 60g, for pig, sheep and goat is 9 ~ 15g.

14.6 Banlangen (Radix Isatidis)

The source is from the root of Isatis tinctoria L., family Cruciferae. The root is dug and collected in autumn, from which silt is removed, dried in the sun and the crude is used.

Medicinal Properties They are bitter in flavor, cold in nature and attributive to the heart and stomach meridians.

Actions Clear away heat and remove toxin, cool the blood and benefit the throat.

Application It is indicated for seasonal febrile diseases with fever, maculae and papules, swollen head due to infection, carbuncle, and sore toxin, etc. Since its action is stronger in clearing away heat and removing toxin, it is usually used for overabundance of heat and good at removing toxin and benefiting throat as well. For exogenous wind and heat with fever, or onset of seasonal febrile diseases with the above syndromes, it is usually combined with Jinyinhua (Flos Lonicerae), Lianqiao (Fructus Forsythiae), and Jingjie (Herba Schizonepetae), etc.; for the treatment of swollen head due to infection with red and swollen face and head, unsmooth throat, usually combined with Xuanshen (Radix Scrophulariae), Lianqiao (Fructus Forsythiae), and Niubangzi (Fructus Arctii), etc., such as Puji Xiaodu Yin (Decoction). It is now often used for infectious diseases of pathogenic factors.

Usage and Dosage It is used in decoction for oral use. The dosage for horse and cattle is 30 ~ 90g, for pig, sheep and goat is 15 ~ 30g.

Notes Contraindicated for deficiency and cold of the spleen and stomach.

14.7 Qinghao (Herba Artemisiae Chinghao)

The source is from the herb of *Artemisia annua* L., family Compositae. The medicinal material is produced in all parts of China, collected in autumn that the flowers are vigorious. The fresh can be used or dried in shade and cut into segments for being used.

Medicinal Properties They are bitter and purgent in flavor, cold in nature and attributive to the liver, gallbladder and kidney meridians.

Actions Clear away asthenic-heat and summer-heat and prevent recurrence of malaria.

Application

a. It is indicated for fever and hectic fever due to yin deficiency. Since it attributes to yin-fen and clear away asthenic-heat, and is accompanied with relieving and dispering action, especially effective in hectic fever due to yin deficiency with anhidrosis. It is usually combined with Yinchaihu (Radix Stellariae), Huhuanglian (Rhizoma Picrorrhizae), Zhimu (Rhizoma Anemarrhenae), and Biejia (Carapax Trionycis), etc., such as Qinggu San (Powder); for the late stage of seasonal febrile diseases with residual heat, fever at night subsiding in the morning, fever being gone but anhidrosis, or prolonged low fever after febrile disease, it is usually used together with Biejia (Carapax Trionycis), Mudanpi (Cortex Moutan Radicis), and Shengdihuang (Radix Rehmanniae), etc., such as Qinghao Biejia Tang (Decoction).

b. It is indicated for summer-heat syndrome or the syndrome with dampness and damp-warm disease with alternating steaming of damp and warm. For summer-heat with fever and perspiration, it can be used together with Xiguacuiyi (Pericarpium Citrulli), Jinyinhua (Flos Lonicerae), Heye (Folium Nelumbinis) and others that clear away summer-heat. For summer-heat syndrome with dampness or damp-warm syndrome, it is combined with Huoxiang (Herba Agastacheis Pogostemonis), Peilan (Herba Eupatorii), Huashi (Talcum) and others that resolve and remove dampness.

c. It is used for malaria with alternating episodes of chills and fever, can prevent recurrence of malaria and can be used alone or combined with

Huangqin (Radix Scutellariae), Huashi (Talcum), and Qingdai (Indigo Naturalis), etc. according to syndrome. Arteannuin is now extracted from the herb and prepared into tablet, injection, etc. for being used so as to improve therapeutic effects further.

Usage and Dosage It is used in decoction for oral use. The dosage for horse and cattle is 15 ~ 60g, for pig, sheep and goat is 6 ~ 15g. It is not suitable to be decocted for a long time, or the fresh can be pounded into juice for medication.

15 Cathartics

Any medicinal herb that can cause diarrhea or lubricate the large intestine, aid in moving the bowels and relieve constipation is known as cathartics.

They have the actions of aiding in moving the bowels, clearing away pathogenic heat and purging the retention of water. Therefore, they are mainly indicated to treat constipation, various interior excess syndromes due to fodder and water retention, and interior excess syndrome due to invasion by heat, etc..

In the light of the difference of the cathartic actions and adaptive conditions, this kind of herbs are classified into three subcategories, that is, purgatives, moistening purgatives and drastic purgatives, of which purgatives and drastic purgatives have the potent actions, especially the latter, while moistening purgatives have moderate actions of lubricating intestines.

Therefore purgatives and drastic purgatives should be used with caution or contraindicated in cases of chronic disease with asthenia of healthy qi, pregnancy, postpartum, and old and weak sick animals. So long as the disease is cured, their administration must be stopped and the overdosage must be avoided. For a case of interior sthenia and deficiency of healthy qi, they should be used together with tonics.

15.1 Dahuang (Radix et Rhizoma Rhei)

The source is from the root and rhizome of *Rheum palmatum* L., *R. tanguticum* Maxim. ex Balf. or *R. officinale* Baill., family Polygonaceae. *Rheum palimatum* L. and *R. tanguticum* Maxim. ex Balf. are also called north Radix et Rhizoma Rhei, which is mainly produced at the provinces of Qinghai, Gansu, etc., *R. officinale* Baill is called South Radix et Rhizoma Rhei, which is mainly produced in Sichuan Province. They are collected in the end of autumn or the next spring before the plants will sprout, the hair and coat being removed, cut into pieces or lumps and then dried in the sun. The crude, or the one steamed or stir-baked with wine, or the one charred can be used for medication.

Medicinal Properties They are bitter in flavor, cold in nature and attributive to the spleen, stomach, large intestine, liver and heart meridians.

Actions Remove stagnation by purgation, clear away heat and purge fire, cool the blood and stop bleeding, remove toxin and promote blood circulation to remove b lood stasis.

Application

a. It is used for stagnation in the passway of the intestine and obstruction of the bowels, an essential medicinal herb to treat the syndrome of stagnation with constipation and especially effective in treatment of constipation due to stagnation of heat. For constipation due to stagnation of sthenic-heat and abdominal pain against pressing, it is usually used together with Mangxiao (Natrii Sulfas), Zhishi (Fructus Aurantii Immaturus) and Houpo (Cortex Magnoliae Officinalis), such as Da Chengqi Tang (Decoction); if the constipation is accompanied with deficiency of qi and blood, Dangshen (Radix Codonopsitis pilosulae), Danggui (Radix Angelicae Sinensis), etc. are added to the above prescription again to support healthy qi and attack pathogenic factors, such as Huanglong Tang (Decoction); for yin deficiency due to stagnation of heat, Shengdihuang (Radix Rehmanniae) and Maimendong (Radix Ophiopogonis), etc. are combined with, such as Zengye Chengqi Tang (Decoction); for constipation due to insufficiency of spleen yang and stagnation of cold, Fuzi (Radix Aconiti Lateralis Praeparata), Ganjiang (Rhizoma Zingiberis), etc. can be also combined with, such as Wenpi Tang (Decoction).

b. It is used for dysentery due to stagnation. For early stage of dysentery of dampness-heat type with abdominal pain and tenesmus, it is usually combined with Huanglian (Rhizoma Coptidis), Muxiang (Radix Aucklandiae), etc. such as Shaoyao Tang (Decoction); for abdominal pain due to fodder retention and unsmooth dysentery, combined with Qingpi (Pericarpium Citri Reticulatae Viride), Muxiang (Radix Aucklandiae), etc.

c. For redness of eye, swollen gums, and oral ulcer, etc. caused by f laring of fire, it can be used whether there is constipation or not usually combined with Huanglian (Rhizoma Coptidis), Huangqin (Radix Scutellariae), and Niuhuang (Calculus Bovis), etc. such as Xiexin Tang (Decoction).

d. It is used for bleeding due to blood-heat manifested as haematemesis and epistaxis or bleeding due to blood stasis causing blood not to attribute to channels, since it can promote blood circulation as well as purge fire and stop bleeding without blood stasis. It can be used alone or together with other medicinal herbs for cooling blood and stopping bleeding.

e. Oral or external use is for pyocutaneous disease due to toxic heat, burn and scald. For carbuncle due to toxic heat, it is usually combined with Jinyinhua (Flos Lonicerae), Pugongying (Herba Taraxaci) and Lianqiao (Fructus Forsythiae), etc.; for intestinal abscess, usually combined with Mudanpi (Cortex Moutan Radicis), Taoren (Semen Persicae), etc., such as Dahuang Mudanpi Tang (Decoction); for burn and scald, can be used alone in powder form, or combined with Diyu (Radix Sanguisorbae) powder, being mixed with sesame oil for external application.

f. It is used for postpartum abdominal pain due to blood stasis, abdominal mass, trauma and syndrome of blood retention. For postpartum abdominal pain due to blood stasis, it can be used together with Taoren (Semen Persicae), etc., such as Xiayuxue Tang (Decoction), for trauma and blood stasis with swelling and pain, used together with Taoren (Semen Persicae), Honghua (Flos Carthami), etc., such as Fuyuan Huoxue Tang (Decoction).

Usage and dosage It is used in decoction for oral use. The dosage for horse and cattle is 18 ~ 45g, for pig, sheep and goat is 6 ~ 12g. Just right amount for external use, and the crude one with stronger purgative action is used for downward discharging. It is later added to decoction or soaked in boiling water for oral use, and is not decocted for a long time. That prepared with wine is suitable for blood stasis because of its better action of circulating blood. The carbonized form is usually used for bleeding syndrome.

15.2 Huomaren (Fructus Cannabis)

The source is from the fruits of *Cannabis sativa* L., family Moraceae. The medicinal material is produced in all parts of China, collected in autumn when the fruits are ripe, dried in the sun and broken. The crude form is used for medication.

Medicinal Properties They are sweet in flavor, bland in nature and attributive to the spleen, large and small intestine meridians.

Actions Moisten the intestine to relieve constipation.

Application

It is used for constipation due to dryness of intestine. For the elderly or pregnant animal that is weak or deficient in the body fluids and blood, it can be combined with Danggui (Radix Angelicae

Sinensis) and Shudihuang (Radix Rehmanniae Praeparata). For constipation due to pathogenic heat injuring yin or constitutional yin deficiency, it can be combined with Dahuang (Radix et Rhizoma Rhei) and Houpo (Cortex Magnoliae Officinalis), such as Maziren Wan (Pill).

Usage and Dosage It is used in decoction for oral use. The dosage for horse and cattle is 120 ~ 180g, for pig, sheep and goat is 12 ~ 30g.

15.3 Gansui (Radix Euphorbiae Kansui)

The source is from the root tuber of *Euphorbia kansui* TN Liou et TP Wang, family Euphorbiaceae. The medicinal material is mainly produced in the areas of Shaanxi, Shanxi, and Henan, etc. dug and collected in the end of autumn or at the early stage of spring, skined, then dried in the sun and prepared with vinegar for being used.

Medicinal Properties They are bitter and sweet in flavor and cold in nature, toxic and attributive to the lung, kidney and large intestine meridians.

Actions Purge the bowels to eliminate the retention of phlegm, induce diuresis to relieve edema.

Application

a. It can be used for edema, tympanites and localized fluids in the chest and hypochondrium but the healthy qi is not weakened since it has violent effect in purging and eliminating the retention of phlegm. It can be powdered singly for oral use or used together with Jing daji (Radix Euporbiae Pekinensis), Yuanhua (Flos Genkwa), etc., such as Shizao Tang (Decoction); for the syndrome resulting from water retention and heat accumulating in the thorax, it is used together with Dahuang (Radix et Rhizoma Rhei) and Mangxiao (Natrii Sulfas), such as Da Xianxiong Tang (Decoction). Now it is mostly used for ascites, exudative pleurisy and intestinal obstruction, etc.

b. For pyocutaneous disease, the powder is mixed with water for external application.

Usage and Dosage It is used in decoction for oral use. The dosage for horse and cattle is 6 ~ 15g, for pig, sheep and goat is 1 ~ 3g. It must be prepared with vinegar for oral use so as to decrease the toxicity.

Notes Contraindicated in pregnant animal or the weak, it is incompatible with Gancao (Radix Glycyrrhizae).

16 Dampness Removing Chinese Medicinal Herbs

Any medicines that remove dampness from the body and treat syndromes mused by dampness are considered as dampness removing Chinese medicinal herbs. Of them, those that take removing wind-dampness and relieving pains due to arthralgia as their main actions are considered as antirheumatic Chinese medicinal herbs. Those that display aromatic properties and have the action to dissolve dampness and to strengthen the function of the spleen fall into dampness resolving aromatic herbs. Those that mainly promote the removal of water and permeate the dampness are considered as medicinal herbs for promoting diuresis and resolving dampness.

They are indicated for wind-cold-damp Bi-syndrome, edema, stranguria, jaundice and syndrome of dampness obstruction in the middle energizer, etc.

Some of them are warm and dry in nature or their permeating and purging properties are violent, and easily consume yin and blood, therefore, they are used with caution in a case with insufficiency of yin, deficiency of blood or consumption of the body fluids.

16.1 Duhuo (Radix Angelicae Pubescentis)

The source is from the root of *Angelica pubescens* Maxim. f. biserrata shan et Yuan, family Umbelliferae. The medicinal material is mainly produced in Sichuan, Hubei, Anhui provinces, etc., dug and collected in the end of autumn and early stage of spring, dried in the sun, cut into pieces and the crude form is used.

Medicinal Properties They are pungent and bitter in flavor, warm in nature and attributive to the liver and bladder meridians.

Actions Expel wind and dampness, stop pain of Bi-syndrome and relieve exogenous pathogenic factors.

Application

a. It is used for Bi-syndrome of wind-cold-dampness type with pain, especially in the back with pain due to dampness. For general arthritis, it is combined with Oianghuo (Rhizoma et Radix Notopterygii) and Qinjiao (Radix Gentianae Macrophyllae); for prolonged Bi-syndrome accompanied with insufficiency of the liver and kidney, and deficiency of qi and blood, combined with Sangjisheng (Ramulus Taxilli), Duzhong (Cortex Eucommiae), Niuxi (Radix Achyranthis Bidentatae) and others that tonify the liver and kidney, such as Duhuo Jisheng Tang (Decoction).

b. For superficial syndrome of wind-cold type with dampness-syndrome, which is manifested as aversion to cold and fever, combined with Qianghuo (Rhizoma et Radix Notopterygii), Fangfeng (Radix Saposhnikoviae) and Jingjie (Herba Schizonepetae).

Usage and Dosage It is used in decoction for oral use. The dosage for horse and cattle is 15 ~ 45g, for pig, sheep and goat is 3 ~ 9g.

Explanation Duhuo (Radix Angelicae Pubescentis) and Qianghuo (Rhizoma et Radix Notopterygii) can all expel wind, dampness and cold, and relieve pain, treat rheumatic arthralgia with pain and superficial syndrome of wind-cold-damp type. But Qianghuo (Rhizoma et Radix Notopterygii) tends to treat superficial syndrome and has strong action of promoting sweating to expel cold, so it is often used to expel wind and dampness and treat Bi-syndrome of the upper part of the body; Duhuo (Radix Angelicae Pubescentis) has weaker action of eliminating exogenous pathogenic factors and works on the side of expelling wind and dampness, and of relieving pain of Bi-syndrome, so it is more on the side of treating pain of Bi-syndrome in the lower part of the body. Both Duhuo (Radix Angelicae Pubescentis) and Qianghuo (Rhizoma et Radix Notopterygii) can combine together to treat general joint pain so that better effect will be achieved.

16.2 Huoxiang (Herba Agastachis)

The source is from the aerial parts of a perennial herb, *Pogostemon cabin* (Blanco) Benth., family Labiatae. The medicinal material is mainly produced in Guangdong Province, cut and collected in summer and autumn, the fresh is cut into segments, or dried in shade for being used.

Medicinal Properties They are pungent in flavor, slightly warm in nature and attributive to the spleen and stomach m meridians.

Actions Eliminate dampness, clear away summer-heat and stop vomiting.

Application

a. For the obstruction of dampness and stagnation of qi in the middle energizer manifested, it is usually combined with Cangzhu (Rhizoma Atractylodis), Houpo (Cortex Magnoliae Officinalis), etc.

b. It is indicated for the early stage of summer-heat-dampness and dampness-warm syndromes of seasonal febrile disease. For affection of exogenous wind-cold in summer and retention of dampness and cold in the interior manifested as aversion to cold, fever, headache, fullness and pain in the chest and epigastrium, vomiting and diarrhea, it is usually combined with Zisuye (Folium Perillae Acutae), and Houpo (Cortex Magnoliae Officinalis), etc., such as Houxiang Zhengqi San (Powder). For the early Stage of damp-febrile disease, used together with Huangqin (Radix Scutellariae), Huashi (Talcum), and Yinchen (Herba Artemisiae Scopariae), etc., such as Ganlu Xiaodu Dan (Pill).

c. It is used for vomiting, and is an essential medicine for the treatment of dampness-retention syndrome with vomiting. It is usually combined with Banxia (Rhizoma Pinelliae), to which Huanglian (Rhizoma Coptidis), Zhuru (Caulis Bambusae in Taeniam), etc. are added again for vomiting being a bit on the heat side.

Usage and dosage It is used in decoction for oral use. The dosage for horse and cattle is 15 ~ 45g, for pig, sheep and goat is 6 ~ 12g. The amount of the fresh is doubled.

16.3 Fuling (Poria)

The source is from the sclerotium of *Poria cocos* (schw) Wolf, family Polyporaceae. The sclerotium mostly parasitizes on the root of Japanese red pine and Pinus massoniana Lamb., family Pinaceae. The medicinal material is mainly produced in the areas of Yunnan, Hubei, and Sichuan, etc., dug and collected from July to September, piled repeatedly, dried in the sun. The crude one is used.

Medicinal Properties They are sweet and bland in flavor, mild in nature, and attributive to the heart, spleen and kidney meridians.

Actions Promote diuresis to resolve dampness from the lower energizer, invigorate the spleen and tranquilize the mind.

Application

a. It is used for edema and dysuria. For all kinds of edema; for that caused by hypofunction of bladder-qi, it is combined with Guizhi (Ramulus Cinnamomi), Zhuling (Polyporus umbellatus), Baizhu (Rhizoma Atractylodis Alba), and Zexie (Rhizoma Alismatis), etc., such as Wuling San (Powder); for that with deficiency of qi, combined with Fangji (Radix Stephaniae Tetrandrae), Huangqi (Radix Astragali); for that with yang-deficency of the spleen and kidney, combined with Fuzi (Radix Aconiti Carmichaeli Praeparata), Ganjiang (Rhizoma Zingiberis), etc., such as Zhenwu Tang (Decoction).

b. For all syndromes of spleen-deficiency, especially effective in spleen-deficiency with dampness, it is combined with Dangshen (Radix Codonopsitis pilosulae) and Baizhu (Rhizoma Atractylodis Alba), such as SijunziTang (Decoction). For retention of fluid due to spleen-deficiency, combined with Guizhi (Ramulus Cinnamomi) and Baizhu (Rhizoma Atractylodis Alba), etc., such as Ling Gui Zhu Gan Tang (Decociton); for diarrhea due to spleen deficiency, combined with Shanyao (Rhizoma Dioscoreae), Baizhu (Rhizoma Atractylodis Alba), and Yiyiren (Semen Coicis), etc..

Usage and Dosage It is used in decoction for oral use. The dosage for horse and cattle is 18 ~ 60g, for pig, sheep and goat is 9 ~ 18g.

17 The Interior Warming Chinese Medicinal Herbs

Any medicinal herb that has the ability to warm up the interior and dispel cold to act on internal conditions is known as the interior warming Chinese medicinal herbs.

The categories of medicinal herbs are warm and heat in nature, and can expel interior cold and invigorate yang-qi. They are indicated for syndrome due to invasion of cold into the interior which affects yang-qi; or cold conditions in the interior due to yang weakness manifested as cold and pain in the chest and epigastric abdomen, vomiting, nausea, diarrhea and dysentery, chillness with cold limbs, and clear and profuse urine, etc.. Some of them can recuperate the depleted yang and rescue the sick animal from collapse, and are indicated for syndrome of yang exhaustion. In addition, some can warm the lung to resolve retention of phlegm, warm the liver to treat colic, lower the adverse-rising qi to relieve hiccup, and can be used for cough due to lung-cold, colic and vomiting due to stomach-cold, cold and pain in the epigastric abdomen.

17.1 Fuzi (Radix Aconiti Carmichalis Praeparata)

The source is from the root of *Aconitum carmichaeli* Debx. family Ranunculaceae. The medicinal material is mainly produced in the areas of Sichuan, Hubei, and Hunan, etc., collected from the last ten-day period of June to the first ten-day period of August, and they are soaked and processed into salty, black, white, bland and soaked aconite.

Medicinal Properties They are pungent and sweet in flavor, hot in nature, toxic and attributive to the heart, kidney and spleen meridians.

Actions Recuperate the depleted yang for resuscitation, supplement fire and strengthen yang, expel cold to relieve pain.

Application

a. It is used for yang exhausion syndrome. It is an essential medicine for the treatment of yang-exhaustion syndrome manifested as clammy perspiration, faint breath, cold clammy limbs, indistinct and faint pulse, it is usually used in combination with Ganjiang (Rhizoma Zingiberis) and Gancao (Radix Glycyrrhizae), such as Sini Tang (Decoction); for exhaustion of qi due to yang-deficiency, it can be combined with Dangshen (Radix Codonopsis pilosulae) to supplement qi to prevent prolapse of qi such as Shen Fu Tang (Decoction).

b. It is used to treat all syndromes of yang deficiency type. For insufficiency of kidney-yang with cold and painful waist and knees, and frequent micturition, it is usually combined with Rougui (Cortex Cinnamomi), Shanzhuyu (Fructus Corni), and Shudihuang (Rhizoma Rehmanniae Praeparatae), etc., such as Yougui Wan (Pill); for insufficiency of spleen-yang and kidney-yang and interior domination of cold and dampness with coldness and pain in epigastric abdomen, loss of appetite and diarrhea, it is combined with Dangshen (Radix Codonopsis pilosulae), Baizhu (Rhizoma Atractylodis Alba), and Ganjiang (Rhizoma Zingiberis), etc.; for yang deficiency resulting in edema, and dysuria, usually used together with Baizhu (Rhizoma Atractylodis Alba), Fuling (Poria) and Shengjiang (Rhizoma Zingiberis Recens), for exogenous affection due to yang-deficiency, combined with Mahuang (Herba Ephedrae), for spontaneous perspiration due to yang-deficiency, combined with Huangqi (Radix Astragali).

c. It is used to treat all pain syndromes of cold type. For arthralgia of wind-cold-dampness type, pain of general joints due to domination of cold and dampness, it is combined with Guizhi (Ramulus Cinnamomi), Baizhu (Rhizoma Atractylodis Alba) and Gancao (Radix Glycyrrhizae); for abdominal pain due to cold accumulation and qi stagnation, combined with Dingxiang (Flos Caryophylli), Gaoliangjiang (Rhizorna Alpiniae Officinarum), etc.

Usage and Dosage It is used in decoction for oral use. The dosage for horse and cattle is 15 ~ 30g, for pig, sheep and goat is 3 ~ 9g. And decocted at first for about a half to one hour until its narcotico-pungent taste is lost when its decoction is tasted by mouth.

Notes Contraindicated in a case with yin-deficiency leading to hyperactivity of yang and pregnant animal because of its pungent, hot, dry and violent properties. It is incompatible with Banxia (Rhizoma Pinelliae), Gualou (Fructus Trichosanthis), Beimu (Bulbus Fritillariae), Bailian (Radix Ampelopsis) and Baiji (Rhizoma Bletillae). It must be soaked for oral use, decocted for a long time, and over dosage and long administration must be avoided.

17.2 Rougui (Cortex Cinnamomi)

The source is from the bark of *Cinnamomum cassia* Presl, family Lauraceae. The medicinal material is mainly produced in the areas of Guangdong, Guangxi, and Yunnan, etc., barked in autumn and then the bark is dried in the shade, cut into pieces or powdered and the crude form is used.

Medicinal Properties They are pungent and sweet in flavor, hot in nature and attributive to the spleen, kidney and heart meridians.

Actions Supplement fire and strengthen yang, expel cold and alleviate pain, warm channels to promote the circulation of the blood.

Application

a. It is used for insufficiency syndromes of kidney-yang. For insufficiency of kidney-yang with chilliness, frequent micturition, and seminal emission, etc., it is usually used in combination with Fuzi (Radix Aconiti Carmichaeli Praeparata), Shudihuang (Rhizoma Rehmanniae Praeparatae), and Shanzhuyu (Fructus Corni), etc., such as Shenqi Wan (Pill); for syndrome with deficiency and weakness of kidney-yang and up-floating of deficiency-yang, manifested as flushed face, dyspnea due to deficiency, the herbal medicine can be used in combination with Fuzi (Radix Aconiti Carmichaeli Praeparata) so as to direct fire to its source.

b. It can be used to treat all pains due to accumulation of cold and stagnation of qi or stasis of the blood. For deficiency and coldness of the spleen and stomach marked by cold pain in epigastric abdomen, it can be used alone or together with Ganjiang (Rhizoma Zingiberis) and Gaoliangjiang (Rhizoma Alpiniae Officinarum); forcolic of cold type with abdominal pain, usually used in combination with Wuzhuyu (Fructus Evodiae), Xiaohuixiang (Fructus Foeniculi).

c. For pudendal carbuncle and pyocutaneous disease of deficienct-cold type that has not been healed for a long time, it is combined with Lujiaojiao (Golla Comus Cervi), Paojiang (Rhizoma Zingiberis Praeparata), and Mahuang (Herba Ephedrae), etc., such as Yanghe Tang (Decoction); or combined with Huangqi (Radix Astragali), Danggui (Radix Angelicae Sinensis), etc., such as Tuoli Huangqi San (Powder).

Usage and Dosage It is used in decoction for oral use. The dosage for horse and cattle is 15 ~ 30g, for pig, sheep and goat is 3 ~ 9g, and decocted later or soaked in water for taking.

Notes Contraindicated for bleeding due to blood-heat, and used with caution for hypermenorrhea and in pregnant animal.

Explanation Rougui (Cortex Cinnamomi) and Guizhi (Ramulus Cinnamomi) are from same plants, the barks of which are used as medicine that is called Rougui (Cortex Cinnamomi), the tender branches of which are used as medicine that is called Guizhi (Ramulus Cinnamomi). The both medicines can expel cold and strengthen yang, Guizhi (Ramulus Cinnamomi) dominates going up to be a bit on the side of expelling cold to relieve exterior syndrome while Rougui (Cortex Cinnamomi) dominates warming the interior to enter the lower energizer, being a bit on the side of warming kidney-yang.

18 Qi Regulating, Chinese Medicinal Herbs

Any medicinal herb that acts to regulate qi-activity, disperse qi stagnation and facilitate qi flow or that takes treatment of syndrome due to abnormal rising of qi as its essential action falls under this category.

Most of these medicinal herbs are fragrant and warm in nature, pungent and bitter in flavor. They all have, to different degrees, the ability to regulate qi, strengthen the spleen, promote qi circulation, stop pain, facilitate qi flow, reverse rebellious qi activities, soothe the liver, disperse the depressed qi or crack qi stagnation and disperse mass, etc.. They are indicated for obstruction of pulmonary qi, failure of dispersion of the liver and the imbalance of function between the spleen and stomach.

Most of them are pungent, warm and fragrant, and may easily damage qi and yin, so one must use them with caution in cases of qi and yin deficiency.

18.1 Qingpi (Pericarpium Citri Reticulatae Viride)

The source is from the immature fruit or its pericarp of *Citrus reticulata* Blanco, family Rutaceae. The fruit or its pericarp is collected during the period of May and June, dried in the sun. The crude or the one stir-baked with vinegar can be used.

Medicinal Properties They are bitter and pungent in flavor, warm in nature and attributive to the liver, gallbladder and stomach meridians.

Actions Soothe the liver to break qi stagnation, eliminate mass and relieve dyspepsia.

Application

a. It is indicated for liver-qi stagnation manifested as breast distending pain. For treatment of hypochondriac pain, it is usually used in combination with Chaihu (Radix Bupleuri), Yujin (Radix Curcumae), etc., for swelling and pain due to breast abscess, usually combined with Gualou (Fructus Trichosanthis), Jinyinhua (Flos Lonicerae), Pugongying (Herba Taraxaci), and Gancao (Radix Glycyrrhizae), etc., for hernia of cold type with abdominal pain, combined with Wuyao (Radix Linderae), Xiaohuixiang (Fructus Foeniculi), and Muxiang (Radix Aucklandiae), etc., to expel cold, regulate qi and relieve pain, such as Tiantai Wuyao San (Powder).

b. For distention of the epigastrium due to indigestion, it is usually combined with Shanzha (Fructus Crataegi), Maiya (Fructus Hordei Germinatus), and Shenqu (Massa Medicata Fermentata), etc., such as Qingpi Wan (Pill).

Usage and Dosage It is used in decoction for oral use. The dosage for horse and cattle is 15 ~ 30g, for pig, sheep and goat is 6 ~ 12g.

Notes Used with caution in the sick animal with qi deficiency since it has violent action to damage qi.

18.2 Xiangfu (Rhizoma Cyperi)

The source is from the rhizome of *Cyperus rotundus* L., family Cyperaceae. The medicinal material is mainly produced in the areas of Guangdong, Henan, and Sichuan, etc., collected during the period of December and October, dried in the sun, the hair being removed by burning. The crude or the one stir-baked with vinegar can be used.

Medicinal Properties They are pungent, slightly bitter and sweet in flavor, mild in nature and attributive to the liver and triple energizer meridians.

Actions Soothe the liver to regulate qi and regulate menstruation to relieve pain.

Application

a. It is used for the syndromes of stagnation of liver-qi with pain in the hypochondrium, distention and pain in epigastric abdomen and pain due to hernia, etc. For pain in the hypochondrium, it can be combined with Chaihu (Radix Bupleuri), Baishaoyao (Radix Paeoniae Alba), Zhike (Fructus Aurantii), etc., such as Chaihu Shugan San (Powder); for liver-qi invading the stomach and accumulation of cold and qi stagnation with cold and pain in epigastrium, combined with Gaoliangjiang (Rhizoma Alpiniae Officinarum) like Liang Fu Wan (Pill); for hernia of cold type with abdominal pain, combined with Xiaohuixiang (Fructus Foeniculi), Wuyao (Radix Linderae), etc..

b. For distension and pain of the breast, especially those caused by stagnation of liver-qi, it is usually combined with Danggui (Radix Angelicae Sinensis), Chuanxiong (Rhizoma Chuanxiong), Baishaoyao (Radix Paeoniae Alba), and Chaihu (Radix Bupleuri), etc., to soothe the liver to remove stagnation and regulate qi and blood.

Usage and Dosage It is used in decoction for oral use. The dosage for horse and cattle is 15 ~ 45g, for pig, sheep and goat is 9 ~ 15g.

19 Fodder Retention Relieving Chinese Medicinal Herbs

Any Chinese medicinal herb that takes promoting digestion and relieving dyspepsia, restoring transportation and transformation of the spleen and stomach as the dominant actions, and meanwhile can treat the syndrome of indigestion is considered as fodder retention relieving Chinese medicinal herbs.

In addition to digesting and relieving accumulation and stagnation, most of them have the actions of increasing appetite and harmonizing the stomach and spleen. They are indicated for indigested syndrome with epigastric and abdominal fullness, belching, acid regurgitation, nausea, vomiting, lack of appetite, and irregular bowel movement such as constipation or diarrhea, etc., which are all due to fodder accumulation.

When they are used, they are usually combined with qi regulating medicinal herbs to promote digestion and relieve dyspepsia, and meanwhile other appropriate medicinal herbs must be combined with separately according to different conditions, such as those for warming the spleen and stomach, clearing away heat or eliminating dampness.

For indigestion due to deficiency of the spleen and stomach, those for nourishing the spleen and stomach must be taken as dominants and the digestive herbs as assistants.

19.1 Shanzha (Fructus Crataegi)

The source is from the fruit of *Crataegus cuneata* Sieb. et Zucc., or C. pinnatifida Bunge, family Rosaceae. The medicinal material is mainly produced in the provinces of Henan, Jiangsu, Zhejiang, and Anhui, etc., collected in the end of autumn and the early stage of winter, dried in the sun. The raw or the stir-baked one can be used.

Medicinal Properties They are sour and sweet in flavor, slightly warm in nature, and attributive to the spleen, stomach and liver meridians.

Actions Digest fodder and dissipate the fodder accumulation, circulate the blood and disperse blood stasis.

Application

a. It is indicated for syndrome of indigestion manifested as diarrhea. For indigestion of fodder, it can be decocted alone or combined with Shenqu (Massa Medicata Fermentata), and Maiya (Fructus Hordei Germanatus), etc., for obvious fullness and pain in epigastrium, it is combined with Muxiang (Radix Aucklandiae), Zhike (Fructus Aurantii), etc., for abdominal pain with diarrhea due to improper diet, the charred one can be ground into powder, which is mixed with water for oral use.

b. It is used for blood-stasis syndrome manifested as postpartum abdominal pain with lochiostasis, hernia with bearing-down distending pain; for postpartum blood-stasis, it is usually combined with Danggui (Radix Angelicae Sinensis), Chuanxiong (Rhizoma Chuanxiong), and Yimucao (Herba Leonuri); for pain due to hernia, combined with Xiaohuixiang (Fructus Foeniculi), Juhe (Semen Citri Reticulatae), etc..

Usage and Dosage It is used in decoction for oral use. The dosage for horse and cattle is 18 ~ 60g, for pig, sheep and goat is 9 ~ 15g. The raw is a bit on the side of promoting digestion and relieving dyspepsia, and the charred one is a bit on the side of relieving diarrhea and dysentery.

19.2 Shenqu (Massa Medicata Fermentata)

This medicinal herb consists of the mixture of fermented powders of wheat flour, Semen Armeniacae Amarum, sprout of Semen Phaseoli, Herba Artmisiae Annuae, Fructus Xanthii, and Herba Polygoni Hydropiperis, etc.. The medicinal material is produced in all parts of China. The crude or the stir-baked is used for medication.

Medicinal Properties They are sweet and pungent in flavor and warm in nature and attributive to the spleen and stomach meridians.

Actions Promote digestion and harmonize the stomach.

Application For dyspepsia with epigastric distention, poor appetite and diarrhea, it is usually combined with Shanzha (Fructus Crataegi), Maiya (Fructus Hordei Germinatus), etc.. Furthermore, it has a little the action of relieving exterior syndrome, so it's more suitable for dyspepsia due to exogenous pathogenic factors.

Usage and Dosage It is used in decoction for oral use. The dosage for horse and cattle is 20 ~ 60g, for pig, sheep and goat is 10 ~ 25g.

19.3 Maiya (Fructus Hordei Germinatus)

The source is from the germinant fruit of *Hordeum vulgare* L., family Gramineae. The medicinal material is produced in all parts of China, the crude or the one stir-baked into yellow is used for medication.

Medicinal Properties They are sweet in flavor, bland in nature and attributive to the spleen, stomach and liver meridians.

Actions Digest the accumulated fodder and harmonize the middle energizer, and help stop lactation.

Application

It is used for fodder accumulation and indigestion, epigastric distress and abdominal distention, especially for helping digest starch fodder, it is usually combined with Shanzha (Fructus Crataegi), Shenqu (Massa Medicata Fermentata), and Jineijin (Endothelium Corneum Gigeriae Galli), etc., for poor transportation and transformation due to deficiency of the spleen and stomach, the tonics can be combined with the herb.

Usage and Dosage It is used in decoction for oral use. The dosage for horse and cattle is 20 ~ 60g, for pig, sheep and goat is 9 ~ 15g.

Notes It is not suitable to be used in breast-feeding period.

20 Hemostatic Chinese Medicinal Herbs

Any medicines that mainly stop various bleeding internally or externally are called hemostatic Chinese medicinal herbs. They separately have the different actions of cooling the blood and stopping bleeding, stopping bleeding by their astringent property, by removing obstructions and by warming channels, etc.. They are mainly indicated for bleeding from all parts of the body, such as hemoptysis, haematemesis, epistaxis, hematuria, bloody stools, and traumatic bleeding.

So, clinically, one must choose proper hemostatic Chinese medicinal herbs that are combined with each other, according to the bleeding causes and specific symptoms and signs, so as to enhance the effects. For example, for bleeding due to blood heat, blood cooling hemostatic herbs should be selected, and combined with those for clearing away heat and cooling the blood; for hyperactivity of yang due to deficiency of yin, those for nourishing yin and suppressing the sthenic yang should be used together; for bleeding due to blood stasis, those for removing blood stasis to stop bleeding should be mainly selected, and used together with those for circulating qi and the blood; for bleeding due to deficiency and cold, those for wanning yang, replenishing qi and strengthening the function of the spleen should be used together; for excessive bleeding followed by exhaustion of qi, those for supplementing primordial qi should be used so as to supplement qi. When one applies the blood cooling hemostatic and astringent hemostatic herbs, he/she must pay attention to whether there is blood stasis or not. If bleeding is mused by remaining of blood stasis, one must add herbs for circulating blood to remove blood stasis according to the conditions, and cannot simply use herbs for relieving bleeding and further cannot choose astringent hemostatic and blood cooling hemostatic herbs so as to prevent blood stasis.

20.1 Daji (Radix Cirsii daponici)

The source is from the root and herb of *Cirsium japonicum* DC., family Compositae. The medicinal material is mainly produced in all parts of China, collected in summer and autumn when the flower opens, and the root is dug in the end of autumn, dried in the sun, cut into segments. The crude or the stir-baked can be used.

Medicinal Properties They are sweet and bitter in flavor, cool in nature and attributive to the heart and liver meridians.

Actions Cool the blood and stop bleeding, remove blood stasis to treat carbuncle.

Application

a. For hemoptysis, epistaxis, and hematuria etc. of blood-heat type, it can be used alone or combined with Xiaoji (Herba Cephalanoplosis segeli), Cebaiye (Cacumen Biotae), etc..

b. For sore and carbuncle, it is effective whether it is used internally or externally, the fresh being best in action. Also it can be combined with Jinyinhua (Flos Lonicerae), Daqingye (Folium Isatidis), and Chishbaoyao (Radix Paeoniae Rubra), etc..

Usage and Dosage It is used in decoction for oral use. The dosage for horse and cattle is 18 ~ 60g, for pig, sheep and goat is 9 ~ 18g.

20.2 Huaihua (Flos Sophorae)

The source is from the flower bud of *Sophora japonica* L., family Leguminosae. The medicinal material is cultivated in most parts of China. The flower bud is collected from June to July, dried in the sun. The crude or stir-baked one can be used.

Medicinal Properties They are bitter in flavor, slightly cold in nature and attributive to the liver and large intestine meridians.

Actions Cool the blood and stop bleeding, clear away liver-heat and lower the fire.

Application

a. It is used for bleeding due to blood-heat, especially for bleeding from the lower part of the body. The carbonized form is usually used. For hematochezia, it is usually combined with Diyu (Radix Sanguisorbae); for hemoptysis and epistaxis, usually used together with Xianhemcao (Herba Agrimoniae), Baimaogen (Rhizoma Imperatae), and Cebaiye (Cacumen Biotae), etc..

b. For conjunctivitis due to flaming-up of the dominant liver-fire, it can be decocted as tea or used together with Xiakucao (Spica Prunellae), Juhua (Flos Chrysanthemi).

Usage and Dosage It is used in decoction for oral use. The dosage for horse and cattle is 15 ~ 30g, for pig, sheep and goat is 6 ~ 12g.

21 Chinese Medicinal Herbs for Invigorating the Blood and Removing Blood Stasis

Chinese medicinal herbs with the main action of facilitating blood flow and removing blood stasis are called medicinal herbs for invigorating the blood and removing blood stasis, in which those with strong power of promoting blood circulation and removing blood stasis are also called drastic medicinal herbs for removing blood stasis.

They are mostly pungent and bitter in flavor and mainly attributive to the liver and heart meridians and xuefen. They can facilitate blood flow and remove blood stasis so as to achieve the therapeutic effects of stimulating the menstrual flow, treating arthralgia and trauma and relieving swelling and pain. They are indicated for unsmooth blood flow and blood stasis manifested as fixed pain; for internal or external lumps, bleeding with darkish blood and purple-dark masses and ecchymosis on the skin and tongue.

There are various causes of forming blood stasis, therefore the causes should be found out by differentiation of syndromes and proper combination of medicinal herbs worked out when they are used. For example, if blood stasis is caused by cold obstruction and qi stagnation, the medicinal herbs warming the interior and dissipating cold should be combined; if by heat invading ying-fen and xue-fen and blood stasis in the interior, those clearing away heat and cooling the blood should be combined; if by obstructive pain due to wind-dampness, wind-dampness eliminating ones should be combined; if by trauma, those that can promote qi circulation and dredge collaterals should be combined; if by abdominal mass, those that resolve phlegm, soften hardness and disperse stagnation should be used together; if the blood stasis syndrome is accompanied by insufficiency of healthy qi, those that supplement qi should be used together.

Since qi and blood are closely related to each other, i. e. free flow of qi leads to free flow of the blood and stagnation of qi results in blood stasis and vice versa, they are usually used in combination with qi regulating medicinal herbs so as to enhance the action of invigorating the blood and removing stagnation.

21.1 Danshen (Radix Salviae Miltiorrhizae)

The source is from the root and rhizome of *Salvia miltiorrhiza* Bunge, family Labiatae. The medicinal herb is produced in most parts of China, mainly in areas of Hebei, Anhui, Jiangsu, and Sichuan, etc., dug and collected in spring and winter, dried in the sun and the crude or the stir-baked with wine is used.

Medicinal Properties Bitter in flavor, slightly cold in nature and attributive to the heart, pericardium and liver meridians.

Actions Promote blood circulation to remove blood stasis, regulate menstruation to relieve pain, cool the blood to relieve carbuncle, and clear away heat from the heart and tranquilize the mind.

Application

a. It is indicated for postpartum abdominal pain resulting from blood stasis, abdominal masses and pain of limbs. Since it is cold in nature, it is especially suitable for syndrome of blood stasis with heat. For animal's syndromes due to blood stasis, it is usually combined with Honghua (Flos Carthami), Taoren (Semen Persicae), and Yimucao (Herba Leonuri), etc., for the pains in the chest and epigastric abdomen resulting from blood stasis and stagnation of qi, combined with Tanxiang (Lignum Santali) and Sharen (Fructus Amomi), such as Danshen Yin (Decoction); for abdominal masses, combined with Sanleng (Rhizoma Sparganii), Ezhu (Rhizoma Zedoariae), Zelan (Herba Lycopi), and Biejia (Carapax Amydae), etc., for trauma with pain due to blood stasis, usually combined with Danggui (Radix Angelicae Sinensis), Honghua (Flos Carthami) and Chuanxiong (Rhizom Chuanxiong); for Bi-syndrome of wind-damp-heat type with red, swollen and painful joints, combined with Rendongteng (Caulis Lonicerae), Chishaoyao (Radix Paeoniae Rubra) and Qinjiao (Radix Gentianae Macrophyllae).

b. For carbuncle with swelling and pain, it is combined with medicinal herbs that clear away heat and eliminate toxic materials, which can help relieve carbuncle. For example, for breast abscess with swelling and pain, combined with Ruxiang (Olibanum), Jinyinhua (Flos Lonicerae) and Lianqiao (Flos Forsythiae), that is, Xiaoru Tang (Decoction).

c. It is used for pathogenic factors invading ying-fen and xue-fen in the seasonal febrile disease; it is usually combined with Shengdihuang (Radix Rehmanniae), Xuanshen (Radix Scrophulariae), and Zhuyexin (the centre of Folium Phyllostachydis Nigrae), such as Qingying Tang. (Decoction)

Usage and Dosage It is used in decoction for oral use. The dosage for horse and cattle is 15 ~ 45g, for pig, sheep and goat is 6 ~ 15g. The stir-baked one with wine can strengthen the action of promoting blood circulation.

Notes It is incompatible with Lilu (Rhizoma et Radix Veratra).

21.2 Honghua (Flos Carthami)

The source is from the flower of *Carthamus tinctorius L.*, *family Compositae. The medicinal material is mainly produced in the areas of Henan*, Hubei, Sichuan, and Zhejiang, etc.. The medicinal material is picked in summer when the flower changes from the yellow into the bright red, dried in shade and the crude is used.

Medicinal Properties They are pungent in flavor, warm in nature and attributive to the heart and liver meridians.

Actions Promote blood circulation to remove blood stasis, promote menstruation and alleviate pain.

Application

a. For blood-stasis syndrome with postpartum abdominal pain and mass, trauma and pain of joints, etc., it is usually combined with Taoren (Semen Persicae), Danggui (Radix Angelicae Sinensis), Chuanxiong (Rhizoma Chuanxiong), and Chishaoyao (Radix Paeoniae Rubra), etc..

b. For darkish skin eruptions due to stagnation of heat and blood stasis, it is combined with Danggui (Radix Angelicae Sinensis), Zicao (Radix Amebiaeseu Lithospermi), and Daqingye (Folium Isatidis), etc., such as Danggui Honghua Yin (Decoction).

Usage and Dosage It is used in decoction for oral use. The dosage for horse and cattle is 10 ~ 30g, for pig, sheep and goat is 3 ~ 6g.

22 Phlegm Resolving, Antitussive and Antiasthmatic Chinese Medicinal Herbs

Any Chinese medicinal herb that helps to dissolve and excrete phlegm out of the lung and takes treatment of phlegm-syndnome as its main action is considered as phlegm resolving Chinese medicinal herbs while any Chinese medicinal herb with the main actions of relieving cough and asthma is called antitussive and antiasthmatic Chinese medicinal herbs. Phlegm resolving herbs concurrently have the action of arresting cough and stopping asthma, and antitussive and antiasthmatic herbs can also concurrently have the effect of dissolving phlegm, so these two kinds of medicinal herbs fall into one chapter, known as phlegm resolving, antitussive and antiasthmatic Chinese medicinal herbs. Phlegm resolving Chinese medicinal herbs is mainly used for cough due to excessive phlegm or asthma due to phlegm-retention unsmooths coughing sputum, etc. Antitussive and antiasthmatic Chinese medicinal herbs are mainly indicated for cough and asthma caused by internal injury and external affection. Syndromes such as goiter, scrofula, yin furuncle and suppurative tissue disease are closely related to phlegm in pathogenesis, therefore, they are also treated by phlegm resolving Chinese medicinal herbs.

External affection and internal damage can all cause cough and excessive sputum, so clinically, besides choosing these kinds of medicinal herbs according to each property of them, one must combine them appropriately with each other according to the pathogenic factors, signs and symptoms. For example, for cough or asthma accompained by exterior syndrome, diaphoretic Chinese medicinal herbs must be combined; for that with internal heat, heat clearing herbs must be combined; for that with internal cold, the interior warming Chinese herbs must be combined; and for that due to asthenia of viscera, tonics must be combined.

22.1 Banxia (Rhizoma Pinelliae)

The source is from rhizome of *Pinellia ternata* (Thunb.) Breit, family Araceae. The medicinal material is produced in all parts of China and mostly in the reaches of the Yangtze River, dug and collected in summer and autumn, prepared by being dried in the sun after the cortex is removed and the fibrous root is got rid of. The crude or the one prepared by being soaked in ginger juice and alum water is used.

Medicinal Properties They are pungent in flavor, warm in nature, toxic and attributive to the spleen, stomach and lung meridians.

Actions Dry dampness and eliminate phlegm, lower the adverse rising qi to stop vomiting, disperse stagnation and lumps and externally disperse swelling and relieve pains.

Application

a. It is used for dampness or cold phlegm syndrome and is an essential medicinal herb to treat dampnessphlegm syndrome. For cough and asthma with plenty of sputum caused by accumulation of phlegm and dampness in the lung, it is usually combined with Jupi (Pericarpium Citri Tangerinae) and Fuling (Poria), such as Erchen Tang (Decoction); for that accompanied by cold, profuse and thin sputum, Xixin (Herba Asari Cum Radice) and Ganjiang (Rhizoma Zingiberis) may be added to the above combination; for that with fever, thick and yellowish sputum, it must be combined with Huangqin (Radix Scutellariae), Zhimu (Rhizoma Anemarrhenae) and Gualou (Fructus Trichosanthis).

b. For vomiting due to the adverse-rising of stomach-qi, it is usually combined with Shengjiang (Rhizoma Zingiberis Recens).

c. It is used for goiter, subcutaneous nodule, large carbuncle and mammary sores. For goiter and subcutaneous nodule, it is used together with Kunbu (Thallus Laminariae seu Eckloniae), Zhebeimu (Bulbus Fritillariae Thunbergii), Haizhao (Sargassum) and others that soften hardness and disperse lumps; for large carbuncle, its crude powder mixed with egg white is applied on the wounded area, which can also be used to treat poisonous snakebite.

Usage and Dosage It is used in decoction for oral use. The dosage for horse and cattle is 15 ~ 45g, for pig, sheep and goat is 3 ~ 10g.

Notes Since it is warm and dry in nature, it is contraindicated or used with caution for dry cough due to yin deficiency, hemorrhagic diseases and heat-phlegm. It is incompatible with Wutou (Rhizoma Aconiti).

Explanation The prepared Banxia is usually taken orally, that is, Jiangbanxia, Rhizoma Pinelliae prepared with ginger and alum, Fabanxia, that prepared with Gancao (Radix Glycyrrhizae) and lime, and Zhulihanxia, that prepared with Zhuli (Succus Bambusae) and alum. Jiangbanxia is effective in lowering the adverse-rising qi to stop vomiting, so it is usually used for vomiting; Fabanxia is effective in eliminating dampness but its warm nature is weaker, so it is usually used for syndrome of dampness-phlegm; Zhulibanxia is cool in nature and can eliminate phlegm-heat, so it is mainly used for phlegm of heat and wind types.

22.2 Xingren (Semen Armeniacae Amarae)

The source is from the mature seed of *Prunus armeniaca* L., or *P. armeniaca* L. var. *ansu Maxim*, family Rosaceae. The medicinal material is produced in the areas of Northeast, North and Northwest China, Xinjiang, and the reaches of the Yangtze River of China, collected in summer when the fruit is ripe and dried in the sun after it is shelled. The crude one is used.

Medicinal Properties They are bitter in flavor, slightly warm in nature, mildly toxic and attributive to the lung and large intestine meridians.

Actions Relieve cough and dyspnea, moisten the intestine and relax the bowels.

Application

a. It is used for cough and dyspnea, is an essential medicine of treating them and can be used to treat all kinds syndromes of cough and dyspnea in combination with other herbs according to syndrome. For cough and dyspnea of wind-cold type, it can be used in combination with Mahuang (Herba Ephedrae) and Gancao (Radix Glycyrrhizae), such as San'ao Tang (Decoction); for cough of wind-heat type, used with Sangye (Folium Mori), Juhua (Flos Chrysanthemi), etc., for that of dryness-heat type, used together with Sangye (Folium Mori), Beimu (Bulbus Fritillariae), and Shashen (Radix Glehniae), etc., for that of lung-heat type, used together with Mahuang (Herba Ephedrae), Shengshigao (Gypsum Fibrosum, unprepared), etc., such as Ma Xing Shi GanTang (Decoction).

b. For constipation due to dryness of intestine, it is usually combined with Homerun (Semen Cannabis), Dangue (Radix Angelica Sinensis), Zhike (Fructus Aurantii) and others, such as Runchang Wan (Pill).

Usage and Dosage It is used in decoction for oral use. The dosage for horse and cattle is 15 ~ 30g, for pig, sheep and goat is 3 ~ 12g.

Notes Since it is mildly toxic, overdosage should be avoided.

23 Chinese Medicinal Herbs for Calmins the Liver to Stop Endogenous Wind

Any medicinal herb that has the actions of calming the liver and suppressing the hyperactive yang or calming the liver-wind, and is indicated for syndrome of liver-yang rising or liver-wind stirring is called Chinese medicinal herbs for calming the liver to stop endogenous wind.

They are mainly attributive to the liver meridian and mostly include animal and mineral medicines such as shell and insect types. They are mainly used for convulsion and epilepsy due to liver-wind stirring inside.

When they are used, one should differently combine them with other medicinal herbs according to the different causes and symptoms. For syndrome of liver-wind stirring inside, which is usually caused by excessive fire-heat and hyperactive liver-yang accompanied by liver-heat, medicinal herbs for clearing away heat, reducing fire and purging liver-heat must be combined. For yin-deficiency with insufficiency of the blood and failure of the liver to nourish resulting in liver-wind stirring inside and liver-yang rising, those for nourishing the kidney and yin or replenishing the blood should be combined.

They should be applied distinctively since most of them are a bit on cold or cool side while some of them are a bit on warm and dry side. For chronic convulsion due to spleen-deficiency, cold or cool medicinal herbs are not suitable, but for yin-deficiency with insufficiency of the blood, the warm-dry medicinal herbs should be used with caution.

Muli (Concha Ostreae)

The source is from the shell of *Ostrea gigas* Thunberg, *O. talienwhanensis* Crosse or *O. rivularis* Gould, family Ostreidae. The medicinal material is produced in the areas along the sea of China, collected in winter and spring and dried in the sun after the shell (meat is removed) is cleaned. The crude or calcined one is used after it is broken.

Medicinal Properties They are salty in flavor, slightly cold in nature and attributive to the liver and kidney meridians.

Actions Calm the liver, suppress the hyperactive yang, soften hardness and disperse the stagnated mass, and astringe, invigorate the kidney to preserve essence.

Application

a. For febrile disease injuring the yin and causing endogenous liver wind manifested as convulsion of limbs, usually combined with Guiban (Ptastrum Testudinis), Biejia (Carapax Amydae), and Dihuang (Radix Rehmanniae), etc., such as Da Dingfeng Zhu (Bolus).

b. For stagnation of phlegm and fire manifested as scrofula, subcutaneous nodule and abdominal mass, it is usually used together with Zhebeimu (Bulbus Fritillariae Thunbergii) and Xuanshen (Radix Scrophulariae), such as Xiaoluo Wan (Bolus). Recently, it is used clinically to treat mass of the hypochondrium, it is usually combined with Danshen (Radix Salviae Miltiorrhizae), Zelan (Herba Lycopi) and Biejia (Carapax Amydae), etc..

c. It is used for all slippery and depletive syndromes. For spontaneous perspiration and night sweat, it may be combined with Huangqi (Radix Astragali) and Mahuanggen (Radix Ephedrae).

Usage and Dosage It is used in decoction for oral use. The dosage for horse and cattle is 18 ~ 40g, for pig, sheep and goat is 6 ~ 12g. It is decocted earlier. The calcined one is used for astringing, invigorating the kidney to preserve essence and controlling acid and the crude one is used for other symptoms.

24 Restoratives

All medicinal herbs for replenishing qi, blood, nourishing yin and yang, improving the functions of internal organs and body immunity, and relieving the various kinds of symptoms of weakness are defined as restoratives, also known as restoratives for reinforcing asthenia or tonics.

Though they complicatedly manifest themselves in clinic, basically, deficiency syndromes are summarized as deficiency of qi, yang, blood, or yin. So correspondingly, restoratives are classified into four categories, restoratives for invigorating qi, reinforcing yang, nourishing yin, and nourishing blood. There are interdependence relationships to exist among qi, blood, yin and yang, so deficiency of yang is often accompanied by deficiency of qi that is inclined to resulting in deficiency of yang, deficiency of qi and deficiency of yang both indicating a hypofunctioning state of human body; deficiency of yin is mostly accompanied by deficiency of blood that will easily lead to deficiency of yin, these two deficiencies indicating a consumption of essence, blood and body fluids. In this way, Chinese medicinal herbs for tonifying qi, yang, blood and yin are usually used according to the mutual promotion principle. In a case with deficiency of qi and blood, or both yin and yang, the restoratives for nourishing both qi and blood, or both yin and yang should be used simultaneously.

Restoratives do not fit for the case with incompletely expelled pathogenic factors, and unshortage of healthy qi so as to prevent pathogenic factors from being removed but make the disease severe. However, in a case with deficiency of healthy qi and uncleared pathogenic factors, the choice of restoratives is recommendable for the purpose of strengthening the body resistance against disease.

Restoratives are usually taken in a long course, and are difficult to digest, so they are often combined with some herbs with actions of strengthening the stomach and spleen to achieve a better effect. They should be used strictly instead of loosely in case some other symptoms arise.

24.1 Dangshen (Radix Codonopsitis Pilosulate)

The medicinal material is from the root of *Codonopsis pilosula* (Franch.) Nannf., and other several species of the same catalogue, family Campanulaceae. It is classified into the wild and cultivated one according to sources, and mostly produced in Shanxi, Shaanxi and Gansu, and harvested in autumn, dried and cut into segments. It is used crudely.

Medicinal Properties They are sweet in flavor, warm in nature, and attributive to the spleen and lung meridians.

Actions Invigorate the stomach and spleen, and benefit qi, promote the production of the body fluids and nourish the blood.

Application

a. It is suitable for deficiency of the spleen with fatigue, loss of appetite, loose stool and syndrome with deficiency of qi and weakness due to various causes. It is usually used together with Baizhu (Rhizoma Atractylodis Alba), Fuling (Poria) and Gancao (Radix Glycyrrhizae).

b. It is suitable for insufficiency of lung-qi with shortness of breath, or cough. It is usually used together with Huangqi (Radix Astragali), Wuweizi (Fructus Schisandrae), such as Bufei Tang (Decoction).

c. It is used for febrile diseases resulting in consumption of both quiz and the body fluids with shortness of breath and thirst. It is usually used together with Maimendong (Radix Ophiopogonis), Wuweizi (Fructus Schisandrae), etc..

Usage and Dosage It is used in decoction for oral use. The dosage for horse and cattle is 18 ~ 60g, for pig, sheep and goat is 6 ~ 12g.

Notes Contraindicated for heat-syndrome, syndrome of deficiency of yin and hyperactivity of yang. It is incompatible with Lilu (Rhizoma et Radix Veratri).

24.2 Huangqi (Radix Astragali)

The source is from the root of *Astragalus membranaceus* (Fisch) Bunge var. *mongholicus* (Bunge) Hsiao and *A. membranaceus* (Fisch.) Bunge, family Leguminosae. It mainly grows in Inner Mongolia, Shanxi, Gansu and Heilongjiang, etc., harvested in spring and autumn, sliced and dried with removal of head and fine roots. It is used crudely or roasted with honey for use.

Medicinal Properties They are sweet in flavor, slightly warm in nature, and attributive to the spleen and lung meridians.

Actions Replenish qi to invigorate yang; benefit the lung to strengthen the body; promote diuresis and relieve edema; relieve skin infection and promote tissue regeneration.

Application

a. It is used for the syndrome with qi-deficiency of the spleen and lung, and visceroptosis with hyposplenic qi manifested as fatigue and weakness due to prolonged illness, it is used together with Dangshen (Radix Codonopsits Pilosulae), such as Shen Qi Gao (Soft extract); for that accompanied by deficiency of yang manifested as chill, fatigue and polyhidrosis, it is used with Fuzi (Radix Aconiti Lateralis Praeparata), such as Qi Fu Gao (Soft extract). For deficiency of both qi and blood, it is used together with Danggui (Radix Angelicae Sinensis). For deficiency of spleen-qi manifested as poor appetite, loose stool or diarrhea, it is used together with Baizhu (Rhizoma Atractylodis Alba), such as Qi Zhu Gao (Soft extract). For descending of spleen-yang, visceroptosis with hyposplenic qi, prolaps of uterus, and prolaps of rectum due to chronic diarrhea, it is used with Dangshen (Radix Codonopsits Pilosulae), Baizhu (Rhizoma Atractylodis Alba), Shengma (Rhizoma Cimicifugae), and Chaihu (Radix Bupleuri), etc., such as Buzhong Yiqi Tang (Decoction). For deficiency of qi resulting in inability to govern the blood manifested as bloody stool, it may be used together with Renshen (Radix Ginseng), Suanzaoren (Semen Ziziphi Spinosae) and Guiyuanrou (Arillus Longan), etc., such as Guipi Tang (Decoction).

b. It is applied for cough due to lung deficiency, superficies-asthenia with profuse sweating, or night sweating. For deficiency of lung qi with cough or shortness of breath or dyspnea, it is frequently used with Ziwan (Radix Asteris), Wuweizi (Fructus Schisandrae), etc., for superficies-asthenia with profuse sweating and suspectility to common cold, it is used with Baizhu (Rhizoma Atractylodis Alba) and Fangfeng (Radix Saposhnikoviae), such as Yupingfeng San (Powder); for night sweating, it can be used with Shengdihuang (Radix Rehmannia), Huangbai (Cortex Phellodendri), etc., such as Danggui Liuhuang Tang (Decoction).

c. It is applied for maltransformation of water and dampness due to deficiency of qi with facial edema, or oliguria. And it is used together every time with Fangji (Radix Stephaniae Tetrandrae), Baizhu (Rhizoma Atractylodis Alba), etc., such as Fangii Huangqi Tang (Decoction). It is remarkably effective for edema from chronic nephritis with presence of protein in urine for a long time.

d. It is applied for unruptured carbuncle or

ruptured but unhealed one due to deficiency of qi and blood. For unruptured carbuncle, it is often used together with Danggui (Radix Angelicae Sinensis), Zaojiaoci (Spina Gleditsiae), etc., such as Tounong San (Powder). For ruptured carbuncle with watery pus and no tendency to healing, it can be used with Danggui (Radix Angelicae Sinensis), Renshen (Radix Ginseng), and Rougui (Cortex Cinnamomi), etc., such as Shiquang Dabu Tang (Decoction).

Usage and Dosage It is used in decoction for oral use. The dosage for horse and cattle is 15 ~ 60g, for pig, sheep and goat is 6 ~ 10g.

Notes Contraindicated in the case with superficial asthenia syndrome with excessive pathogenic factors, stagnation of qi and dampness, indigestion, hyperactivity of yang due to deficient yin, onset of carbuncle at the early stage or one ruptured but unhealed for a long time with excessive heat-toxin.

24.3 Bajitian (Radix Morindae Officinalis)

The source is from the root of *Morinda officinalis* How, family Rubiceae. The medicinal material is mainly produced in Guangdong, Guangxi, and Fujian, etc., available all year round. The one after being dried, steamed, and its wood centre removed is named as Bajirou. It is cut into fragments and dried, used crudely or soaked in a salty solution.

Medicinal Properties It is sweet and pungent in flavor, slightly warm in nature, and attributive to the kidney and liver meridians.

Actions Nourish kidney-yang, strengthen tendons and bones, and eliminate wind and dampness.

Application

a. It is applied for deficiency of kidney-yang manifested as impotence, sterility, or cold pain in lower abdomen. In the case of impotence and sterility, it is used with Dangshen (Radix Codonpsitis Pilosulae), Shanyao (Rhizoma Dioscoreae), and Roucongrong (Herba Cistanchis), etc., for cold pain in lower abdomen, used with Gaoliangjiang (Rhizoma Alpiniae Officinarum), Rougui (Cortex Cinnamomi), Wuzhuyu (Fructus Evodiae), etc., such as Baji Wan (Pill).

b. For deficiency of kidney-yang accompanied by lumbago or fatigue, it is used with Bixie (Rhizoma Dioscoreae), Duzhong (Cortex Eucommiae), etc., such as Jingang Wan (Pill).

Usage and Dosage 3 ~ 9 g is used in decoction.

Notes It is not suitable for the case with deficiency of yin causing hyperactivity of yang or for that with damp heat.

24.4 Danggui (Radix Angelicae Sinensis)

The source is from the root of *Angelica sinensis* (Oliv.) Diels, family Umbelliferae. The medicinal material is mainly produced in the areas of Gansu, Shaanxi, etc., and that produced in Min County of Gansu is the best in quality. The root is dug at the end of autumn, whose fibrous root is removed, dried with mild fire, cut into pieces and used crudely or stir-baked with wine for medication.

Medicinal Properties They are sweet and pungent in flavor, warm in nature and attributive to the liver, heart and spleen meridians.

Actions Nourish the blood, promote blood circulation, relieve pain and moisten the intestine.

Application

a. It is indicated for blood deficiency syndromes such as sallow complexion, pale lips, and whitish nails, it is usually used together with Huangqi (Radix Astragali), such as Buxue Tang (Decoction).

b. It is used for various pains due to blood stasis or Bi-syndrome with pain due to wind-damp. For pains due to blood stasis in limbs, it may be combined with Danshen (Radix Salviae Miltiorrhizae), Ruxiang (Olibanum), and Moyao (Myrrha), etc., such as Huoluo Xiaoling Dan (Bolus); for swelling and pains due to trauma, usually combined with Dahuang (Radix et Rhizoma Rhei), Taoren (Semen Persicae), and Honghua (Flos Carthami), etc., such as Fuyuan Huoxue Tang (Decoction); for pains in shoulder and arms due to wind-dampness, usually combined with Qianghuo (Rhizoma et Notopterygii), Fangfeng (Radix Saposhnikoviae), and Jianghuang (Rhizoma Curcumae Longae), etc., such as Juanbi Tang (Decoction). It is also used together with Guizhi (Ramulus Cinnamomi), Baishaoyao (Radix Paeoniae Alba) and Shengjiang (Rhizoma Zingiberis Recens) in treating abdominal pain due to deficiency and cold of the middle-energizer.

c. It is used for large carbuncle and pyocutaneous disease. For the early stage of those with redness, swelling and pain, it is combined with Jinyinhua (Flos Lonicerae), Chishaoyao (Radix Paeoniae Rubra), and Tianhuafen (Radix Trichosanthis), etc., such as Xianfang Huoming Yin (Decoction); for the middle age of that with pus before rupture, it should be combined with Zaojiaoci (Spina Gleditsiae), Huangqi (Radix Astragali), and Shudihuang (Radix Rehmanniae Praeparata), etc., such as Tounong San (Powder); for that unhealed after rupture due to deficiency of qi and blood, and with running pus, it is usually combined with Dangshen (Radix Codonpsitis Pilosulae), Huangqi (Radix Astragali), and Shudihuang (Radix Rehmanniae Praeparata), etc., such as Shiquan Dabu Tang (Decoction).

d. It is used in sick animals with constipation due to blood deficiency and dryness of the intestine, especially for those with weakness due to prolonged illness, the aged and animal with postpartum blood deficiency and constipation due to insufficiency of the body fluids, it is usually combined with Shengheshouwu (Radix Polygoni Multifiori, unprepared), Huomaren (Fructus Cannabis), Roucongrong (Herba Cistanches), etc..

In addition, 5% of Danggui Injection is injected in the acupoints such as Feishu (BL 13) and Tanzhong (CV17) to treat chronic bronchitis.

Usage and Dosage It is used in decoction for oral use. The dosage for horse and cattle is 15 ~ 60g, for pig, sheep and goat is 10 ~ 15g.

24.5 Shudihuang (Radix Rehmanniae Praeparata)

The source is from the root of Rehmannia glutinosa Libosch., family Scrophulariaceae. The medicinal material is mainly produced in the areas of Henan, Hebei, and Inner Mongolia, etc.. After dried, it is prepared by steaming with wine, Sharen (Fructus Amomi) and Jupi (Pericarpium Citri Tangerinae), and drying itself repeatedly until it becomes black internally and externally, has the quality of softness and stickiness. It is cut into slices for medication.

Medicinal Properties They are sweet in flavor, slightly warm in nature and attributive to the liver and kidney meridians.

Actions Enrich the blood and nourish yin, and supplement the essence.

Application

a. It is indicated for blood deficiency manifested as sallow complexion; it is usually combined with Danggui (Radix Angelicae Sinensis), Baishaoyao (Radix Paeoniae Alba), Chuanxiong (Rhizoma Chuanxiong), etc., such as Siwu Tang (Decoction).

b. For deficiency of liver-yin and kidney-yin is usually used together with Shania (Rhizome Discourage), Shanzhuyu (Fructus Corni), etc., such as Liuwei Dihuang Wan (Bolus). For insufficiency of kidney yin resulting in preponderant deficiency-fire with low fever and obvious night sweat, used together with Zhimu (Rhizoma Anemarrhenae), Huangbai (Cortex Phellodendri), Guiban (Plastrum Testudinis) and others that can nourish yin to clear fire, such as Da Buyin Wan (Bolus).

c. For insufficiency of the essence and blood is usually used with Heshouwu (Radix Polygoni Muttifiori), Niizhenzi (Fructus Ligustri Lucidi),

Hanliancao (Herba Ecliptae), and Shanzhuyu (Fructus Corni), etc.

Usage and Dosage It is used in decoction for oral use. The dosage for horse and cattle is 200 ~ 400g, for pig, sheep and goat is 50 ~ 100g.

24.6 Maimendong (Radix Ophiopogonis)

The source is from the root tuber of Ophiopogon japonicus (Thunb.) Ker-Gawl., family Liliaceae. The medicinal material is produced in all parts of China. It is dug and collected in summer, the fibrous roots being removed, cleaned, dried and used crudely.

Medicinal Properties They are sweet and slightly bitter in flavor, slightly cold in nature and attributive to the heart, lung and stomach meridians.

Actions Nourish yin and moisten the lung, benefit the stomach and regenerate the body fluids.

Application

a. The medicinal herb is indicated for dry cough and sticky sputum due to lung-dryness and cough and hemoptysis due to asthenia of viscera. In treating pathogenic warm-dry invading the lung resulting in dry cough, it is usually combined with Sangye (Folium Mori), Xingren (Semen Armeniacae Amarae) and Shengshigao (Gypsum Fibrosum), such as Qingzhao Jiufei Tang (Decoction); in treating insufficiency of lung-yin manifested as cough and hemoptysis due to asthenia of viscera, usually combined with Tianmendong (Radix Asparagi), such as Erdong Gao (Soft extract).

b. It is used for yin-syndromes of the stomach and intestine. For insufficiency of stomach-yin with dry tongue and thirst, it is usually combined with Shashen (Radix Adenophorae), Shengdihuang (Radix Rehmanniae), Yuzhu (Rhizoma Polygonati Odorati), etc., such as Yiwei Tang (Decoction); for constipation due to dryness of the intestine, may be combined with Shengdihuang (Radix Rehmannia) and Xuanshen (Radix Scrophulariae) to compose a formula that is Zengye Tang (Decoction).

Usage and Dosage It is used in decoction for oral use. The dosage for horse and cattle is 15 ~ 45g, for pig, sheep and goat is 2 ~ 18g.

Notes It is not suitable for cough due to exogenous wind and cold, and obstruction of phlegm-dampness in the lung.

25 Astringent Chinese Medicinal Herbs

Chinese medicinal herbs with the main actions of inducing astringency are Astringent Chinese Medicinal Herbs.

Astringents have a "closing" effect on the body to prevent further discharge from the body, therefore, they are not suitable in the case with exogenous pathogenic factors that are not removed or interior stagnation of any dampness and domination of sthenia pathogenic factors.

25.1 Fuxiaomai (Fructus Tritici Levis)

The source is from the blighted caryopsis of Triticum aestivum L., family Gramineae. The medicinal material is produced all over China. The wheat is washed; the floated one is picked out, and then dried in the sun. The raw or stir-baked one is used for medication.

Medicinal Properties They are sweet in flavor, cool in nature, and attributive to the heart meridian.

Actions Replenish qi, clear away heat and arrest sweating.

Application

a. It is used for spontaneous perspiration and night sweat. It can be used alone to treat night sweat and persistent deficient sweating. For persistent spontaneous perspiration due to weakness of the body, it is usually used together with Muli (Concha Ostreae), Mahuanggen (Radix Ephedrae) and Huangqi (Radix Astragali) to compose a formula, that is, Muli San (Powder).

b. In treating hectic fever and overstrain-fever, it is often used together with the medicinal herbs that nourish yin and clear asthenia-heat such as Shengdihuang (Radix Rehmanniae, unprepared), Maimendong (Radix Ophiopogonis), and Digupi (Cortex Lycii Radicis).

Usage and Dosage It is used in decoction for oral use. The dosage for horse and cattle is 30 ~ 120g, for pig, sheep and goat is 12 ~ 18g. Or stir-fried until burnt and then ground into powder for oral use.

Lecture Three

Science of Prescriptions of the Traditional Chinese Veterinary Materia Medica

Science of Prescriptions is a subject dealing with treatment and the theories of compatibility of prescriptions as well as the clinical application. It has a close relationship with all other clinical branches, linking up the basic theories with clinical practices.

A prescription is composed of selected drugs and suitable doses based on syndrome differentiation for etiology and the composition of therapies in accordance with the principle of formulating a prescription. It serves as a chief means of treating diseases clinically.

Likewise, the application of prescriptions also went through the process from experience to theory. At first people chose prescriptions and drugs in the light of syndrome only. But with the increasing number of prescriptions and long-term clinical practices, they gradually summarized some regularity of functions of the prescriptions, thus the therapeutic theory came into being. By the therapeutic method is meant a guiding principle of treatment for different syndromes. It is established on the basis of differentiating syndromes for etiology and selecting treatment. For instance, heat-clearing therapy is used for treating interior-heat syndrome; the interior-warming therapy for treating interior-cold syndrome, and the activating blood to resolve stasis therapy is used for the syndrome of blood stasis. Once the therapeutic theory is developed, it will be looked upon as an important principle and theoretical basis to guide people in the application of extant prescriptions or creation of new ones. If a disease is marked by aversion to cold, fever, headache and aching body, anhidrosis and asthma, thin and whitish tongue fur, superficial and tense pulse, it may be determined as exterior-cold syndrome caused by exogenous wind-cold through comprehensive analysis of the four diagnostic methods. It should be treated by pungent and warm prescriptions for relieving exterior syndrome in accordance with the principles of diaphoretic therapy for exterior syndromes and drugs of warm nature for cold syndromes. In this case, the veterinary may either adopt a set prescription (Mahuang Tang) with modification, or draw up, by him, a pungent and warm prescription for relieving exterior syndrome with proper drugs in the light of the principle of formulating a prescription. And the sick animals will get recovered after taking the drugs that are decocted as required. Thus it can be seen that the therapeutic method is the theoretical basis for formulating a prescription and the prescription is the concrete embodiment of the former, namely "prescriptions come from and are guided by therapeutic methods". The two aspects interrelate closely with each other, constituting the two important links in the process of treatment based on syndrome differentiation.

1 Compositions and Modification of the Prescriptions

Apart from very few single drugs of all the prescriptions used clinically, the great majority of them are compound drugs consisting of two or more drugs. The reasons are that the action of a single drug is usually limited, and some of them may produce certain side effects or even toxicity. But when several drugs are applied together, ensuring a full play of their advantages and inhibiting the disadvantages, they will display their superiority over a single drug in the treatment of diseases. This can be illustrated in the following three aspects. Firstly, drugs of similar action, if used simultaneously, can strengthen the therapeutic effect for serious diseases. For example, the synergism of Mangxiao (Natrii Sulphas) and Dahuang (Radix et Rhizoma Rhei) can enhance the therapeutic effects of eliminating pathogenic factors by purgation in the treatment of serious heat accumulation syndrome, e. g. Da Chengqi Tang. Secondly, drugs of different actions in combination can broaden the therapeutic scope in the treatment of complex conditions. For example, Dangshen (Radix Codonopsitis Pilosulae) capable of reinforcing qi and Maimendong (Radix Ophiopogonis) of nourishing yin in combination has the action of reinforcing both qi and yin for deficiency of both qi and yin, e. g. Shengmai San. Thirdly, drastic or noxious drugs may be applied with some drugs capable of reducing or removing their side effect or toxicity so that they are not likely to generate or at least produce less damage to the body resistance or toxic reaction. For instance, Gansui (*Euphorbia kanscu* Lioumss) has the function of eliminating retained fluid, but is drastic and toxic in property. If Dazao (Fructus Ziziphi Jujubae) are added to it, they can alleviate its side effect, e. g. Shizao Tang. Hence, it's obvious that a rational and appropriate compatibility of the drugs contributes to the full play of the drugs' efficacy and helps reduce to its maximum or get rid of the toxicity and side effect of drugs. That's why compound prescriptions are so widely used. To achieve the above requirement so that the made prescriptions can meet the clinical syndromes to the greatest extent, it is imperative that the drugs be chosen and prescriptions made with flexibility to suit the specific syndromes under the guidance of the formulation principle.

1.1 Composition of Prescriptions

How to make prescriptions with different drugs based on their rational compatibility? Besides an accurate differentiation of syndromes, establishment of therapeutic method and appropriate choice of drugs and doses, it is necessary to follow the peculiar principle of monarch (jun), minister (chen), adjuvant (zuo) and guide (shi) drugs in a prescription.

Monarch drug: Being an essential ingredient in a prescription, it plays a leading curative role aiming at the cause or the main syndrome of a disease.

Minister drug: It helps strengthen the curative effect of the monarch drug.

Adjuvant drug: It refers to: ①the ingredient to cooperate with the monarch and minister drugs to strengthen the therapeutic effects or treat the accompanying diseases or syndromes; ②the ingredient to inhibit the drastic effects or toxicity of the monarch and minister drugs; ③the ingredient to possess the properties and flavor opposite to those of the monarch drug, but play supplementing effect in the treatment when serious diseases due to excessive pathogenic factors make sick animals refuse the drug.

Guiding drug: It refers to: ①the ingredient leading the other drugs in the prescription to the affected part; and ②the ingredient regulating the properties of other drugs in the prescription.

Take Mahuang Tang for example to explain the above principle of formulation. This recipe is composed of Mahuang (Herba Ephedrae) 30g, Guizhi (Ramulus Cinnamomi) 30g, Xingren (Semen Armeniacae Amarae) 30g, and Gancao (Radix Glycyrrhizae) 21g. It is used to treat exterior excess syndrome due to affection of exogenous pathogenic wind-cold, marked by aversion to cold, fever, anhidrosis, asthma, thin and whitish fur, superficial and tense pulse. The syndrome herein is caused by exogenous wind-cold. Its chief syndrome is attack of wind-cold on the superficies, the ac companying one being the obstruction of the lung-qi. Thus the therapy for expelling cold to relieve exterior syndrome and facilitating the flow of lung-qi to relieve asthma should be employed. The first two ingredients of this recipe are both pungent in flavor and warm in property, capable of expelling cold to relieve the exterior syndrome. But Mahuang (Herba Ephedrae) bears a drastic efficacy with a large dosage, and is thus used as monarch drug to deal with the cause of the disease and the chief syndrome. Guizhi (Ramulus Cinnamomi) helps Mahuang (Herba Epheadrae) induce sweating to disperse cold and expel exterior pathogenic factors and functions as minister drug. Xingren (Semen Armeniacae Amarae) acts as adjuvant drug with the function of descending adversely rising qi, stopping cough and relieving asthma, and is specially supposed to treat accompanying symptoms. The last ingredient in this recipe can mediate drug properties on the one hand, belonging to a mediating drug of guiding drugs. On the other hand, it is sweet in flavor and mild in property, and can alleviate excessive diaphoretic effects induced by the first two ingredients that are pungent and warm in nature, so it is concurrently the adjuvant drug. Here is the outline of the formulation of the above recipe:

Mahuang Tang

Monarch drug—Mahuang (Herba Ephedrae): Relieving exterior syndrome by means of diaphoresis, activating the flow of the lung-qi to relieve asthma

Minister drug—Guizhi (Ramulus Cinnamomi): Assisting the monarch drug to induce diaphoresis, relieve exterior syndrome and expel cold

Adjuvant drug—Xingren (Semen Armeniacae Amarae): Helping the monarch drug to promote the flow of lung-qi and stop asthma

Guiding drug—Gancao (Radix Glycyrrhizae): Mediating the drug properties and preventing the impairment of vital qi due to excessive diaphoresis

The above principle of forming a prescription shows that drugs in a prescription have their respective importance of effect in the order of the monarch, minister, adjuvant and guiding drugs. Besides, the drugs in a recipe are related to one another — the monarch and minister drugs cooperate with each other, the adjuvant drug coordinates or inhibits the monarch and minister drugs, ensuring the optimum effect of the recipe by means of a supplementary or opposite relationship. Moreover, not every recipe comprises invariably the monarch, minister, adjuvant and guiding drugs, nor does each of the four ingredients play a single role in a recipe. So the composition of monarch, minister, adjuvant and guiding drugs depends on therapeutic requirements. Although monarch drug is indispensable to a recipe, it does not necessarily follow that the rest three should be included in the recipe. If the monarch drug has adequate potency, minister drug will not be included in the recipe. If the first two ingredients bear no toxic and drastic property, adjuvant drug is not needed. If the drugs meant to treat the chief syndrome can come to the affected part, no guiding drug will be involved. Some minister drugs may have the function of the adjuvant drug concurrently, so do some adjuvant drugs of the guiding drug, e. g., Gancao (Radix Glycyrrhizae) in Mahuang Tang is a guiding drug in itself, but has a concurrent function with the adjuvant drug. Therefore, this principle should never be applied mechanically.

The drugs in a prescription in accordance with the compatible principle have their special effect and action. They interact with and inhibit one another. In this way, they form an organic whole with rigorous compatibility, thereby producing significant therapeutic effect.

1.2 Modification of a Prescription

In the clinical application of a set prescription, it is necessary to modify it flexibly under the guidance of the compatible principle and in accordance with the condition of illness, constitution, and age, sex of the sick animals and the occurring season, climate. Only, when the principle is unified with flexibility can the prescription tally with the syndromes and the expected therapeutic goal be achieved. In other words, it is necessary to follow the principle but not adhere to the originally established prescription mechanically. The modification of a prescription includes the following three ways:

1.2.1 Modification of Drugs

This can be subdivided into two. One is the modification of the adjuvant drug, by which is meant subtracting some unsuitable drugs in the original prescription or adding some necessary drugs that are not in the original prescription to meet the need of treatment of present accompanying symptoms. But it should be carried out on the condition that the present chief syndrome is identical with that of the original prescription, but the accompanying symptoms are different. Since the adjuvant drug plays a minor role in a recipe, its modification is not likely to bring about a radical change of the original potency. This is also known as "the modification in the light of the symptoms". For instance, Si Junzi Tang is chiefly designed for qi-deficiency syndrome of the spleen and stomach, marked by short breath, lack of strength, poor appetite, loose stool, pale tongue with whitish fur, thready and weak pulse. It is composed of Dangshen (Radix Codonopsis Pilosulae), Baizhu (Rhizoma Atractylodis Alba), Fuling (*Poria cocos* (Schw.) Wolf) and Zhigancao (Radix Glycyrrhizae Praeparatae), and functions to replenish qi and invigorate the spleen. If the above symptoms are accompanied by oppressed sensation over the abdominal distention and obstructed flow of qi, which results from dysfunction of spleen-qi, Chenpi (Pericarpium Citri Reticulatae) can be added to the above recipe to promote the flow of qi and relieve distention. This gives rise to another prescription entitled Yigong San.

The other refers to the modification of monarch and minister drugs or the change of the monarch drug and its compatibility by adding to or subtracting other drugs from the original prescription. Consequently radical changes will take place in terms of the potency. For example, when Guizhi (Ramulus Cinnamomi) in Mahuang Tang is substituted for Shigao (Gypsum Fibrosum), it is called Mahuang Xingren Gancao Shigao Tang. The former takes Mahuang (Herba Ephedrae) as the monarch drug, which is combined with Guizhi (Ramulus Cinnamomi) to treat exterior-excess syndrome caused by wind-cold through inducing diaphoresis and expelling cold. While the latter take Mahuang (Herba Ephedrae) and Shigao (Gypsum Fibrosum) together as the monarch drug to remove heat from the lung to relieve asthma. Thereby, it is effective for curing cough with asthma due to lung-heat. It can be seen that though the two are different only in one ingredient, because of the change of the relationship between monarch drug and its compatibility, the main action of the prescription changes accordingly. In that case, the prescription for relieving exterior syndrome with pungent and warm drugs is changed consequently into one for relieving exterior syndrome with pungent and cool drugs.

In the clinical application of set prescriptions, the above modifications can be conducted according to the actual requirements. Generally, "the modification in the light of the symptoms" is easy to master and conforms to the thought of the veterinary, and is thus commonly used in clinical treatment.

1.2.2 Modification of Dosage

It refers to increasing or decreasing the dosage of a drug in an established prescription without any change in its ingredients so as to change its potency or even its compatibility as well as its action and indication. For instance, Sini Tang and Tongmai Sini Tang are both composed of Fuzi (Radix Aconiti Lateralis), Ganjiang (Rhizoma Zingiberis) and Zhigancao (Radix Glycyrrhizae Praeparatae), acting respectively as the monarch, minister and adjuvant drugs. Because the former contains less of the first two ingredients than the latter, and functions to recuperate depleted yang and rescue sick animals from collapse, it is used to treat the syndrome due to excessive yin and deficient yang marked by cold limbs, aversion to cold, lying in curl, dyspeptic diarrhea, deep faint and thready pulse. In the latter, however, both of the first two ingredients are increased. So it functions to warm the interior to recuperate depleted yang and promote blood circulation. It is mainly applied to treating the syndrome due to excessive yin repelling yang marked by cold limbs, no aversion to cold, flushed mouth and tongue, dyspeptic diarrhea and extremely faint or even depleting pulse. Take Xiao chengqi Tang and Houpo Sanwu Tang for another example, both are composed of Dahuang (Radix et Rhizoma Rhei), Zhishi (Fructus Aurantii Immaturus) and Houpo (Cortex Magnoliae Qfficinalis). In the former prescription, Dahuang (Radix et Rhizoma Rhei) is in large dosage and used as the monarch drug, and Zhishi (Fructus Aurantii Immaturus) as the minister drug. Houpo (Cortex Magnoliae Qfficinalis) is in half the dosage of Dahuang (Radix et Rhizoma Rhei) and serves as the adjuvant drug. This recipe, with the function of purging away heat and relieving constipation, is indicated for mild case with excess syndrome of the Yangming fu organs. In the latter, however, Houpo (Cortex Magnoliae Officinalis) serves as the monarch drug with a larger dosage, Zhishi (Fructus Aurantii Immaturus) as the minister with the dose larger than that of the former prescription, and Dahuang (Radix et Rhizoma Rhei) as the adjuvant drug in half the dosage of Houpo (Cortex Magnoliae Officinalis). The latter prescription functions to promote the flow of qi and relieve constipation and is mainly used to treat constipation due to stagnation of qi

It can be seen from the above that Sini Tang and Tongmai Sini Tang, though different in the doses of drugs, have the same compatibility. They differ only in their power of actions and pathologic conditions. However, because of modification of the quantity of the ingredients, the compatible relationships of the two prescriptions are changed, thus leading to the change of their actions and indications.

2 Common Forms of Prescriptions

The forms of prescription refer to the definite patterns processed, after drugs are made up into a prescription in line with certain compatibility, according to the condition of the sick animals, the properties of the drugs and the route of administration as well. An appropriate form is indispensable to the action of a prescription and the efficacy of the drugs.

Here are the commonly used forms as follows:

2.1 Decoction

It refers to the medicinal solution obtained by removing the dregs after soaking and decocting the prepared herbal pieces in water for a period of time. It is chiefly for oral administration, e.g., Mahuang Tang, Guizhi Tang, etc.. Besides it can be applied externally for washing and steaming. It bears the characteristics of being easily absorbed and producing quick curative effects. It is especially suitable for serious or unsteady cases. And it can be modified according to the changes of disease. It can meet the need of treatment based on syndrome differentiation, so it is one of the forms most widely used in clinical practice. Nevertheless it has its disadvantages. Firstly, it needs to be taken in large amount, and the effective compositions of some drugs are difficult to extract and easy to volatilize. Secondly, it takes a longer time to boil the drugs and thus it is not beneficial to the rescue work of critical sick animals. Thirdly, it is inconvenient to carry and difficult to take for its bitter taste.

2.2 Powder

The drugs are ground into fine powder and well mixed for oral use and external use. Generally the powder for oral administration is usually ground into fine powder to be taken orally with warm boiled water, such as Qili San, Xingjun San, and so on. Some drugs may be ground into coarse powder to be boiled with water and then the residue is removed to get the liquid for oral use. So they got the name of "boiled powder", e. g. Yinqiao San, Baidu San, etc.. The powder for external use is generally applied to or sprinkled on a sore or affected part of the body. Jinhuang San and Shengji San are such examples. But some are for eye droppings or throat insufflation, e. g. Babao Yanyao, Bingpeng San, etc.. Powder is prepared easily and absorbed quickly, stable in property, insusceptible of going bad, and convenient to use and carry with less drugs used.

2.3 Pill

It refers to the round solid preparation of drugs obtained by grinding drugs into fine powder or extracting medicinal materials and then mixing with excipient. Compared with decoction, pills have the advantages of slow absorption, lasting potency, small dosage, and are convenient to carry and take because of their small volumes. Therefore they are commonly adopted to treat chronic diseases or diseases with deficiency syndromes, e. g. Liuwei Dihuang Wan, Xiangsha Liujunzi Wan, etc. Sometimes they are also used for chronic treatment with drugs drastic in properties, e. g. Shizao Wan, Didang Wan, etc.. They may also contain drugs aromatic in flavor, and unsuitable to be boiled in water, e. g. Angong Niuhuang Wan, Suhexiang Wan, etc.. The pills or boluses commonly used are listed as follows:

2.4 Tablet

It refers to a flat preparation made by pressing the mixture of fine powder or extract of drugs with some excipients, small in size and accurate in dosage, convenient administration.

2.5 Injection

It refers to either the aseptic solution, or the aseptic suspension or the aseptic powder for preparing liquid, which has gone through the process of extracting, refining and preparing drugs. It is used for subcutaneous, muscular or intravenous injections, with the characteristics of being accurate in dosage, quick in action and not influenced by the digestive system.

All the forms mentioned above have their respective features and need to be applied clinically in the light of the pathological conditions and the drug properties in a recipe. In addition, such forms as capsule, moxa, press preparation, enema and spray are also widely used. At present the forms of Chinese Patent medicines have amounted to roughly 60 varieties. Moreover many traditional products have got new forms through renovation, which improve clinical effect as well.

3 Prescriptions In Common Use

3.1 *Mahuang Tang* (*Ephedra Decoction*)

Source: Shanghan Lun (Treatise on Exogenous Febrile Diseases).

Ingredients:

No. 1 Mahuang (Herba Ephedrae) 30 g;

No. 2 Guizhi (Ramulus Cinnamomi) 30 g;

No. 3 Xingren (Semen Armeniacae Amarae) 30 g;

No. 4 Zhigancao (Radix Glycyrrhizae Praeparatae) 21 g.

Administration: Decoct the above drugs in water for oral use.

Actions: Inducing perspiration to relieve exterior pathogenic factors, dispersing the lung to relieve asthma.

Clinical Application: This recipe is for exterior-excess syndrome due to exogenous wind-cold, marked by aversion to cold, fever, dyspnea without perspiration, thin and whitish tongue fur, and superficial and tense pulse. It is applicable to common cold, flu, acute bronchitis, bronchial asthma and other diseases which chiefly manifest aversion to cold without perspiration, cough and dyspnea. If complicated with pathogenic dampness marked by arthralgia, heaviness of the body, add Baizhu (Rhizoma Atractylodis Alba) to eliminate dampness, thus making a new one entitled Mahuang Jiazhu Tang. In case of mild aversion to cold but chiefly asthma, subtract Guizhi (Ramulus Cinnamomi) to make a recipe entitled San'ao Tang specially for dispersing the lung to relieve asthma. In case of serious aversion to cold without perspiration and general aching body accompanied by dysphoria due to interior heat, double the dosage of Mahuang (Herba Ephedrae) to enhance the potency of inducing perspiration for eliminating the pathogenic factors and add Shigao (Gypsum Fibrosum) to clear away the interior heat, called Da Qinglong Tang.

Elucidation: The syndrome is due to attack of wind-cold on the exterior, obstruction of defensive qi, stagnation of interstitial space, and failure of pulmonary qi to disperse. Inducing perspiration to expel pathogenic factors from the exterior and dispersing the lung to relieve asthma should treat it. In this recipe, ingredient No. 1 acts as the monarch drug. It is capable of inducing sweating to dispel exogenous pathogenic factors and dispersing the lung to relieve asthma. Ingredient No. 2, as the minister drug, is capable of dispersing pathogenic cold by warming the meridians. Ingredient No. 3 capable of relieving stagnant lung-qi and No. 1 serve as adjuvant drug to enhance the effect of relieving cough and asthma. The last ingredient serves as the guiding drug, capable of mediating drug properties and invigorating qi and enriching mid-energizer so as to prevent ingredients No. 1 and No. 2 from inducing excessive sweating to impair the vital-qi.

Cautions: The recipe is contraindicated for the exterior syndrome due to wind-cold with sweating because it is drastic in inducing sweating. So sick animals with general debility, blood deficiency and serious interior heat should use it with great caution.

3.2 *Yinqiao San* (*Powder of Lonicera and Forsythis*)

Source: Wenbing Tiaobian (Detailed Analysis of Seasonal Febrile Diseases).

Ingredients:

No. 1 Jinyinhua (Flos Lonicerae) 45 g;

No. 2 Lianqiao (Fructus Forsythiae) 45 g;

No. 3 Xianlugen (Rhizoma Phragmitis) 45g;

No. 4 Jiegeng (Radix Platycodi) 30 g;

No. 5 Bohe (Herba Menthae) 30 g;

No. 6 Niubangzi (Fructus Arctii) 30 g;

No. 7 Zhuye (Herba Lophatheri) 30 g;

No. 8 Jingjiesui (Spica Schizonepetae) 30 g;

No. 9 Dandouchi (Semen Sojae Praeparatum) 30 g;

No. 10 Shenggancao (Radix Glycyrrhizae Recens) 21 g.

Administration: These ingredients are decocted for oral administration.

Actions: Relieving exogenous pathogenic factors with drugs pungent in flavor and cool in property, clearing away heat and eliminating toxin.

Clinical Application: This recipe is indicated for seasonal febrile diseases in the early stage and severe syndrome of exterior heat. The usual symptoms are fever without perspiration, or hindered perspiration, slight aversion to wind-cold, thirst, cough, red tip of the tongue, thin whitish or yellowish fur, superficial and rapid pulse. It is applicable to flu, acute tonsillitis, epidemic encephalitis B, epidemic cerebrospinal meningitis, and parotitis, etc., which belong to the exterior syndrome of wind-heat and are marked by fever, and no perspiration. In case of excessive heat marked by high fever with thirst, add Shigao (Gypsum Fibrosum), Huangqin (Radix Scutellariae) and Daqingye (Folium Isatidis) to clear away heat and purge fire.

Elucidation: Affection by wind-heat leads to stagnation of defensive qi and failure of the lung in purifying and descending function, for which the method of relieving exterior syndrome by expelling wind, clearing away heat and toxin should be adopted. Ingredients No. 1 and No. 2 in larger amount act as monarch drug, capable of clearing away heat and eliminating toxin as well as expelling exterior pathogenic factors. Ingredients No. 5 and No. 6 pungent in flavor and cool in property, No. 8 and No. 9 pungent in flavor and warm in property, act together as minister drug, enhancing the effect of expelling pathogenic factors, and involving no impairment of the body fluid. Ingredient No. 7 and Lugen (Rhizoma Phragmitis) can clear away heat and promote the production of body fluid, and No. 4 disperses the lung to relieve cough, the three constituting adjuvant drug. The last ingredient acts as guiding drug, capable of clearing away heat and toxin, mediating drug property, and being good at soothing the throat together with No. 4.

Cautions: It is not applicable to exterior syndrome due to wind-heat with aversion to cold, absence of perspiration, but a mild fever without thirst.

3.3 *Baihu Tang* (*White Tiger Decoction*)

Source: Shanghan Lun (Treatise on Exogenous Febrile Diseases).

Ingredients:

No. 1 Shigao (Gypsum Fibrosum) 180g;

No. 2 Zhimu (Rhizoma Anemarrhenae) 60g;

No. 3 Zhigancao (Radix Glycyrrhizae Praeparatae) 45g;

No. 4 Jingmi (Semen Oryzae Sativae) 30g.

Administration: Decoct the above drugs till the rice is well done and take the decoction several times after removal of the residue.

Actions: Clearing away heat and promoting the production of body fluid.

Clinical Application: This recipe is indicated for yangming meridian syndrome, marked by high fever, polydipsia, profuse perspiration, aversion to heat, full forceful pulse. It is applicable to infectious diseases such as lobar pneumonia, epidemic encephalitis B, epidemic hemorrhagic fever, and gingivitis with the four characteristics of this syndrome (high fever, polydipsia, profuse perspiration, and full large pulse). In case of impairment of qi and body fluid due to excessive heat marked by full large but weak pulse, thirst not quenched with drink, add Dangshen (Radix Codonopsis) to replenish qi and promote the production of body fluid. In case of intense heat in both qi-fen and xue-fen in combination with skin rashes, add Shengdihuang (Radix Rehmanniae) and Shuiniujiao (Cornu Bubali) to clear away heat and cool the blood. In case of wind stirring up due to excessive heat concomitant with tic of limbs, add Lingyangjiao (Cornu Antelopis) and Gouteng (Ramulus Uncariae cum Uncis) to relieve convulsion and spasm. In case of interior heat complicated with dampness, such as damp-warm syndrome with heat prevailing over dampness, or arthritis with reddish, swelling and painful joints, add Cangzhu (Rhizoma Atractylodis) to dry pathogenic dampness, constituting another recipe entitled Baihu plus Cangzhu Tang.

Elucidation: This syndrome is the result of excessive heat in qi-fen scorching the body fluid, and should be treated by clearing away heat and promoting the production of body fluid. Ingredient No. 1, pungent and sweet in flavor and extremely cold in property, acts as monarch drug, capable of purging away heat without impairing body fluid. Ingredient No. 2, which is bitter in flavor and cold moist in property, serves as minister drug, strengthening the action of the monarch drug and promoting the production of body fluid as well. Ingredients No. 3 and No. 4 function as adjuvant drug that can not only reinforce the stomach and protect body fluid, but also prevent the stomach from being injured by the extremely cold drug as ingredient No. 1. Besides, No. 3 also plays the role of guiding drug, mediating drug properties.

Cautions: This must not be applied to sick animals with full large but weak pulse when pressed and pale tongue, which are attributable to fever due to blood deficiency.

3.4 *Huanglian Jiedu Tang* (*Decoction of Coptis for Detoxification*)

Source: Cui's Prescriptions from Waitai Miyao (Clandestine Medical Essentials from Imperial Library).

Ingredients:

No. 1 Huanglian (Rhizoma Coptidis) 45g;

No. 2 Huangqin (Radix Scutellariae) 60g;

No. 3 Huangbai (Cortex Phellodendri) 30g;

No. 4 Zhizi (Fructus Gardeniae) 45g.

Administration: Decoct the above drugs for oral use.

Actions: Purging pathogenic fire and toxin.

Clinical Application: This recipe is indicated for excess of fire, toxin and heat in the triple energizer, marked by high fever, thirst for drink, or dysentery with fever, or sores and carbuncles, red tongue with yellowish fur, rapid and forceful pulse. It is applicable to septicemia, pyosepticemia, dysentery, pneumonia, urinary infection, epidemic cerebrospinal meningitis、epidemic encephalitis B, and other infectious diseases that are attributable to syndrome due to intense fire and toxin. If concomitant with constipation, add Dahuang (Ratix et Rhizoma Rhei) to promote defecation and cause heat and toxin to descend. If concomitant with hematemesis, epistaxis and skin rashes, add Shengdihuang (Radix Rehmanniae), Xuanshen (Radix Scrophulariae) and Mudanpi (Cortex Moutan Radicis) to clear away heat and cool blood. In case of jaundice, add Yinchenhao (Herba Artemisiae Scopariae) and Dahuang (Radix et Rhizoma Rhei) to eliminate dampness and jaundice.

Elucidation: The syndrome is the result of heat and toxin accumulating in the three energizer, and should be treated with drugs of extremely cold nature and capable of purging fire and toxin. Ingredient No. 1 specializes in purging away heart-fire and fire in the middle-energizer, acting as monarch drug. Ingredients No. 2, No. 3 and No. 4 function to purge fire respectively in the upper-energizer, the lower-energizer, and triple energizer, and together serve as minister and adjuvant drugs. All the ingredients are extremely bitter in flavor and cold in property, thus can achieve a better effect of expelling fire and toxin when combined.

Cautions: This recipe mostly comprises drugs of bitter and cold nature. Since drugs of bitter and dry nature and those of bitter and cold nature are apt to impair respectively yin and the stomach, this recipe is not applicable to sick animals who do not manifest syndrome of excessive accumulation of fire and toxin. Besides this recipe should be abandoned immediately after it creates the expected result. It must not be applied to sick animals suffering from a serious impairment of yin with deep-red tongue without fur.

3.5 *Lizhong Tang* (*Decoction for the Function of Middle Energizer*)

Source: Shanghan Lun (Treatise on Exogenous Febrile Diseases).

Ingredients:

No. 1 Dangshen (Radix Codonopsis Pilosulae) 60g;

No. 2 Ganjiang (Rhizoma Zingiberis) 60g;

No. 3 Zhigancao (Radix Glycyrrhizae praeparatae) 30g;

No. 4 Baizhu (Rhizoma Atractylodis Alba) 45g.

Administration: Decoction for oral administration.

Actions: Warming the middle-energizer to dispel cold and replenishing qi to invigorate the spleen.

Clinical Application: This recipe is used to treat deficiency-cold syndrome of the spleen and stomach, marked by abdominal pain, preference for warmth and for pressure, aversion to cold, cold limbs, poor appetite, vomiting, diarrhea, or bleeding due to yang deficiency with little dark blood, pale tongue with whitish fur, deep and thready pulse. It is applicable to such diseases characterized by vomiting, diarrhea, cold and pain as acute or chronic gastroenteritis, gastro-duodenal ulcer, etc., which pertain to deficiency-cold in the stomach and spleen. In case of severe cold syndrome, add the dosage of ingredient No. 2 or add Fuzi (Radix Aconiti Carmichaeli Praeparata) to enhance the effect of warming the middle-energizer to dispel cold, which constitutes another recipe known as Fuzi Lizhong Wan. When used in the treatment of bleeding due to yang deficiency, substitute No. 2 with Paojiang (Rhizoma Zingiberis Praeparatae). In case of excessive deficiency syndrome, add No. 1 to replenish qi and reinforce the spleen. In case of severe diarrhea, add the dosage of No. 4 to invigorate the spleen and arrest diarrhea. In case of severe vomiting, add Wuzhuyu (Fructus Evodiae) and Shengjiang (Rhizoma Zingiberis Recens) to warm the stomach and stop vomiting. In case of dysfunction of the spleen in transportation due to deficiency, and phlegm generated from dampness, add Banxia (Rhizoma Pinelliae) and Fuling (Poria) to resolve phlegm, which constitutes another recipe, entitled Lizhong Huatan Wan.

Elucidation: The syndrome results from yang-deficiency of middle-energizer, and deficiency-cold of the spleen and stomach, leading to dysfunction in digestion and transformation. It should be treated by warming the middle-energizer to dispel cold and replenishing qi to invigorate the spleen. Ingredient No. 2 acts as monarch drug to restore yang of the spleen and stomach by means of warming the middle-energizer to dispel cold. Ingredient No. 1 is used as minister drug for supplementing qi, and strengthening the spleen to promote transportation and transformation. Ingredient No. 4 serves as adjuvant drug, used together with ingredients No. 1 and No. 2 for replenishing qi, strengthening the spleen and eliminating dampness. Ingredient No. 3, sweet and mild in nature, helps other drugs in the recipe reinforce the spleen and stomach, and mediates other drugs properties, playing the part of guiding drug.

3.6 *Da Chengqi Tang* (*Decoction for Potent Purgation*)

Source: Shanghan Lun (Treatise on Exogenous Febrile Diseases).

Ingredients:

No. 1 Dahuang (Radix et Rhizoma Rhei) 60g;

No. 2 Houpo (Cortex Magnoliae Officinalis) 90g;

No. 3 Zhishi (Fructus Aurantii Immaturus) 60g;

No. 4 Mangxiao (Natrii Sulphas) 120g.

Administration: No. 2 and No. 3 are to be decocted prior to No. 1 and the decoction is to be taken after removing the residue and infusing No. 4 in it.

Actions: Drastic purgation for eliminating accumulated heat.

Clinical Application: This recipe is used to treat excessive-heat syndrome of yangming-fu organ, manifested as constipation, abdominal pain with tenderness, afternoon fever, thirst with great desire to drink, reddened and prickled tongue with dry yellowish fur or dry blackish fur with fissures, deep and forceful pulse. It is applicable to acute simple ileus, adhesive ileus, acute cholecystitis, acute pancreatitis as well as symptoms of high fever, convulsion, and mania with constipation, yellowish fur and forceful pulse as seen in the course of febrile diseases. In case of consumption of qi due to excessive heat with lassitude, add Dangshen (Radix Codonopsis Pilosulae) to replenish qi and prevent exhaustion of qi induced by drastic purgation. If the accumulating heat impairs yin manifesting extreme thirst with great desire to drink, dry reddened tongue with little fur, add Xuanshen (Radix Scrophulariae) and Shengdihuang (Radix Rehmanniae) to nourish yin for the production of body fluid and to moisten dryness for relieving constipation.

Elucidation: The syndrome results from pathogenic heat accumulating in the large intestine, leading to obstructed flow of fu-qi. This should be treated by eliminating heat to relax the bowels. Ingredient No. 1 acts as monarch drug to eliminate heat and stagnated fodder and stool in the gastrointestinal tract, and relieve constipation. No. 4 possesses the effect of moistening dryness and softening mass, and is used as minister drug to reinforce the effect of the monarch drug. Both No. 3 and No. 2 are adjuvant drugs, capable of promoting flow of qi and relieving fullness and helping the monarch and minister drugs remove the stagnated fodder and stool to accelerate discharge of heat and mass. The four drugs in compatibility generates a drastic efficacy in purging heat and mass. Subtract No. 4 and Xiao Chengqi Tang is formed, which has a mild efficacy. Replace No. 2 and No. 3 with Gancao (Radix Glycyrrhizae), and Tiaowei Chengqi Tang is obtained for laxation of heat and mass. The above three share the name of "San Chengqi Tang" and the selective application of them is made clinically in the light of the degrees of severity.

Cautions: It must be applied with great care to those with excess syndrome of yangming-fu organ but without excessive heat, or those with constitutional deficiency of qi and yin.

3.7 *Xiao Chaihu Tang* (*Minor Decoction of Bupleurum*)

Source: Shanghan Lun (Treatise on Exogenous Febrile Diseases).

Ingredients:

No. 1 Chaihu (Radix Bupleuri) 45g;

No. 2 Huangqin (Radix Scutellariae) 45g;

No. 3 Dangshen (Radix Codonopsitis Pilosulae) 30g;

No. 4 Zhigancao (Radix Glycyrrhizae Praeparatae) 20g;

No. 5 Banxia (Rhizoma Pinelliae) 45g;

No. 6 Shengjiang (Rhizoma Zingiberis Recens) 30g;

No. 7 Dazao (Fructus Ziziphi Jujubae) 20pcs.

Administration: Decoct the above drugs in water for oral application.

Actions: Treating shaoyang disease by mediation.

Clinical Application: This recipe is used to treat febrile diseases in shaoyang meridian, marked by alternate attacks of chills and fever, thin and whitish fur and taut pulse. It is applicable to common cold, influenza, chronic hepatitis, acute or chronic cholecystitis, pleuritis, pyelonephritis, postpartum infection, gastric ulcer and other diseases, which manifest the above symptoms and pertain to febrile disease in shaoyang meridian. In case of dysphoria without vomiting, which is indicative of heat accumulating in the chest, replace ingredients No. 3 and No. 5 with Gualou (Fructus Trichosanthis) to clear away heat and regulate qi for relieving chest fullness. In case of body fluid impaired by heat with manifestation of thirst, replace No. 5 with Tianhuafen (Radix Trichosanthis) to produce body fluid for quenching thirst. If the spleen is affected by the stagnated liver-qi with abdominal pain, replace No. 2 with Baishaoyao (Radix Paeoniae Alba) to nourish the liver for relieving spasm and pain.

Elucidation: The syndrome is due to pathogenic factors invading shaoyang meridian and fighting against healthy qi in between the exterior and interior. Ingredient No. 1, as monarch drug, dispels the pathogenic factors located in the half exterior and restores the functional activity of shaoyang meridian. Ingredient No. 2 acts as minister drug to purge away heat located in the half interior. They are primary drugs for treating shaoyang disease, one for dispelling pathogenic factor, and the other clearing away the accumulated heat. Ingredients No. 5 and No. 6 can regulate the stomach and relieve vomiting, while ingredients No. 3, No. 4 and No. 7 supplement qi and strengthen body resistance to prevent the pathogenic factors from getting into the interior. These five ingredients play the role of adjuvant drug. Ingredient No. 4 coordinates the actions of various drugs in the recipe and also serves as guiding drug.

Cautions: This recipe should be used with great caution for sick animals with shaoyang disease but manifesting d eficiency of yin and blood because Ingredient No. 1 possesses lifting and dispersing action and ingredients No. 2 and No. 5 are of dry nature.

3.8 *Sijunzi Tang* (*Decoction of Four Noble Ingredients*)

Source: Taiping Huimin Hejiju Fang (Benevolent Prescriptions from Taiping Pharmaceutical Bureau).

Ingredients:

No. 1 Dangshen (Radix Codonopsitis Pilosulae) 45g;

No. 2 Baizhu (Rhizoma Atractylodis Alba) 45g;

No. 3 Fuling (Poria) 45g;

No. 4 Zhigancao (Radix Glycyrrhizae Praeparatae) 30g.

Administration: Decoct the above drugs in water for oral application.

Actions: Replenishing qi and strengthening the spleen.

Clinical Application: This recipe is used to treat qi deficiency syndrome of the spleen and stomach marked by pale mouth and tongue, shortness of breath and lassitude, poor appetite, watery stool, pale tongue with whitish fur, thready and weak pulse. It is applicable t o chronic gastritis, gastroduodenal ulcer and other diseases, which have the above symptoms and pertain to qi deficiency of the spleen and stomach. If concomitant with qi stagnation, add Chenpi (Pericarpiurn Citri Reticulatae) to promote the flow of qi, which forms another recipe entitled Yigong San. If concomitant with phlegm and dampness retaining in the interior and cough with profuse sputum, add Chenpi (Pericarpium Citri Reticulatae) and Banxia (Rhizoma Pinelliae) to remove phlegm by drying dampness, which constitute another recipe entitled Liujunzi Tang. If concomitant with cold-dampness retaining in the middle-energizer, functional disorder of the stomach-qi and obstructed circulation of qi, add Muxiang (Radix Aucklandiae), Sharen (Fructus Amomi), Chenpi (Pericarpium Citri Reticulatae) and Banxia (Rhizoma Pinelliae) to promote the flow of qi of the middle-energizer and regulate the stomach to relieve vomiting. This forms a new recipe named Xiangsha Liujunzi Tang.

Elucidation: The syndrome is due to qi deficiency of the spleen and stomach, leading to disorder of transportation and transformation. It should be treated by means of replenishing qi and strengthening the spleen. Ingredient No. 1 possesses the effect of invigorating qi, and acts as monarch drug to strengthen the spleen and nourish the stomach. Used as minister drug, No. 2 is effective in invigorating the spleen and eliminating dampness. Ingredient No. 3 serves as adjuvant drug to eliminate dampness and strengthen the spleen. The last ingredient acts to invigorate qi, reinforce the middle-energizer and mediate the Properties of other drugs as guiding drug. The combination of ingredients No. 1, No. 2 and No. 4 brings about the effects of invigorating qi, reinforcing the spleen and stomach, while that of ingredients No. 2 and No. 3 invigorating the spleen and eliminating dampness to promote transportation and transformation. The four ingredients in compatibility constitute the recipe. For reinforcing the spleen and stomach with sweat and warm drugs, which has the characteristics of reinforcement without stagnancy and warming without dryness.

3.9 *Siwu Tang* (*Decoction of Four Ingredients*)

Source: Xianshouli Shangxuduan Mifang (Clandestine Prescriptions for Waunds and Bonesetting Handed down by the Fairy).

Ingredients:

No. 1 Shudihuang (Radix Rehmanniae Praeparatae) 45g;

No. 2 Danggui (Radix Angelicae Sinensis) 45g;

No. 3 Baishaoyao (Radix Paeoniae Alba) 45g;

No. 4 Chuanxiong (Rhizoma Ligustici Chuanxing) 30g.

Administration: Decoct the above drugs in water for oral application.

Actions: Nourishing and regulating blood.

Clinical Application: This recipe is indicated for syndrome of deficiency and stagnation of nutrient qi and blood, marked by pale tongue and thready pulse. It is applicable to chronic dermatosis, chronic eczema, urticaria, orthopedic and traumatic diseases, and others, which pertain to deficiency and stagnation of nutrient qi and blood. If concomitant with deficiency of qi manifested as lassitude and shortness of breath, add Dangshen (Radix Codonopsitis Pilosulae) and Huangqi (Radix Astragali) to supplement qi for generating blood, which forms another recipe entitled Shengyu Tang. In case of dominant blood stasis manifested as sharp abdominal pain during menstruation, replace the third ingredient with Chishaoyao (Radix Paeoniae Rubra) and add Taoren (Semen Perisicae) and Honghua (Flos Carthami) to enhance the action of promoting blood circulation to remove blood stasis, which is known as Taohong Siwu Tang. In case of blood deficiency complicated with cold syndrome manifesting abdominal pain with preference for warmth, add Rougui (Cortex Cinnamomi), Paojiang (Rhizoma Zingiberis Praeparatae) and Wuzhuyu (Fructus Evodiae) to warm and dredge the vessels. In case of blood deficiency with heat syndrome manifesting dry mouth and throat, replace the first ingredient with Shengdihuang (Radix Rehmanniae) and add Huangqin (Radix) and Mudanpi (Cortex Moutan Radicis) to clear away heat and cool blood.

Elucidation: The syndrome results from deficiency of nutrient qi and blood, and obstructed blood circulation. It should be treated by nourishing and regulating blood. The first ingredient is capable of nourishing yin and blood and used as monarch drug. Ingredient No. 2 acts as minister drug with effects of nourishing blood and the liver and regulating menstruation. The last two ingredients function as adjuvant drug, the former nourishing blood and the liver and the latter promoting circulation of both qi and blood. Ingredients No. 1 and No. 3 are designed for nourishing blood, while No. 2 and No. 4 for nourishing and regulating blood. The compatibility of this recipe aims to nourish blood without inducing stasis, promote blood circulation without impairing it. Therefore they also have the function of regulating menstruation.

3.10 *Pingwei San* (*Powder for Regulating Stomach Function*)

Source: Taiping Huimin Hejiju Fang (Benevolent Prescriptions from Taiping Pharmaceutical Bureau).

Ingredients:

No. 1 Cangzhu (Rhizoma Atractylodis) 45g;

No. 2 Houpo (Cortex Magnoliae Officinalis) 45g;

No. 3 Chenpi (Pericarpium Citri Reticulatae) 45g;

No. 4 Zhigancao (Radix Glycyrrhizae Praeparatae) 30g.

Administration: Decocted with 30g of fresh ginger and 30 pieces of Chinese dates in water for oral application.

Actions: Drying dampness and invigorating the spleen, promoting the flow of qi and regulating the stomach.

Clinical Application: This recipe is indicated for dampness retention in the spleen and stomach, marked by anorexia, vomiting, belching and regurgitation, thick greasy whitish fur, slow pulse. It is applicable to chronic gastritis, dysfunction of digestive tract, gastro-duodenal ulcer, and others that have chief manifestations as abdominal fullness and greasy whitish fur and are ascribed to dampness retention in the spleen and stomach. In case of heat transformed from dampness with greasy yellowish fur, add Huanglian (Rhizoma Coptidis) and Huangqin (Radix Scutellariae) to clear away heat-dampness. If concomitant with interior cold syndrome manifesting cold sensation in the abdomen, loose stool, and aversion to cold with preference for warmth, add Ganjiang (Rhizome Zingiberis) and Cookout (Semen Alpena Katsumadai) to warm and dispel cold-dampness. If concomitant with disharmony between the spleen and stomach manifesting poor digestion, add Shenqu (Massa Fermentata Medicinalis), Shanzha (Fructus Crataegi) and Maiya (Fructus Hordei Germinatus) to promote digestion. In case of diarrhea due to excessive dampness, add Fuling (Poria) and Zexie (Rhizoma Alismatis) to promote diuresis and relieve diarrhea.

Elucidation: The syndrome is caused by dampness in the middle energizer, disorder of qi function, dysfunction of the spleen and stomach. Drying dampness and reinforcing the spleen, promoting flow of qi and regulating the stomach, should treat it. With large dosage, No. 1 is used as monarch drug to eliminate dampness and activating the spleen. No. 2 has the function of eliminating dampness and promoting the circulation of qi and functions as minister drug, which can not only enhance the effect of the monarch drug but also relieve abdominal fullness. As adjuvant drug, No. 3 promotes the flow of qi and regulates the stomach. It assists No. 2 in promoting the flow of qi to relieve abdominal fullness and strengthens the effect of No. 1 in eliminating dampness and regulating the middle-energizer. No. 4 serves as guiding drug, mediating the properties of other drugs. Fresh ginger and Chinese dates are meant respectively to regulate the stomach and reinforce the spleen so that the effect of regulating the spleen and stomach is further enhanced.

Cautions: This recipe is contraindicated for those with deficiency of the spleen and stomach or usual deficiency of yin.

3.11 *Wuling San (Powder of Five Drugs Containing Poria)*

Source: Shanghan Lun (Treatise on Exogenous Febrile Diseases).

Ingredients:

No. 1 Zhuling (*Polyporus umbellatus* (Pers.) Fr.) 45g;

No. 2 Zexie (Rhizoma Alismatis) 45g;

No. 3 Baizhu (Rhizoma Atractylodis) 30g;

No. 4 Fuling (Poria) 45g

No. 5 Guizhi (Ramulus Cinnamomi) 30g.

Administrations: Decoct the above drugs in water for oral application.

Actions: Inducing diuresis, eliminating dampness, warming yang and promoting qi function.

Clinical Application: This recipe is indicated for syndrome of fluid retention in the interior, marked by edema, diarrhea, dysuria, and pale tongue with glossy whitish fur, soft pulse. It is applicable to edema from nephritis, acute enteritis, uroschesis, and others, which pertain to fluid-retention and dampness in the interiror. In case of excessive retention of fluid, Wupi San (Chenpi, Pericarpium Citri Reticulatae), Fulingpi (Cortex Sclerotii Poriae), Shengjiangpi (Exocarpium Zingiberis Recentis), Sangbaipi (Cortex Mori Radicis) and Dafupi (Pericarpium Arecae) may be used simultaneously to enhance the effect of inducing diuresis and relieving swelling.

Elucidation: The syndrome is caused by dysfunction of the spleen in transportation and transformation, leading to retention of fluid in the interiror and dysfunction of qi transformation of the bladder. Inducing diuresis and warming yang to promote qi function should treat it. With a large dosage, No. 2 is effective in inducing diuresis and used as monarch drug. No. 4 and No. 1 are sweet and mild in nature and function as minister drugs, which can promote diuresis and enhance the effect of the monarch drug in relieving swelling. No. 3 in combination with No. 4 futher strengthens the function of reinforcing the spleen and inducing diuresis, while No. 5 promotes the function of the bladder in excretion of urine. No. 3 and No. 5 together play the role of adjuvant drugs. Originally this recipe was used for treating retention of fluid manifesting dysfunction of the bladder in excretion of urine and bradyuria, which are caused by the unrelieved exogenous pathogenic factors retaining in taiyang meridian. Thus it can be inferred that No. 5 also bears the effect of relieving exterior syndrome and eliminating pathogenic factors. This recipe is mainly to invigorate the spleen and induce diuresis and, thus, fit for retention of fluid due to dysfunction of the spleen.

Cautions: Warm in nature as this recipe is, it should not be used for the case of heat transformed from water-dampness.

3.12 *Erchen Tang* (*Erchen Decoction*)

Source: Taiping Huimin Hejiju Fang (Benevolent Prescriptions from Taiping Pharmaceutical Bureau).

Ingredients:

No. 1 Banxia (Rhizoma Pinelliae) 30g;

No. 2 Jupi (Pericarpium Citri Reticulatae) 30g;

No. 3 Fuling (Poria) 30g;

No. 4 Zhigancao (Radix Glycyrrhizae Praeparatae) 24g.

Administration: Decoct the above drugs in water for oral application.

Actions: Drying dampness, resolving phlegm, promoting the flow of qi and regulating the middle-energizer.

Clinical Application: This recipe is indicated for cough or vomiting due to phlegm-dampness, marked by cough with much whitish sputum which is easy to expectorate, vomiting, greasy whitish fur and smooth pulse. It is applicable to such diseases with much whitish sputum as chronic bronchitis, pulmonary emphysema, chronic gastritis, vomiting of pregnancy and neurotic vomiting, and others that are caused by phlegm-dampness. It is a basic recipe for treating phlegm syndrome and can be modified to treat various kinds of such syndrome. In case of heat-phlegm, add Huangqin (Radix Scutellariae) and Danxing (*Arisaema cum* Bile) to clear away heat and remove phlegm. In case of cold phlegm, add Ganjiang (Rhizoma Zingiberis) and Xixin (Herba Asari cum Radice) to warm and dispel cold phlegm. In case of wind phlegm, add Tiannanxing (Rhizoma Arisaematis) and Zhuli (Succus Phyllostachydis Henonis) to subdue wind and remove phlegm. In case of phlegm from indigestion, add Laifuzi (Semen Raphani) and Shenqu (Massa Fermentata Medicinalis) to promote digestion and remove phlegm.

Elucidation: The syndrome is caused by dysfunction of the spleen in transportation and transformation and disorder of qi activity, leading to accumulation of dampness, which is then transformed into phlegm. This should be treated by drying dampness, resolving phlegm and regulating the circulation of qi and the middle-energizer. Pungent, warm and dry in nature, No. 1 acts as monarch drug, which is effective in stopping cough by drying dampness and eliminating phlegm, and capable of suppressing the adversely ascendant qi and regulating the stomach so as to relieve vomiting.

Used as minister drug, No. 2 can promote the effect of No. 1 as well as promote the flow of qi to eliminate phlegm. No. 3 acts as adjuvant drug, and can reinforce the spleen and promote diuresis to root out "the source of producing phlegm", and achieve the effect of treating both the superficial symptoms and the root cause of the disease when in combination with the monarch and minister drugs. No. 4 serves as guiding drug, which bears the function of regulating the spleen and mediating drug properties. Fresh ginger is meant to assist No. 1 and No. 2 in their action of promoting the flow of qi, removing phlegm, regulating the stomach and relieving vomiting as well as to reduce the toxic effect of No. 1. Wumei (Fructus Mume) is supposed to preserve the lung-qi so that phlegm is removed without body resistance impaired. This compatibility focuses on eliminating dampness and removing phlegm but has also the effect of reinforcing the spleen and regulating the flow of qi, which is a reflection of the basic principle in treating phlegm syndrome. That is why this recipe is universal for various kinds of such syndrome. It is advisable to use the old ones in the selection of No. 1 and No. 2 so as to reduce the dry and dispersive action of both.

Cautions: Warm and dry as this recipe is, it is contraindicated for dryness syndrome of the lung due to yin deficiency or hemoptysis.

Lecture Four

The Science of Veterinary Acupuncture and Moxibustion of the Traditional Chinese Veterinary Medicine

Veterinary acupuncture and moxibustion, one of the therapeutic methods in traditional Chinese veterinary medicine (TCVM) with a history of several thousand years, developed in the ancient times and has contributed much to the healthcare and medical treatment for animals in China.

1 General Introduction to Acupoints

Acupoints are the spots where qi and blood from the viscera and meridians effuse and infuse in the body surface. The name of acupoints indicates that the basic characteristics: the spots where qi and blood from the viscera and meridians effuse and infuse; acupoints are usually located in the interstices in the thick muscles or between tendons and bones.

1.1 Classification of acupoints

Acupoints can usually be classified into meridian acupoints, extraordinary acupoints and Ashi points according to the characteristics of acupoints.

1.1.1 Meridian acupoints

Meridian acupoints, also known as acupoints of the fourteen meridians, refer to the acupoints located on the twelve meridians as well as the governor and conception vessels.

1.1.2 Extraordinary acupoints

Extraordinary acupoints refer to the acupoints not included in the acupoints of the fourteen meridians.

1.1.3 Ashi points

Ashi points actually refer to tenderness spots. Such points are marked by no fixed location, no pertaining meridians and no names.

1.2 Functions of acupoints

The functions of acupoints are marked by proximal curative effect, distal curative effect and special curative effect.

1.3 Methods for locating acupoints

Accurate location of acupoints is prerequisite to the treatment of disease with veterinary acupuncture and moxibustion therapy. However, accurate location of acupoints depends on proper selection of acupoints.

The methods commonly used to locate acupoints are anatomical landmarks, finger measurement, simple location and searching acupoints.

1.3.1 Anatomical landmark

Anatomical landmarks include fixed landmarks and moving landmarks.

1.3.2 Simple locations

The ancient doctors according to their clinical experience developed simple location. For example, Baihui.

1.3.3 Searching acupoints

Searching acupoints means that the doctor presses around the acupoint to decide its exact location. The acupoints are usually located in the bone spaces, muscular interstices and depression, pressing around is helpful for finding such spaces and interstices. Acupoints usually reflect pathological changes. Under pathological conditions, searching such tenderness points for needling is often satisfactorily effective.

2 Needles

The round-sharp needles, filiform needles, triangular needles, wide needles, and fire needles are usually used in clinical treatment. Most of them are made of stainless steel. (Figs. 4 - 1).

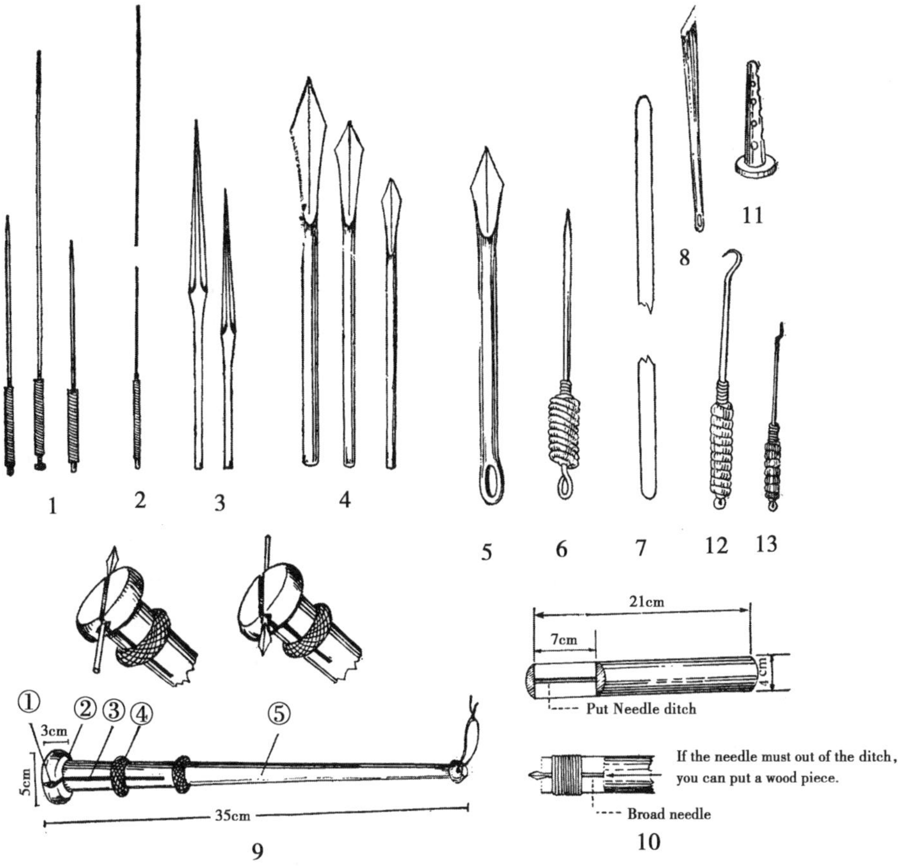

Fig. 4 - 1 Needles

1 ~ 2. filiform needles; 3. triangular needles; 4. wide needles; 5. piercing Huang needle; 6. the fire needle; 7. gas releasing needle; 8. brow needles; 9. needle holder (①needle hammer; ②needle stick; ③saw kerf; ④hoop; ⑤hammer handle); 10. needle stick; 11. standpipe; 12. yu tang hook; 13. san wan needles

3 Manipulating Methods for the Veterinary Acupuncture and Moxibustion

3.1 Safekeeping

The animals for veterinary acupuncture and moxibustion therapy must be safe kept well and truly at first. The veterinary acupuncturist should pay attention to the animals' expression and amicably deal with the animals in order to relax them and enable them to cooperate in the treatment.

3.2 Sterilization

Sterilization includes sterilization of the needles, selected region for needling, and the hands of the veterinary acupuncturist.

3.3 Needling methods

3.3.1 Needling with filiform needles and round-sharp needles

3.3.1.1 Insertion

The needle usually should be held with the right hand known as the puncturing hand. The left hand, known as the pressing hand, pushes firmly against the area close to the acupoint or presses the needle body from both sides to assist the right hand. The needle should be inserted coordinately with the help of both hands. The following are some of the commonly used methods of insertion in clinical treatment.

Nailing insertion of the needle (Inserting the needle aided by the pressure of the finger of the pressing hand): Press beside the acupoint with the nail of the thumb or the index finger of the left hand, hold the needle with the right hand and keep the needle tip closely against the nail, and then insert the needle into the acupoint (Fig. 4 – 2.1). This method is suitable for puncturing with short needles.

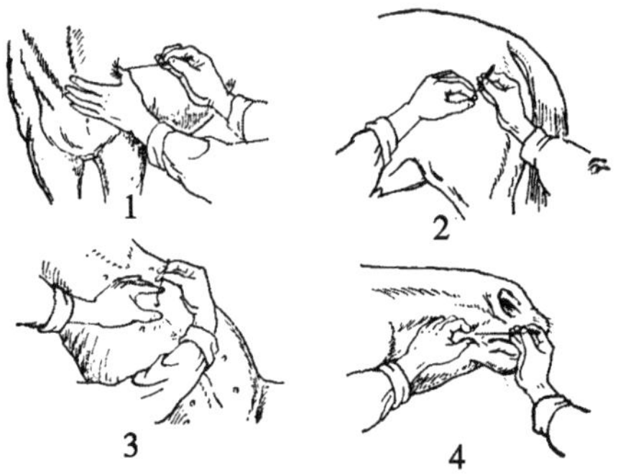

Fig. 4 – 2 Insertion

Holding insertion of the needle (Inserting the needle with the help of the puncturing and pressing hands): Hold the needle tip with sterilized dry cotton balls held by the thumb and the index finger of the left hand, keep the needle tip on the skin surface of the acupoint. Then insert the needle into the skin with both hands (Fig. 4 – 2.2). This method is suitable for puncturing with long needles.

Relaxed insertion of the needle (Inserting the needle with the fingers stretching the skin): Stretch the skin where the acupoint is located with the thumb and the index finger of the left hand, hold the needle with the right hand and then insert it into the area between the two fingers (Fig. 4 – 2.3). This method is suitable for puncturing the acupoints located on the regions with loose skin.

Lifting and pinching insertion of the needle (Inserting the needle by pinching the skin): Pinch the skin up around the acupoint with the thumb and index finger of the left hand; insert the needle into the acupoint with the right hand (Fig. 4 – 2.4). This method is suitable for puncturing the acupoints where the muscles are thin.

3.3.1.2 Angle and depth of insertion

Different angles and depth in needling the same acupoint may puncture different tissues; produce varied needling sensation and therapeutic effects. To different acupoints, appropriate angle and depth are selected according to the location of the acupoints. Therefore, angle and depth are especially important in the process of needle insertion. Correct angle and depth are helpful in producing desired therapeutic effects and preventing needling accidents.

Angle of insertion: The angle formed by the needle and the skin surface is usually classified into three kinds:

Perpendicular insertion: Perpendicular insertion, in which the needle is inserted perpendicularly, means that there is an angle of 90° formed between the needle and the skin surface. This method is applicable to most acupoints on the body.

Oblique insertion: The needle is inserted obliquely to form an angle of approximately 45° between the needle and the skin surface. This method is used for needling the acupoints close to the important viscera or tissues, or the acupoints, which are not suitable for perpendicular and deep insertion.

Horizontal insertion: The needle is inserted transversely to form an angle of about 15° between the needle and the skin surface, also known as transverse insertion. This method is applicable to the areas where the muscle is thin.

Depth of needle insertion: In clinical treatment, the depth of insertion mostly depends upon the location of the acupoints, the constitution of the sick animal and the pathological conditions.

Basic manipulation techniques:

Twirling-rotating: After the needle is inserted to the desired depth, the needle is twirled and rotated backward and forward with the thumb, index and middle fingers of the right hand.

Lifting-thrusting: After the needle is inserted to a certain depth, the needle is lifted and thrusted perpendicularly and continuously.

Twirling-rotating and lifting-thrusting are the two basic manipulations and can be used individually or in combination. The amplitude of twirling and the scope of lifting-thrusting as well as the frequency and duration of manipulation depend upon the sick animal's constitution, pathological conditions and the acupoints to be needled.

3. 3. 1. 3 Retention and withdrawal of the needle

Retention: Retention means to hold the needle in the acupoint after the use of needling manipulations. For sick animals with a slow and weak needling sensation, retention of the needle may strengthen needling effect and induce needling sensation. Whether the needles should be retained or not and the duration of retention depend on the sick animals' conditions, For common diseases, the needles can be withdrawn or be retained for 10-20 minutes after the application of needling manipulations. But for some special diseases, the time for retaining the needle may be appropriately prolonged. At the same time, manipulations may be performed at intervals in order to improve the therapeutic effects.

Withdrawal: For the withdrawal of the needle, press the skin around the acupoint slightly with sterilized dry cotton balls held by the left hand, rotate the needle handle gently and lift it slowly to the subcutaneous level with the right hand, then withdraw it quickly and press the punctured acupoint with sterilized dry cotton balls to prevent bleeding. After the treatment, the veterinary acupuncturist should count the number of the needles to make sure that all the needles are withdrawn.

3. 3. 1. 4 Management of possible accidents

Stuck needle: When stuck needle happens, the veterinary acupuncturist may feel tense and unsmooth beneath the needle and difficult to twirl, rotate, lift or thrust the needle. Then the needle should not be twirled and rotated again. The methods used to cope with such accidents vary according to the conditions of the sick animal. If stuck needle is caused by nervousness and excessive contraction of the local muscles, the veterinary acupuncturist should appropriately prolong the retention of the needle, or press the local region gently, or insert another needle near the stuck needle. If it is caused by excessive rotation to one direction, then rotation of the needle to the opposite direction slight lifting and thrusting will solve the problem.

Bent needle: When the needle is bent, twirling and rotating manipulation should in no case be applied. The needle may be removed and withdrawn slowly by following the direction of bending. The following methods can be taken to avoid bending the needle: During needling, the veterinary acupuncturist should manipulate the needle gently, avoiding forceful manipulation. And during the retention period, the needle handle shall in no case be impacted or pressed by external force.

Broken needle: When the needle is the broken, veterinary acupuncturist should keep calm. If the broken part protrudes over the skin, it should be removed with forceps. If the broken part is kept at the same level of the skin or a little depressed, the skin around the needle is pressed perpendicularly with the thumb and the index finger of the left hand in order to expose the broken end, which is then removed with forceps. If it is completely sunken into the skin, surgical treatment should be resorted to. The following methods can be taken to prevent breaking of the needle: The quality of the needle is inspected carefully prior to the treatment; manipulation should be performed gently and slightly, avoiding forceful manipulation lest the needle be broken.

Hematoma: Hematoma refers to swelling pain caused by subcutaneous hemorrhage around the area needled. If the local region is cyanotic or painful after the withdrawal of the needle, the needled region should be immediately sterilized with dry cotton balls for a while to stop bleeding. If hematoma is mild, it will disappear automatically. If the local swelling and pain is serious and the area with cyanosis is large, cold compress can be used to stop immediate bleeding. After bleeding is stopped, hot compress or pressure is performed slightly and gently to help disperse the hematoma.

Pneumothorax: On puncturing the acupoints located on the supraclavicular fossa, chest, back, axilla, and hypochondriac region, deep insertion may lead to pneumothorax due to the injury of the pleura and lung. The manifestations are sudden chest distress, pectoralgia, and short breath. In severe cases, there may exhibit dyspnea, cyanosis of the lips, sweating and drop of blood pressure. Physical examination may find hyperresonance in percussing the chest, attenuation or disappearance of esicular respiration, or shift of the trachea to the healthy side. X-ray can diagnose the degree of pneumothorax. If it is mild, the sick animal may take some antitussive and antiseptic. The sick animal should be treated under careful inspection. In severe cases, emergency measures should be employed at once.

3.3.2 Needling methods of the three-edged needles and the wide needles

The three-edged needle is the needle shaped with a triangular head and a sharp tip, known as "lance needle" in the ancient times. It is used to prick superficial vein for bloodletting.

The wide needles have a lance head and a sharp cutting edge, and they are divided into 3 kinds of needles. They are used for bloodletting at horses and cattle's.

3.3.3 Needling method of the fire needles

The fire needle method is that the acupoints are pricked with the special needles which have been burnt to hot, and it is a traditional technology used very abroad in veterinary clinic. Because of the local muscle burn at pricking, the stimulating at acupoints can last longer.

It has been proved by practice that the fire needle method has a special effect for the treatment of asthenia-cold syndrome, such as rheumatic and the chronic disease at waist and limbs.

3.3.4 Moxibustion methods

Moxibustion is a therapy used to treat and prevent diseases by applying burning moxa to stimulate the animal body. The material used for moxibustion is mainly Chinese mugwort leaf, which is fragrant and easy to be ignited. It is processed into mugwort wool for clinical use.

3.3.5 Electro-acupuncture

Electro-acupuncture is a kind of therapy by which the needle is attached to a trace pulse current after it is inserted to the selected acupoint for the purpose of producing synthetic effect of electric and needling stimulation. Different kinds of electric instruments are used. Most of them are composed of semiconductor elements and adopt oscillators. They generate low-frequency impulse current, which is close to bioelectric current in the animal body.

3.3.6 Acupoint injection

Acupoint injection, also known as "water-needling" or "aqupuncture", is a therapy used to inject the medicated solution into an acupoint or a point with positive reaction to produce a synthetic effect of acupuncture and medicaments.

4 Selection of Acupoints and Compatibility of Acupoints

The therapeutic effects of veterinary acupuncture and moxibustion are induced through stimulating the selected acupoints by means of acupuncture or moxibustion. Therefore the selection and combination of acupoints play an important role in the treatment with veterinary acupuncture and moxibustion. Correct selection and combination of acupoints is key to the curative effects of veterinary acupuncture and moxibustion therapy.

4.1 Methods for selecting acupoints

The main principles for selecting acupoints are to select along the meridians and according to the pathological conditions.

4.1.1 Selection of proximal acupoints

This is a method to select acupoints located on the adjacent area of the disease because acupoints can be needled to treat the diseases around them or near them.

4.1.2 Selection of distal acupoints

It means to select acupoints distal to the affected area. The acupoints selected in this case are usually located below the elbows and knees. This method for selecting acupoints is based on the distribution of meridians and collaterals. For example, Housanli is selected to treat diarrhea.

4.1.3 Selection of contralateral acupoints

It is a method used to select the acupoints located opposite to the affected side, known as "to select the acupoints on the fight side to treat diseases on the left, and vice versa". For example, the acupoints located on the right ankle are selected to treat the sprain of the left ankle.

4.1.4 Selection of acupoints according to the symptoms

This method is used to select acupoints for the treatment of syndromes involving the whole body, the clinical manifestations of which are not limited to a certain local area according to the indication of acupoints and the theory of syndrome differentiation. For example, Dazhui is selected to treat high fever; Tianmen is selected to treat cold; Feishu or Feipan is selected to treat chogh; Zhigou (TE 6) is selected to treat constipation.

4.2 Methods for the compatibility of acupoints

The prescription used in acupuncture treatment is composed of over two acupoints, which are combined in the light of the syndromes. The purposes of combining acupoints are to strengthen the essential and comprehensive therapeutic effects. Therefore there are two ways to combine acupoints: one is to combine the acupoints with the same or similar indication and the other is to combine the corresponding acupoints according to pathological conditions or differentiation of syndromes.

5 Main Factors Affecting the Curative Effects of Veterinary Acupuncture and Moxibustion

There are many factors, which may affect the curative effect of veterinary acupuncture and moxibustion including every aspect in the procedure of treatment concerning both the veterinary acupuncturist and the sick animal. Among these factors, proper treatment, correct selection of acupoints and performance are the most important ones.

6 Horse Acupuncture

6.1 Head Region

see Fig. 4-3, Fig. 4-4, Fig. 4-5.

There are 36 points on the head of the horse. The most commonly used points are points 01, 06, 07, 09, 10 17, 21, and 35.

01. DA-FENG-MEN. Great wind gate: GV-24.

LO: This point has 3 locations. The main location is on the dorsal midline at the rostral end of the mane. The bilateral auxiliary points are 3cm ventro-lateral to the main point.

AN: Main acupoint on external sagittal crest at caudal end of parietal bones. Axillary acupoint On bifurcation of external sagittai crest (temporal lines) 3cm ventral and iateral to the main point. Musclesofaunculares dorsales, superficial temporal a. and v., and auriculopalpebral n are distributed at the point.

ME: Angular insertion toward the mane along the subcutaneous tissue with filiform needle for 3cm deep. For fire needling, 2cm deep. Moxibustion or cauterization point.

IN: Teranus, encephalitis, encephalomyelitis encephaledema.

#13. SAN-JIANG. Three rivers.

LO: On the angular vein about 3cm ventral to the medial canthus.

AN: Distribution of the angular and v, and the tacial n.

ME: Lowering the head which engorges the vein. Hemoacupuncture with a triangular or a small wide needle 1cm deep.

IN: Enterospasm, gastric distension, indigestion, keratitis, periodical ophthalmia, and conjunctivitis.

#29. SUO-KOU. Locking mouth.

LO: 6cm dorsal and caudal to anguli oris on the outer edge of orbicularis oris.

AN: Innervated by buccal a and v, and maxillary or buccal n.

ME: Puncture with a filiform needle dorsocaudally towards the point #30. Or puncture with a fire needle 3cm deep, or cauterization.

IN: Tetanus, facial paralysis.

#34. TONG-GUAN. Passing pass.

LO: On the ventral surface of the tongue, on sublingual vein lateral to frenulum linguae.

AN: Sublingual a and v distributed.

ME: Hemo-acupunclure 1cm deep. Wash with cold water

IN: Glossitis and stomatitis, hard swollen tongue, stomach heat, sunstroke.

#35. ER-JIAN. Ear tip.

LO: The vein at the tip of the ear.

AN: On the convex surface of the ear at the conjunction of the medial, middle and lateral branches of greater auricularis v. Along the point are caudal auricular a and n.

ME: Hemo-acupuncture.

IN: Enterospasm, heat stroke, common cold.

6.2 Trunk and Tail

There are 50 points in the trunk and tail of the horse. The most commonly used points are numbers 39, 40, 43, 59, 62, 68, 69, 73 & 78.

#39. JIU-WEI. Cervical 9 points.

LO: On each side of the neck, a group of 9 acupoints form an arching chain. The most cranial point is about 3cm caudal to the point #38, and 3.5cm ventral to the mane. The caudal most point is 4.5cm cranial to the cranial angle of the scapula, and 5cm ventral to the mane. The remaining 7 points are equidistantly located between the most cranial and caudal points.

AN: All points are located along the ventral border of the cervical rhomboideus. In cmnio-caudal orientation, the 1st and 2nd points reach m obliquus capitis caudalis. The 3rd and 4th reach m splcnius and m semispinalis capitis. From the 5th to the 8th points, all are in m splenius. All points arc supplied by the deep cervical a and v, and the dorsal br. of the cervical n.

ME: Filiform needling 4.5 ~ 6cm deep: fire needling 2.5 ~ 3cm.

IN: Rheumatism of the cervical muscles.

#40. JING-MAI. Jugular vein.

LO and AN: 6cm caudal to the mandibular angle on the jugular vein. The point is at the junction of the upper and middle 1/3 of the jugular groove. Bilateral.

ME: Raise the head of the horse. Apply pressure to lower part of the jugular vein to engorge the vein. Hemo-acupuncture the vein at the point with a wide needle or a three-edged needle.

IN: Sunstroke, toxication, congestion of the lung, laminitis urticaria, cerebral hyperemia, glossitis and stomatitis.

#43. DA-ZHUI. Large vertebra; GV-14.

LO: In the depression along the midline between C7 and T1 vertebrae. Single point.

AN: Under the skin are the ligamentum supraspinale and lig. nuchae. Supplied by the deep cervical a and v, and the dorsal branches of the 8th cervical and the 1st thoracic n.

ME: Filiform needle punctures forwards and downwards for 6 ~ 9cm deep.

IN: Coryza, fever, cough, rheumatism of the loin and back, convulsion.

59. PI-SHU. Spleen association point; BL-19.

LO: In the 15th intercostal space between muscles of longissimus thoracis and iliocostalis thoracis. Bilateral.

AN: Under the skin is m. latissimus dorsi. Supplied by the intercostal a and v, and the dorsal br. of the thoracic n.

ME: Perpendicular insertion with round-sharp or fire needle 2 ~ 3cm deep. Filiform needle inserts upwards or downwards for 3 ~ 4. 5cm.

IN: Enterospasm, intestinal constipation, indigestion, gaseous bowel, diarrhea, excessive salivation due to cold of stomach.

#62. GUAN-YUAN-SHU. Asso. pt. of enclosed original energy: BL-21; 26.

LO: In the depression caudal to the last rib between the muscles of the longissmus and iliocostalis. Bilateral.

AN: Supplied by the last intercostal a, v, and n.

ME: Perpendicular insertion with round-sharp or fire needle 2 ~ 3cm deep. Filiform needle inserts upwards or downwards for 3 ~ 4. 5cm.

IN: Constipation, enteritis, gastroenteritis, abdominal pain, indigestion, dilatation of stomach, gaseous bowel.

#68. BAI-HUI. Hundred meetings; GV-20.

LO: In the depression between the spinous processes of the last lumbar and the first sacml vertebrae. Single point.

AN: Under the skin is the lig. supraspinale. Supplied by the lumbar a and v, and dorsal br. of the last lumbar n.

ME: Perpendicular insertion. Fire or round-sharp needle 3 ~ 4. 5cm, filiform needle 6 ~ 7. 5cm.

IN: Rheumatism of the hind quarter, arthritis of hip, paralysis of the hind quarter, overexertion, tetanus, colic, gaseous bowel, dilatation of stomach, diarrhea, sprain of the loin and hip.

#69. SHEN-SHU. Kidney asso. pt. ; BL-23.

LO: Six cm lateral to the point Bai-hui (# 68). Bilateral.

AN: Under the skin is m gluteus medius. Supplied by the cranial gluteal a, v and n.

ME: Perpendicular insertion. Fire or round-sharp needle 3 ~ 4. 5cm; filiform needle 6cm. Horizontal insertion may reach point #70 or #71.

IN: Rheumatism in the loin and hip region, lumbar paralysis

#73. DAI-MAI. Thoracic vein.

LO: On the thoracic wall about 6cm caudal to the processus anconaeus on the 7th rib. The point is at the external thoracic vein. Bilateral.

AN: Accompanied by the external thoracic and the lateral thoracic n.

ME: Hemo-acupuncture with wide needle I cm deep.

IN: Enterospasm, sunstroke, gastroenteritis.

#78. QIAN-SHU. Flank asso. pt.

LO: In the right flank 22cm ventral to the dorsal midline and 4. 5cm caudal to the last rib.

ME: Make a surgical incision at the point and insert a gas releasing needle into the intestinal tract to release the gas.

IN: Gaseous distension of the bowel.

#82. HOU-HAI. Caudal sea: GV-l.

LO: in the depression between the anus and the ventral base of the tail. Single point.

AN: Between the coccygeal muscle and the sphincter ani. Supplied by the middle coccygeai a and v, and the caudal rectal n.

ME: Puncture forwards and upwards. Fire or round-sharp needle 9cm: filiform needle 12 ~ 18cm.

IN: Constipation, swelling of anus, dysentery, rectal paralysis, tenesmus.

6.3 Thoracic Limb

Thirty-two acupoints were identified in this region. The most frequently used points are: #87, #94, #103, #104, #113, and #l16.

#87. BO-JIAN. Shoulder tip SI-14.

LO: In the depression just cranial to the cranial angle of scapula. Bilateral.

AN: Under the skin is the muscle of the cervical trapezius. Between m rhomboideus cervicis and m serratus ventralis cervicis. Supplied by the dorsal scapular a and v, dorsal br of the cervical n, and the suprascapular n.

ME: Insert needle slightly backwards and downwards. Needle reaches tile medial surface of scapula. Round-sharp or fire needle 3 ~ 5cm: filiform needle 10 ~ 12cm.

IN: Rheumatism of the thoracic limb, paralysis of tile suprascapular n.

#94. QIANG-FENG. Robbing wind.

LO: The large depression 15cm caudoventral to the shoulder joint. Bilateral.

AN: At the junction of m deltoideus and m triceps brachii. Between long head and lateral head of the triceps. Supplied by brs of the caudal circumflex humeral a and v, and an axiilaris and radialis.

ME: Perpendicular insertion. Round-sharp needle 3 ~ 4cm, filiform needle 8 ~ 10cm.

IN: Rheumatism, paralysis, sprain, and arthritis of the thoracic limb. Myositis of sternobrachiocephalicus, radial paralysis.

#103. XIONG-TANG. Thoracic vein.

LO and AN: Cranial and dorsal to axilla and medial to m biceps brachii under the skin is cephalic vein. Bilateral.

ME: Engorge the vein by holding the head high. Hemo-acupuncture 1cm deep.

IN: Acute arthritis of the shoulder and elbow joints, myositis of sternobrachiocephalicus, laminitis, rheumatism in the shoulder and arm region, sprain shoulder.

#104. ZHOU-SHU. Elbow association point.

LO: In the depression between the tuber of olecmnon and the lateral epicondylc, of humerus. Bilateral.

AN: Under the skin is the lateral head of m triceps brachii. Supplied by the deep brachial a and v, and the radial n.

ME: Perpendicular insertion. Fire or round-sharp needle 3 ~ 4cm; filiform needle 6cm.

IN: Arthritis, swelling, and rheumatism of the elbow joint, shoulder paralysis.

#113. QIAN-CHAN-WAN. Thoracic fetlock.

LO and AN: On both the caudomedial and caudolateral of the fetlock. On the medial and lateral palmar veins slightly proximal to the fetlock. Each limb has 2 points.

ME: Hemo-acupuncture 1cm deep.

IN: Sprain, contusion and arthritis of the fetlock joint, flexor tenositis, and tendosynovitis.

#116. QIAN-TI-TOU. Toe of hoof.

LO: Two to 3cm lateral to the dorsal median line of the toe. 1cm dorsal to the periople or the junction between the coronary border of the hoof anti tile skin. Bilateral.

AN: Under the skin is the coronary vcneous plexus.

ME: Hemo-acupuncture 1cm deep.

IN: Constipation, enterospasm, laminitis, painful and swollen hoof, sprain of the fetlock.

6.4 Pelvic Limb

There are 36 acupoints on the pelvic limb of

the horse. The most commonly used points are #119, #124, #127, #128, #132, #133, #139, #148, and #151.

119. BA-SHAN. Attach to mountain: BL-48/53.

LO: In the depression midpoint between the greater trochanter of femur and the point Bai-hui (#68). Bilateral.

AN: Under the skin are gluteai fascia, and mm gluteus superficialis and gluteus medius. Supplied by the cranial gluteal a, v and n.

ME: Perpendicular insertion. Round-sharp or fire needle 3 ~4. 5cm; filiform needle 10 ~ 12cm.

IN: Rheumatism, arthritis of hip, myositis o: m biceps femoris, sciatic paralysis, sprain of loin and croup.

#124. HUAN-TIAO. Ring cmniad; GB-30.

LO: In the depression 6cm cranial to the greater trochanter of femur. Bilateral.

AN: Between m gluteus superficialis and m tensor fasciae latae. Supplied by the cranial gluteus a, v and n, and the femoral n.

ME: Perpendicular insertion. Round-sharp or fire needle 3 ~4. 5cm; filiform needle 6 ~7. 5cm.

IN: Swelling and pain in the loin and hip, pelvic limb paralysis myositis of thigh muscles.

#127. DA-KUA. Greater trochanter.

LO: Six cm cranioventral to the greater trochanter of femur. Bilateral.

AN: In the depression between m tensor fasciae latae and gluteus superficialis. Supplied by the gluteus cranialis a, v and n.

ME: Angular insertion caudoventrally. Fire or round-sharp needle 3 ~4. 5cm; filifrom needle 6 ~ 8cm.

IN: Rheumatism of the hind quarters, arthritis of the hip joint, pelvic limb paralysis, myositis of biceps femoris, pain and sprain in the hip and croup.

#128. XIAO-KUA. Third trochanter.

LO: In the dcprcssion 3. 5cm caudoventral to the 3rd trochanter. Bilateral.

AN: in the muscular groove of m biceps femoris. Supplied by the caudal femoral a and v, and the tibial and fibular n.

ME: Perpendicular insertion. Round-sharp or fire needle 3 ~4. 5cm; filiform needle 6 ~8cm.

IN: Same as #127.

#131. HUI-YANG. Meeting of Yang.

LO: In the muscle groove 6cm craniolateral to the root of tile tail. Bilateral.

AN: On the most proximal end of tile muscle groove of mm biceps femoris and semitendinosus Supplied by the gluteus cauldalis a. v and n.

ME and IN: Same as #128.

#132. XIE-Qi. Evil qi

LO: In the muscle groove of biceps femoris and semitendinosus, and horizontal to the anus. Bilateral.

AN: Supplied by the same a, v and n as #131.

ME: Perpendicular insertion. Round-sharp or fire needle 4. 5cm; filiform needle 6 ~ 8cm

IN: Rheumatism, arthritis, pain or swelling in the pelvic limb. Myositis of biceps fcmoris. semimembranosus or semitendinosus.

#133. HAN-GOU. Weat groove; BL--51.

LO: In the same muscle groove of biceps femoris and semi tendinosus, 7cm distal to # 132. Bilateral.

AN: Supplied by the deep femoral il and n, and the tibial n.

ME and IN: Same as # 132.

#139. HOU-SAN-LI. Pelvic 3 miles: ST-36.

LO: Dorsolateral surface of leg. 7. 5cm distill to the ventral border of the patella. In the depression equidistance to the tubcrositv of the tibia and the head of the fibula. Bilateral.

AN: In the muscle groove of m extensor digitorum Iongus and m extensor digitorum lateralis Supplied by the cranial tibial a and v, and the fibular n.

ME: Angular insertion downwards. Round-sharp or fire, needle 2 ~ 4cm; filiform needle

4.5 ~6cm.

IN: Indigestion, constipation, dysentery, rheumatism of the pelvic limb, paralysis of tibial or fibular nerves.

#148. HOU-CHAN-WAN. Pelvic fetlock.

Counter point of the thoracic limbs #113.

#149. LAO-TANG. Proximal sesamoid of pelvic limb.

Counter point ofthe thoracic limbs #114.

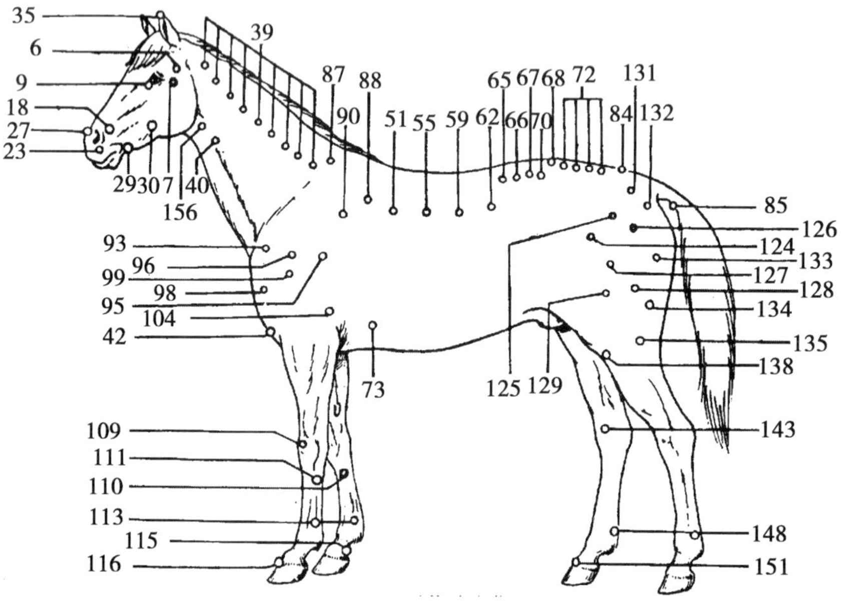

Fig. 4 –3 Acupoints and body surface of the horse

6. SHANG-GUAN 7. XIA-GUAN 9. JING-MING 18. XUE-TANG 23. JIANG-YA 27. CHOU-JIN 29. SUO-KOU 30. KAI-GUAN 35. ER-JIAN 39. JIU-WEI 40. JING-MAI 42. CHUAN-HUANG 51. FEI-ZHI-SHU 55. WEI-SHU 59. PI-SHU 62. GUAN-YUAN-SHU 65. YAO-QIAN 66. YAO-ZHONG 67. YAO-HOU 68. BAI-HUI 70. SHEN-PANG 72. BA-LIAO 73. DAI-MAI 84. WEI-GEN 85. WEI-BEN 87. BO-JIAN 88. BO-LAN 90. FEI-PAN 93. JIOAN-JING 95. CHONG-TIAN 96. JIAN-ZHEN 98. JIAN-YU 99. JIAN-WAI-YU 104. ZHOU-SHU 109. XI-YAN 110. XI-MAI 111. ZHAN-JIN 113. QIAN-CHAN-WAN 115. QIAN-TI-MEN 116. QIAN-TI-TOU 124. HUAN-TIAO 125. HUAN-ZHONG 126. HUAN-HOU 127. DA-KUA 128. XIAO-KUA 129. HOU-FU-TU 131. HUI-YANG 132. XIE-QI 133. HAN-GOU 134. YANG-WA 135. QIAN-SHEN 138. LUE-CAO 143. QU-CHI 148. HOU-CHAN-WAN 151. HOU-TI-TOU 156. HOU-MEN

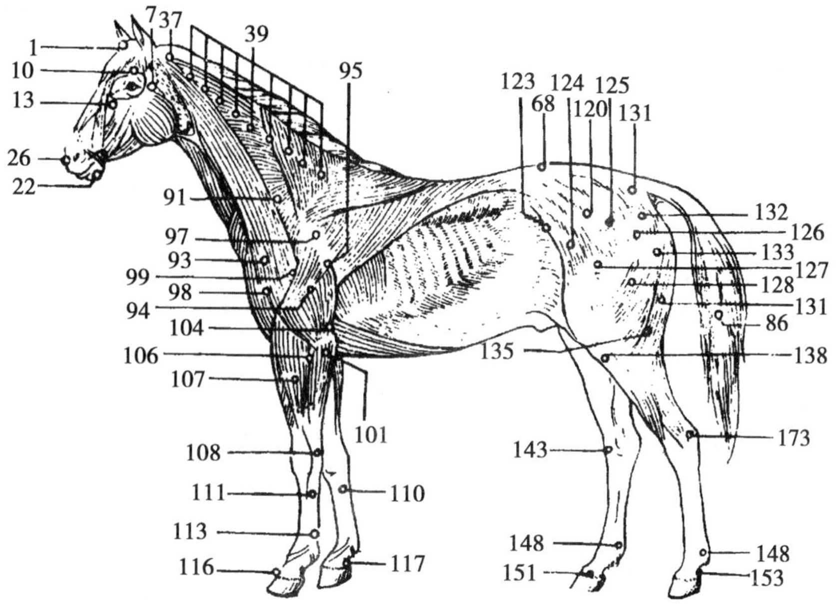

Fig. 4 – 4 Acupoints and Musculature of the horse

1. DA-FENG-MEN 7. XIA-GUAN 10. JING-SHU 13. SAN-JIANG 22. CHENG-JIANG 26. WAI-CHUN-YIN 37. FENG-MEN 39. JIU-WEI 94. QIANG-FENG 95. CHONG-TIAN 97. TIAN-ZHONG 98. JIAN-YU 99. JIAN-WAI-YU 101. CHENG-DENG 104. ZHOU-SHU 106. CHENG-ZHONG 107. QIAN-SAN-LI 108. GUO-LIANG 110. XI-MAI 111. ZHAN-JIN 113. QIAN-CHAN-WAN 116. QIAN-TI-TOU 117. QIAN-JIU 120. LU-GU 123. JU-LIAO 124. HUAN-TIAO 125. HUAN-ZHONG 126. HUAN-HOU 127. DA-KUA 128. XIAO-KUA 131. HUI-YANG 132. XIE-QI 133. HAN-GOU 134. YANG-WA 135. QIAN-SHEN 138. LUE-CAO 143. QU-CHI 148. HOU-CHAN-WAN 151. HOU-TI-TOU 153. HOU-JIU 173. KUNLUN

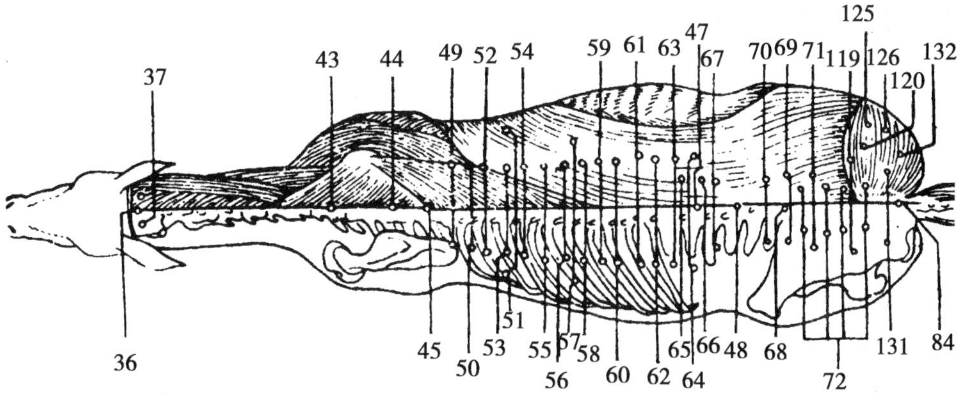

Fig. 4 – 5 Acupoints and Musculature of the horse

36. TIAN-MEN 37. FENG-MEN 43. DA-ZHUI 44. QI-JIA 45. SAN-CHUAN 47. MING-MEN 48. YANG-GUAN 49. JUE-YIN-SHU 50. DU-SHU 51. FEI-ZHI-SHU 52. FEI-SHU 53. GE-SHU 54. DAN-SHU 55. WEI-SHU 56. GAN-ZHI-SHU 57. GAN-SHU 58. SAN-JIAO-SHU 59. PI-SHU 60. QI-HAI-SHU 61. DA-CHANG-SHU 62. GUAN-YUAN-SHU 63. XIAO-CHANG-SHU 64. PANG-GUAN-SHU 65. YAO-QIAN 66. YAO-ZHONG 67. YAO-HOU 68. BAI-HUI 69. SHEN-SHU 70. SHEN-PANG 71. SHEN-JIAO 72. BA-LIAO 84. WEI-GEN 119. BA-SHAN 120. LU-GU 125. HUAN-ZHONG 126. HUAN-HOU 131. HUI-YANG 132. XIE-QI

7 Bovine Acupoints

7.1 Head Region

There are 17 points on the head of the ox. The most commonly used points are #01, #03, #06, #09, #10, #11, #16, and #17.

#01. TIAN-MEN. Heavenly gate.

LO: On the dorsal midline between the two ear bases. Shaking the head up and down, there is a depression felt ill the above site. Single point.

AN: in the depression between the occipital crest and the alae atlantis. Under the skin is the ligamentum nuchae. Supplied by the cranial br of the occipital a and n.

ME: Angular insertion backwards and downwards. Fire or round-sharp needle 3cm; filiform needle 3 ~ 6cm.

IN: Coryza, cold, epilepsy, tetanus, syncope, cerebral vessel congestion.

#02. LONG-HUI. Dragon meeting.

LO: Midpoint between the 2 lateral canthuses on the dorsal midline. Single point.

AN: Under the skin is the frontal sinus. Supplied by the frontal il, v and n.

ME: Cauterization or moxibustion.

IN: Coryza, cold, congestion of cerebral vessel, syncope, epilepsy.

#03. ER-JIAN. Ear tip.

LO and AN: On the convex surface of the ear 3cm from the ear tip. At the 3 brs of the greater auricular vein. Three points on each side.

ME: Hemo-acupuncture.

IN: Sunstroke, coryza, cold, colic fever, toxic symptoms, and abdominal pain.

#04. ER-GEN. Ear base.

LO: in the depression ventrocaudal to the ear base.

AN: Between the ear base and tile cranial border of the alae atlantis. Under the skin is m obliquus capitis cranialis. Supplied by the caudal auricular a and v, and the caudal auricular br of the facial n.

ME: Angular insertion inwards and downwards. Wide or fire needle 1 ~ 1.5cm; round-sharp or filiform needle 3 ~ 6cm.

IN: Cold, coryza, exhaustion, fatigue, rheumatism of the cervical muscles.

#05. TONG-TIAN. Communicate with heaven.

LO: On the dorsal midline 6 ~ 8 above the line of medial canthus. Single point.

AN: Under the skin is the dorsal wall of the frontal sinus. Supplied by the frontal a, v and n.

ME: ttorizontal insertion upwards. Fire needle 2 ~ 3cm. Cauterization.

IN: Same as #01.

#06. SHAN-GEN. Mountain base.

LO: There are 3 loci of this point. All are found on tile dorsal surface of the planum nasolabiale of the nose. File main point is on the midline at the dorsal most portion of the planum. The 2 auxiliary points lire located lateral to the main point and dorsal to the dorsal angle of nostril.

AN: Under the skin are m dilator naris apicalis and m dilator naris medialis. Supplied by the maxillary labial and caudal lateral nasal a, the dorsal nasal v, and the buccal n.

ME: Hemo-acupuncture 1cm. Filiform needle 3 ~ 4.5cm.

IN: Sunstroke, cold, coryza, cough, colic pain, syncope.

#07. BI-ZONG. Middle of nose.

LO: At the center between the two ventral angles of the nostril. Single point.

AN: On the planum nasolabiale. Under the skin is m dilator naris apicalis. Supplied by the lateral nasal a and vi and the dorsal nasal n.

ME: Hemo-acupuncture 1cm.

IN: Indigestion, fever, loss of appetite, lip swelling, epistaxis.

#08. CHENG-JIANG. Receiving saliva.

LO: On the midline at the veternal edge of the lower lip.

AN: Under the skin is m orbicularis oris. Supplied by the mental a, v and n.

ME: Hemo-acupuncture 1cm.

IN: Stomach and intestinal disturbances, laryngitis, stomatitis, fever, heat in the 5 organs.

#09. JING-SHU. Eye association point.

LO: At the midpoint of the upper eyelid above the globe.

AN: Just ventral to the supraorbital process of the frontal bone. Supplied by the supraorbital a, lind the frontal n.

ME: Press the eyeball downwards, puncture between the supraorbital process and the globe. Filiform needle 3 ~5cm. Hemo-acupuncture the corresponding conjuntiva.

IN: Syncope, swelling of the upper eyelid, liver meridian problem.

#10. JING-MING. Clear eye.

LO: On the lower eyelid at the junction of the medial and middle 1/3 of the line between the two canthuses.

AN: Between the globe and the lacrimal bone. Supplied by the zygomatic a, v and n.

ME: Press the eyeball upwards. Filiform needle insertion inwards and downwards 3cm. Hemo-acupunture the corresponding conjuntiva.

IN: Conjunctivitis, keratitis, fever of liver meridian.

#11. TAI-YANG. Great yang.

LO: In the temporal fossa 3cm caudal to the lateral canthus.

AN: Under the skin are the superficial temporal a and v, and the deep temporal n. and the facial n.

ME: Perpendicular insertion filiform needle 3 ~6cm. Hemo-acupuncture 1 ~2cm.

IN: Sunstroke, cold, coryze, conjunctivitis, keratitis.

#12. SAN-JIANG. Three streams.

LO and AN: Four and a half cm ventromedial to the medial canthus on the angular vein of the eye. Supplied by the malar a. and the trochlear n.

ME: Hemo-acupuncture 1cm.

IN: Swelling of the eyelid, colic, indigestion.

#13. BI-SHU. Nose association point.

LO and AN: At the 2 sides of the dorsal nose 4.5cm dorsal to the dorsal angles of tile nostril, midpoint between the dorsal angle and the nasoincisive notch. Supplied bv the dorsal nasal a and v, and the lateral nasal n.

ME: Hemo-acupuncture 1.5cm.

IN: Coryza, cold, cough, nasal swelling and pain.

#14. KAI-GUAN. Open-close.

LO: On the cheek along the rostral border of the masseter muscle between the upper and lower 3rd premolars.

AN: Supplied by the facial a and v, and the ventral buccal n.

ME: Angular insertion upwards and backwards. Round-sharp 2 ~ 3cm: filiform needle 3 ~4.5cm.

IN: Facial paralysis, trismus, swelling in the buccai region.

#15. BAO-SAI. cmbrace mandible.

LO: Seven and a half cm dorsocaudai to the point #14.

AN: In the masseter muscle between the last upper and lower molars. Supplied by the masseter a and v, and the dorsal buccal n.

ME and IN: Same as #14.

#16. TONG-GUAN. Passing pass.

LO and AN: On the sublingual vein of the ventral surface of the tongue. Lateral to the rostral rim of the frenulum linguae.

ME: Hemo-acupuncture 1cm.

IN: Loss of appetite, sunstroke, laryngitis, laryngopharyngitis, hard-swollen tongue. Acupuncture to the point during spring and fall promotes the animals' health.

#17. SHUN-QI. Orderly qi.

LO and AN: The opening of the incisive duct lateral to the incisive papilla. Supplied by the greater palatine a, v and n.

ME: Special treatment. Passing a smooth, slender and flexible twig into the point 18 ~211cm.

IN: Ruminal tympany, cold, coryza, ocular discharge.

7.2 Trunk and Tail

There are 37 points in the region of trunk and tail. The most commonly used points arc #20, #22, #24, #31, #35, #36, #39, #40, #50, #51, #52, and #53.

#18. FENG-MEN. Wind gate.

LO: In the depression 6cm ventral to the point #04.

AN: Between the glandula parotis and the cranial border of the alae atlantis. Under the skin is m obliquus capitis cranialis. Supplied by the occipitalis a and v, and the dorsal br of the cervical m.

ME: Angular insertion inwards and downwards. Fire or round-sharp needle 3cm; filiform needle 4.5cm.

IN: Cold, coryza, rheumatism of the cervical muscle, tetanus.

#20. JING-MAI. Jugalar vein.

LO and AN: On the jugular vein 6cm caudal to the prominentia laryngis.

ME: ttemo-acupuncture 1cm.

IN: Acute poisoning, sunstroke, pneumonia, congestion (if cerebral vessel, heat in the five organs.

#21. DAN-TIAN. Red field.

LO: On the dorsal midline between the processus spinosus of the TI, lind T2 vertebrae. Single point.

AN: Under the skin are the interspinai muscle and Jig. Supplied by the costocervical a and v, and the dorsal br of the 1st thoracic n.

ME: Angular insertion forwards and downwards. Wide, fire or round-sharp needle 3cm, filitoml needle 6cm.

IN: Sunstroke, exhaustion, rheumatism of thoracic limb, shoulder problems.

#22. QI-JIA. Withers.

LO: On the dorsal midline between processus spinosus of the T4 and T5 vertebrae. At the intersection of the dorsal extension of the spine of the scapula and the midline of the back. Single point.

AN: Similar to #21, but is innervated by the dorsal br of the 4th thoracic n.

ME: Same as #21.

IN: Rheumatism of the thoracic limb, cough due to heat in the lung, swelling and pain in the shoulder and scapula, displacement of scapula.

#23. SAN-CHUAN. Three rivers.

LO: On the dorsal midline between the processus spinosus of the T5 and T6 vertebrae. Single point.

AN: Under the skin are the interspinal muscle and the interspinous lig. Supplied by the intercostal a and v, and the dorsal br of the 5th thoracic n.

ME: Same as #21.

IN: Same as # 21, diarrhea, abdominal pain.

#24. SU-QI. Reviving qi.

LO: On the dorsal midline between the processus spinosus of the T8 and T9 vertebrae. Single point.

AN: Similar to #23, but is innervated by the 8th intercostal n.

ME: Same as #21.

IN: Cough, asthma, heat in the lung.

#25. AN-FU. Calm and happy.

LO: On the dorsal midline between the processus spinosus of the T10 and T11 vertebrae. Single point.

AN: Under the skin are the supraspinous lig and the interspinal muscle and lig. Supplied by the intercostal a and v, and the dorsal br of the 10th intercostal n.

ME: Same as #21.

IN: Diarrhea, rheumatism, pneumonia, lung congestion, bronchitis, heat in the lung.

#26. TIAN-PING. Heavenly peace.

LO: On the dorsal midline between the processus spinosus of the T13 and L1 vertebrae. Single point.

AN: Similar to #25, but ix innervated by the last intercostal n.

ME: Perpendicular insertion. Wide, fire or round-sharp needle 3cm, filiform needle 4. 5cm.

IN: Anuria, epistaxis, hematuria, hemafecia, other bleeding problems, enteritis.

#27. HOU-DAN-TIAN. Caudal red field.

LO: On the dorsal midline between the processus spinosus of the El and L2 vertebrae. Single point.

AN: Similar to #25, but is supplied by the 1st and 2nd lumbar a and v, and the dorsal br of the 1st lumbar n.

ME: Same as #26.

IN: Indigestion, anuria, loss of appetite, lumbar rheumatism, swelling and pain in the loin and hip region.

#28. MING-MEN. Life gate.

LO: On the dorsal midline between the processus spinosus of the L2 and L3 vertebrae. Single point.

AN: Similar to #26. Supplied by the 2nd lumbar a and v, and the dorsal br of the 2nd lumbar n.

ME: Same as #26.

IN: Same as #27, retained placenta.

#29. AN-SHEN. Calming kidney.

LO: On the dorsal midline between the precessus spinosus of the L3 and L4 vertebrae. Single point.

AN: Similar to #26. Supplied by the 3rd lumbar a, v and n.

ME: Same as #26.

IN: Same as #28.

#30. YAO-ZHONG. Middle lumber.

LO: Six cm lateral to #29.

AN: Between the transverse processes of the L3 and LA vertebrae. Under the skin is m longissimus dorsi. Supplied by the 3rd lumbar a, v and n.

ME: Same as #26.

IN: Rheumatism of the loin and hip, pain in the same area, cystitis.

#31. BAI-HUI. Hundred meetings.

LO: On the dorsal midline between the processus spinosus of the L6 and SI vertebrae. Single point.

AN: Similar to #25. Supplied by the last lumbar a, v and n.

ME: Perpendicular insertion. Wide, fire or round-sharp needle 3 ~4. 5cm, filiform needle 6 ~ 9cm.

IN: Rheumatism or sprain of the loin and hip, ruminal tympany, difficiency of the kidney.

#32. SHEN-SHU. Kidney association point.

LO: In the muscle groove 8cm lateral to the point #31.

AN: Under the skin is m gluteus medius. Supplied by the cranial gluteal a, v and n.

ME: Perpendicular insertion. Wide, fire or round-sharp needle 3 CT: filiform needle 4. 5cm.

IN: Rheumatism or sprain of the loin and hip.

#33. DAI-MAI. Thoracic vein.

LO and AN: On the lateral thoracic vein 10cm caudal to the olecranon at the 7th rib.

ME: Hemo-acupuncture 1cm.

IN: Cold, dysentery, sunstroke, enteritis, abdominal pain.

#34. LIU-MAI. Six channels.

LO: About 10cm lateral to the dorsal midline in the 10th to 12th intercostal spaces. Three on each side.

AN: In the muscle groove of m longissimus dorsi and imliocostalis thoracis. Supplied by the corresponding intercostal a, v and n.

ME: Angular insertion inwards and backwards. Wide, fire or round-sharp needle 3cm: filiform needle 6cm.

IN: Constipation, diarrhea, loss of appetite, indigestion gaseous distension.

#35. PI-SHU. Spleen association point.

This is the cranial most point of # 34.

#36. GUAN-YUAN-SHU. Association point of enclosed original energy.

LO: Caudal to the last rib and cranial to the tip of processus transversi of LI vertebra.

AN: In the muscle groove between' m longissimus dorsi and m iliocostalis. Supplied by the last intercostal a, v and n.

ME: Puncture inwards and downwards. Wide, fire or round-sharp needle 3cm; filiform needle 4.5cm, or puncture toward the vertebral column 6 ~ 9cm.

IN: Same as #34.

#37. DU-JIAO. Ruminal corner.

This is the left side of the point #36.

#38. SHI-ZHANG. Ruminal tympany.

LO: On the left side of the thoracic wall 18cm from the dorsal midline at the 1 lth intercostal space. Single point.

AN: Supplied by the intercostal a and v, and the dorsal br of the thorac'ic n.

ME: Puncture downwards. Wide or filiform needle 9cm. Allow the nnedle to enter the rumen.

IN: Same as #34.

#39. TONG-QIAO. Reaching key sites.

LO and AN: Cranial to #34 in the muscle groove of m longissimus dorsi and m iliocostalis thoracis. In the 6th to 9th intercostal spaces, 4 points on each side. Supplied by the intercostal a, v and n.

ME: Perpendicular insertion. Wide, fire or round-sharp needle 3cm; filiform needle 4.5 ~ 6cm.

IN: Lung problems cough, exhaustion, rheumatism of the local region.

#40. QIAN-SHU. Flank association point.

LO: In the left side of the flank at the deepest part of the paralumbar fossa. Single point.

AN: Supplied by the deep circumflex iliac a and v, and the superficial br of the 1st and 2nd lumbar n.

ME: Special treatment. Puncture inwards and downwards with gas releasing needle 6 ~ 9cm. Let the gas out slowly.

IN: Bloat.

#41. DI-MING. Dripping clear water.

LO and AN: On the superficial cranial epigastric (subcutaneous abdominal) vein 15cm cranial to the umbilicus. At the site of the milk well.

ME: Hemo-acupuncture 1cm. For draining, insert the draining tube to the vicinity of the point where swelling occurs.

IN: Anuria, ascites, abdominal subcutaneous swelling, inflammation of the front quater of udder.

#42. YANG-MING. Yang clear.

LO: In the depression lateral to the base of each teat.

AN: Supplied by tile mammary a and v, and the spermaticus externus n.

ME: Angular insertion upwards and inwards. Wide needle 1 ~ 2cm.

IN: Anuria, inflammation of the mammary gland.

#43. HAI-MEN. Gate of sea.

LO: Three cm lateral to the umbilicus.

AN: Supplied by the caudal abdominal a and v, and the ventral br of the last intercostal n.

ME: Puncture upwards with a wide needle 1cm, or puncture the swollen area repeatedly, or insert the draining tube.

IN: Anuria, ascite, edema in the ventral abdomen.

#44. YIN-SHU. Associate point of genitalia.

LO: On the mid-sagittal plane dorsal to the scrotum or the vulva. Single point.

AN: In male, under the skin are the fascia and the corpus spongiosum penis. Supplied by the external pudendal n and v, and the caudal scrotal n. In female are m constrictor valva and m sphinctor ani externus. Supplied by the pudendal a and v, and the internal pudendal n.

ME: Perpendicular insertion. Filiform, fire or

round-sharp needle 1 ~ 2cm.

IN: Prolapse of vagina or uterus, scrotitis, orchitis.

#45. CHUAN-HUANG. Draining point.

LO: On the mid-sagittal plane cranial to the sternum. Single pt.

AN: Supplied by the transverse cervical and the brachiocephalicus a, ventral br of the cervical n, and the cranial thoracic n.

ME: Puncture with a piercing t luang needle from one side to the other side of the point.

IN: Subcutaneous edema or swelling in the thoracic region.

#46. KAI-FENG. Opening wind.

LO: One vertebra cranial to # 47. Single point.

AN: Supplied by the caudal gluteus a, v and n.

ME: Angular insertion forwards. Wide needle I cm; filiform needle 2cm.

IN: Sunstroke, anuria, rheumatism.

#47. WEI-GEN. Tail base.

LO: On the dorsal midline. At the mobile intervertebral joint of the tail base. Single point.

AN: In the depression between the processus spinosus of the S5 and Cdl vertebrae. Supplied by the caudal gluteus a, v and n.

ME: Perpendicular insertion. Wide. fire or round-sharp needle 1 ~ 2cm: filiform needle 3cm.

IN: Constipation, diarrhea, prolapse of the anus or uterus, fever.

#48. WEI-JIE. Tail segment.

LO: One vertebra caudal to # 47. Single point.

AN: In the depression between the processus spinosus of the Cdl and Cd2 vertebrae. Supplied by the dorsolateral caudal a and v, and the caudal dorsal plexus of nerve.

ME: Perpendicular insertion. Wide or round-sharp needle 1cm, filiform needle 1 ~ 1. 5cm.

IN: Same as #47.

#49. WEI-GAN. Tail trunk.

LO: Two vertebrae caudal lo #47. In the depression between the processus spinosus of the Cd2 and Cd3. Single point.

AN and ME: Same as #48.

IN: Anuria, problems of urinary system.

#50. WEI-BEN. Tail vein: Figs 6 ~ 10. 25.

LO and AN: On the median caudal vein 7. 5cm from the base of tail. Single point.

ME: ttemo-acupuncture 1cm.

IN: Constipation, rheumatism of pelvic limb, abdominal pain, tail paralysis.

#51. WEI-JIAN. Tail tip.

LO: At the tip of the tail.

AN: Supplied by the caudal a and v, and the caudal neural plexus.

ME: Hemo-acupuncture or conventional acupuncture.

IN: Shock, exhaustion, intoxication, fever.

#52. HOU-HAI. Caudal sea.

LO: in tile depression between the anus and the ventral tail base. Single point.

AN: Under the skin are m coccygeus and m sphincter ani externa.

ME: Puncture forwards and upwards. Fire or round-sharp needle 3 ~ 4. 5cm: filiform needle 5 ~ 10cm.

IN: Dysentery, constipation, prolapse of anus, indigestion.

#53. GANG-TUO. Anal prolapse.

LO: Two cm lateral to the anus.

AN: Supplied by the pudendal n, caudal rectal n, and the medial pudendal a and v.

ME: Angular insertion forwards and downwards. Round-sharp or filiform needle 3-5cm. Electroacupuncture. Aqua- acupuncture.

IN: Prolapse of anus or rectum.

#54. YIN-TUO. Vaginal prolapse.

LO: Two cm lateral to the vulva.

AN: Supplied by the pudendal n, and brs of pudendal a and v.

ME: Angular insertion forwards and downwards. Filiform needle 4 ~ 8cm. Electro- or aqua-acupuncture.

IN: Prolapse of vagina or uterus.

7.3 Thoracic limb

There are 19 points in this region. The most frequently used points are #56, #58, #61, #65, #69, #70, and #71.

#55. XUAN-TANG. Magnificent hall.

LO: 3 to 6cm lateral to the space between the processus spinosus of the T4 and T5 vertebrae.

AN: Under the skin are m trapezius and m rhomboideus and the transversal spinal fascia. Supported by the dorsal br of the cervical n, and the costocervical a and v.

ME: Puncture inwards and downwards medjal to the scapular cartilage. Wide, fire or round-sharp needle 9cm: filiform needle 10 ~ 15cm.

IN: Sprain, swelling or pain in the shoulder area, myositis.

#56. BO-JIAN. Shoulder tip.

LO: Ill tile depression just cranial to the cranial angle of the scapula.

AN: tinder the skill arc Ill trapezius and m rhomboideus and serratus ventralis. Supplied by the deep cervical it and v, and the accessory and the cervical and thoracic spinal n.

ME: Angular insertion. Wide, round-sharp or fire needle 6cm; filiform needle 10 ~ 15cm. The needle reaches the space between the spapula and the trunk.

IN: Rheumatism of thoracic limb, sprain and swelling of scapula.

#57. FEI-MEN. Lung gate.

LO: On tile cranial border of the scapula 8 cnl ventral to #56.

AN: Reaching m cervical serratus ventralis. Supplied by the superficial cervical a and v, and the dorsal br of the cervical n.

ME: Angular insertion inwards and downwards. Traditional (wide round-sharp and fire) needles 3cm: filiform needle 6 ~ 9cm.

IN: Rhcumatisnl of the thoracic limb, paralysis, sprain or pain of the scapular region, cough.

#58. BO-LAN. Shoulder post.

LO: in the depression just caudal to the caudal angle of the scapula.

AN: Undcr tile skin are m latissimus dorsi and m scrratus vcntralis. Supplied by the intercostal a and v, and the dorsal br of tile thoracic n.

ME: Angular insertion inwards and downwards, Traditional needles 3cm; filiform needle 6 ~ 9cm.

IN: Same as #56.

#59. FEI-PAN. Hugging lung.

LO: At the intersection of the upper and middle 1/3 of the caudal scapular border.

AN: Under the skin is the long head of m triceps brachii. Supplied by the subscapular a and v, and tile intercostal and the radial n.

ME: Angular insertion forwards and inwards. Traditional needles 3cm, filiform needle 4.5cm.

IN: Same as #57.

#60. JIAN-JING. Shoulder well.

LO: At the shoulder joint, Dorsal to the lateral tuberosity of the humerus.

AN: In tile depression between insertions of m supraspinatus and m infraspinatus. Supplied by the curcumflex humeral a and v, and the suprascapula n.

ME: Angular insertion inwards lind downwards. Traditional needles 3 ~ 4.5cm, filiform needle 6 ~ 9cm.

IN: Rheumatism of thoracic limb, paralysis, swelling or pain of the shoulder region, arthritis, nlvositis.

#61. QIANG-FENG. Robbing wind.

LO: In the depression 8cm caudoventral to the shoulder joint.

AN: At the junction of the caudal border of deltoideus and the lateral head and long head of m triceps brachii. Supplied by the radial n, the subscapula a and v, and the deep brachial a and v.

ME: Perpendicular insertion. Traditional needles 3 ~4. 5cm: filiform needle 6cm.

IN: Same as #60.

#62. CHONG-TIAN. Rushing to heaven.

LO: In the depression 6cm dorsocaudal to #61.

AN: Under the skin are m deltoideus and m triceps brachii. Supplied by the dorsal thoracic a and v, and the radial n.

ME: Perpendicular insertion. Traditional needles 3 ~4. 5cm; filiform needle 6cm.

IN: Same as #60.

#63. JIAN-WAI-YU. Transport point of shoulder externa.

LO: In the depression caudal to the shoulder joint.

AN: Caudodorsal to the lateral tuberosity of the humerus. Supplied by the caudal circumflex humeral a and v, and the axillary n.

ME and IN: Same as #62.

#64. ZHOU-SHU. Elbow association point.

LO: In the depression just cranial to the tip of the processus anconeus of the ulna.

AN: Between the insertions of the lateral and long head of m triceps brachii. Supplied by the deep brachial a and v, and the radial n.

ME: Angular insertion inwards and downwards. Traditional needles 3cm, filiform needle 4. 5cm.

IN: Sprain, swelling of elbow, rheumatism, paralysis of the area.

#65. XIONG-TANG. Thoracic vein.

LO and AN: On the cephalic vein. Lateral to the sternum and cranial to the axilla.

ME: Hemo-acupuncture 1cm.

IN: Acute arthritis of the shoulder and elbow joints, sprain, swelling or pain of the shoulder area, sunstroke, heat in the heart and lumg.

#66. JIA-QI. Axilla.

LO: In the center of the axilla between the lateral surface of the trunk and the medial surface of the arm.

AN: Under the skin are m subscapularis and m serratus ventralis. Supplied by the bracial a and v, and the bracial plexus n.

ME: Special treatment. Make an incision at the point and insert a gas releasing needle in a direction towards point #61. After removing the needle, swing the limb several times.

IN: Sprain of the thoracic limb.

#67. WAN-HOU. Behind the knee.

LO: In the center of the caudal carpus.

AN: In the depression between the accessory carpal bone and the tendon or the superficial digital flexor. Supplied by the palmer br of the cranial interosseous a, and the ulnar n.

ME: Perpendicular insertion 1. 5 ~2. 5cm.

IN: Sprain or swelling of the carpal area, rheumatism of the thoracic limb.

#68. XI-YAN. Carpus.

LO: In the depression at the ventral part of the dorsal surface of the carpal joint.

AN: Between the tendons of m extensor carpi radialis and m extensor digitalis communis. Supplied by the dorsal carpal a and v, and the radial n.

ME: Perpendicular insertion 1cm.

IN: Swelling or pain in the carpal region, carpitis.

#69. QIAN-CHAN-WAN. Thoracic fetlock:

LO and AN: Two points on each limb. In the depression 3cm proximal to the dew claws, one medial to the 2nd and the other lateral to the 5th digits. On the medial and lateral palmar digital v.

ME: Hemo-acupuncture 1. 5cm.

IN: Arthritis of the carpus, tendonitis tendovaginitis, sprain or contusion of fetlock joint.

#70. YOUNG-QUAN. Flushing spring.

LO: In the center of the dorsal surface of the foot 3cm proximal to the junction of the 3rd and the 4th digits.

AN: Supplied by the coronary a and v and tile dorsal digital n.

ME: Hemo-acupuncture 1 ~1. 5cm

IN: Sprain. or pain in the fetlock joint, inflammation of the coronary corium, swelling of the hoof.

#71. QIAN-TI-TOU. Toe of hoot.

LO and AN: Two points on each limb. On the dorsal surface of the 3rd and the 4th digits. At the junction between the coronary border of the hoof and the skin. Under the point is the coronary venous plexus.

ME: Hemo-acupuncture 1cm.

IN: Sprain or pain re. the fetlock area, laminitis, constipation, abdominal pain, cold.

#72. QIAN-DENG-ZHAN. Thoracic lantern.

LO: in the depression between the dew claws and the 3rd and 4th digits. Two points on each limb.

AN: Along the point are the dorsal digital a. and the ventral digital n.

ME: Hemo-acupuncture the palmar digital vein 1cm deep.

IN: Sunstroke, laminitis.

#73. QIAN-TI-MEN. Heel of hoof.

LO and AM: Two points on each limb. On the caudal and proximal border of the hoof bulb. Under the point is the coronary venous plexus of the hoof.

ME: Hemo-acupuncture 1cm.

IN: Fatigue, exhaustion, laminitis.

7.4 Pelvic Limb

There are 17 points in the region of pelvic limb. The most frequently used points are #81, #82, #8, 3, #84, #86 and #87.

#74. QI-MEN. Gate of qi.

LO: Nine cm lateral to the dorsal midlinc and caudal to the wing of the ilium.

AN: The point ix at the intersection between the cranial border of the glutco-biceps musctc and the line between the coxae tuberculum and the ischial tuberosity. Supplied by the sciatic n, anid the brs of the cranial glutcal a and v.

ME: Perpendicular insertion. Fire or round-sharp needle 3cm: filiform nccdlc 6cm.

IN: Rheumatism of the pelvic limb, infertility.

#75. JU-LIAO. At the hip.

LO: In the depression caudovcntral to the tuber coxae.

AN: Under the skin arc m tensor fascia lata and m quadriceps femoris. Supplied by the curcumfJcxa ilium profunda a and v, and the femoral n.

ME: Perpendicular insertion. Fire or round-sharp needle 3 ~ 4.5cm; filiform needle 6cm.

IN: Rheumatism, swelling or para in the loin and croup, paralysis of the pelvic limb, infertility.

#76. DA-KUA. Greater trochanter.

LO: in the depression 9 ~ 12cm dorsal to the greater trochanter of the femur.

AN: Under the skin ix m gluteus medius. Supplied by the glutcus cranialis a and v, and the ischiadicus n.

ME: Same as #75.

IN: Rheumatism or paralysis of the pelvic limb, sprain or pain of the hip area.

#77. XIAO-KUA. Lesser trochanter.

LO: In the depression 6cm ventral to the greater trochanter of the femur.

AN: Under the skin are m tensor lata and m biceps, Supphed by the deep and the caudal femoral a and v, and the sciatic n.

ME and IN: Same as #76.

#78. DA-ZHU. AN. Coxo-femoral Bruit.

LO: in the depression 6cm cranial to the greater trochanter of the femur.

AN: Under the skin are m tensor fascia lata, quadriceps femoris, and gluteus medius. Supplied by the cranial gluteus n, and the circumtlcxa ilium superficialis a and v

ME and IN: Same as #76.

#79. XIE-QI. Evil qi.

LO: Nine cm dorsocaudal to the greater trochantcr.

AN: in the muscle groove of mm biceps femoris and semitcndinosus. Supplied by the glutcus caudalis a, v and n.

ME: Perpendicular insertion. Fire or round-sharp needle 3 ~4. 5cm; filiform needle 6cm.

IN: Rheumatism of the pelvic limb, swelling or pain in the croup and loin region.

#80. YANG-WA. Face up tile.

LO: In the muscle groove of the biceps femoris, 12cm dital to #79.

AN: Supplied by the deep femoral a and v, and the tibial n.

ME and IN: Same as #79.

#81. SHEN-TANG. Kidney hall.

LO and AN: On the saphenous vein 9cm distal to the skin fold of the thigh.

ME: Hemo-acupuncture 1cm.

IN: Orchitis, scrotitis, sprain or swelling in the pelvic limb.

#82. LUE-CAO. Stifle.

LO: In the depression slightly ventrolateral to the patella.

AN: Between the lateral and middle patellar lig. Supplied by the poplitcal a and v and the tibial n.

ME: Angular insertion backwards and upwards. Fire or round-sharp needle 3 ~4. 5cm.

IN: Sprain of stifle joint, rheumatism of the pelvic limit.

#83. YANG-LING. Yang grave.

LO: In the depression 12cm caudal to the patella, and caudodorsal to the lateral condyle of the tibia.

AN: Under the skin is m biceps femoris. Supplied by the caudal femoral a and v, the caudal femoral cutaneous n, and the tibial n.

ME: Perpendicular insertion. Fire or round-sharp needle 3cm; filiform needle 4 ~6cm.

IN: Rheumatism, pain or swelling in the stifle region, paralysis of the pelvic limb.

#84. HOU-SAN-LI. Pelvic 3 miles.

LO: On the lateral surface of the leg, in the muscle groove 9cm distal and lateral to point #82.

AN: In the muscle groove of the long digital extensor, and the lateral digital extensor, distal to the head of the fibula. Supplied by tile cranial tibial a and v, and the fibular n.

ME: Angular insertion inwards and backwards. Filiform needle 6 ~7. 5cm.

IN: Indigestion. diarrhea, constipation, rheumatism or paralysis of tile pelvic limb.

#85. QU-CHI. Pond on the curve.

LO and AN: On the dorsal metatarsal vein at the dorsal surface of the tarsal joint, lateral to the 2nd and 3rd tarsal bones.

ME: Hemo-acupuncture 1cm.

IN: Swelling or pain of tarsal joint, rheumatism of the area.

#86. HOU-CHAN-WAN. Pelvic fetlock.

LO and AN: In the depression 3cm proximal to the dewclaws.

ME: Hemo-acupuncture 1. 5cm.

IN: Laminitis, sprain or pain in the tarsal region.

#87. DI-SUI. Dripping water.

LO, AN and ME: Same as #70.

IN: Cold, sunstroke, sprain or pain of the foot, inflammation of the coronary corium, laminitis.

#88. HOU-TI-TOU. Toe of pelvic hoof.

Corresponding to point #71 of the thoracic limb.

#89. HOU-DENG-ZHAN. Pelvic lantern.

Corresponding to point #72 of the thoracic limb.

#90. HOU-TI-MEN. Heel of pelvic limb.

Corresponding to point #73 of the thoracic limb.

see Fig. 4 - 6, Fig. 4 - 7, Fig. 4 - 8, Fig. 4 -9 and Fig. 4 - 10.

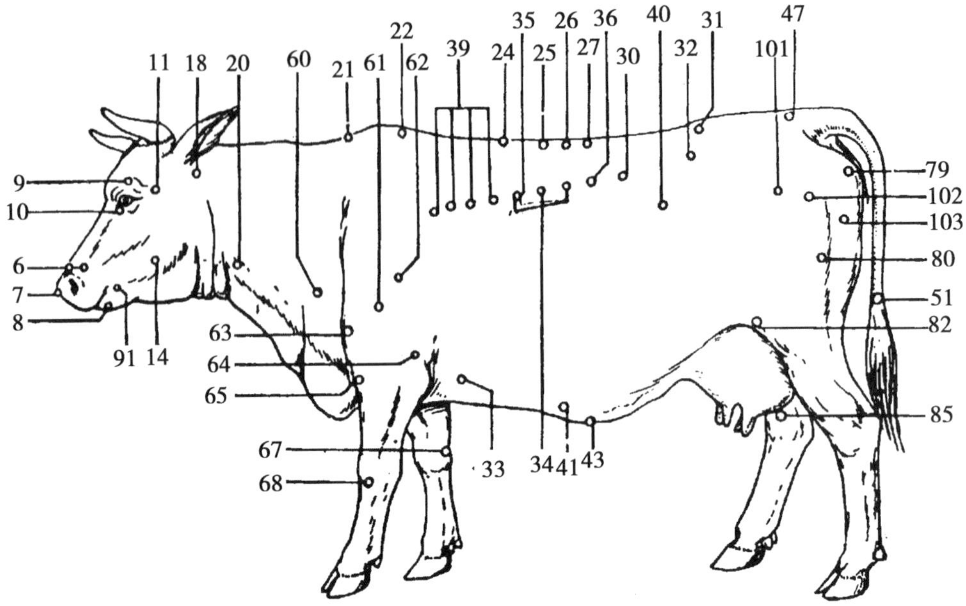

Fig. 4 –6 Acupoints and body surface of the Cattle

6. SHANG-GUAN 7. BI-ZHONG 8. CHENG-JIANG 9. JING-SHU 10. JING-MING 11. TAI-YANG 14. KAI-GUAN 18. FENG-MEN 20. JING-MAI 21. DAN-TIAN 22. QI-JIA. 24. SU-QI 25. AN-FU 26. TIAN-PING 27. HOU-DAN-TIAN 30. YAO-ZHONG 31. BAI-HUI 32. SHEN-SHU 33. DAI-MAI 34. LIU-MAI 35. PI-SHU 36. GUAN-YUAN-SHU 39. TONG-QIAO 40. QIAN-SHU 41. DI-MING 43. HAI-MEN 47. WEI-GEN 51. WEI-JIAN 60. JIAN-JING 61. QIANG-FENG 62. CHONG-TIAN 63. JIAN-WAI-YU 64. ZHOU-SHU 65. XIONG-TANG 67. WAN-HOU 68. XI-YAN 79. XIE-QI 80. YANG-WA. 82. LUE-CAO 85. QU-CHI 91. SUO-KUO 101. HUAN-ZHONG 102. HUAN-HOU 103. HAN-GOU

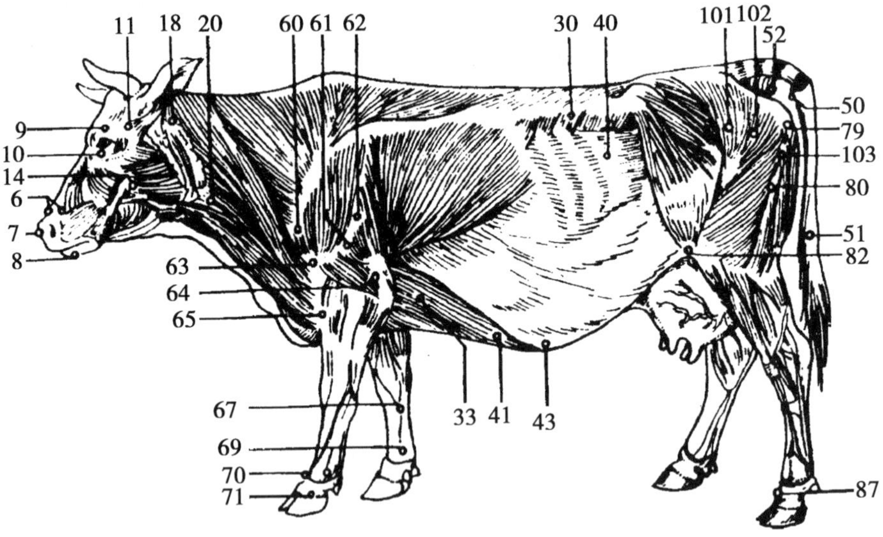

Fig. 4 –7 Acupoints and Musculature of the Cattle

6. SHANG-GUAN 7. BI-ZHONG 8. CHENG-JIANG 9. JING-SHU 10. JING-MING 11. TAI-YANG 14. KAI-GUAN 18. FENG-MEN 20. JING-MAI 30. YAO-ZHONG 31. BAI-HUI 32. SHEN-SHU 33. DAI-MAI 40. QIAN-SHU 41. DI-MING 43. HAI-MEN 50. WEI-BEN 51. WEI-JIAN 52. HOU-HAI 60. JIAN-JING 61. QIANG-FENG 62. CHONG-TIAN 63. JIAN-WAI-YU 64. ZHOU-SHU 65. XIONG-TANG 67. WAN-HOU 69. QIAN-CHAN-WAN 70. YONG-QUAN 71. QIAN-TI-TOU 79. XIE-QI 80. YANG-WA. 82. LUE-CAO 87. DI-SHUI 101. HUAN-ZHONG 102. HUAN-HOU 103. HAN-GOU

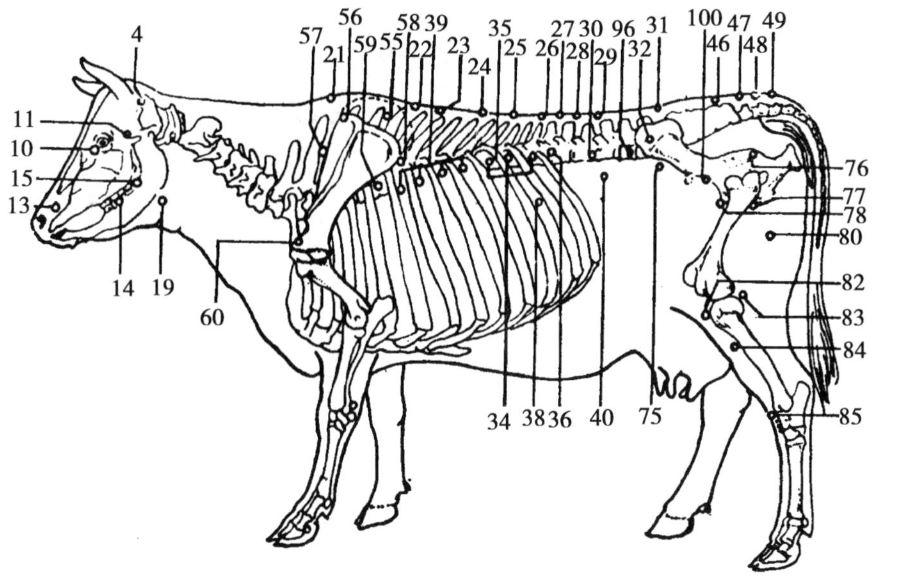

Fig. 4 – 8 Acupoints and Skeleton of the Cattle

1. ER-GEN 10. JING-MING 11. TAI-YANG 13. BI-SHU 14. KAI-GUAN 15. BAO-SAI 19. HOU-MEN 21. DAN-TIAN 22. QI-JIA. 23. SAN-CHUAN 24. SU-QI 25. AN-FU 26. TIAN-PING 27. HOU-DAN-TIAN 28. MING-MEN 29. AN-SHEN 30. YAO-ZHONG 31. BAI-HUI 32. SHEN-SHU 34. LIU-MAI 35. PI-SHU 36. GUAN-YUAN-SHU 38. SHI-ZHANG 39. TONG-QIAO 40. QIAN-SHU 46. KAI-FENG 47. WEI-GEN 48. WEI-JIE 49. WEI-GAN 55. XUAN-TANG 56. BO-JIAN 57. FEI-MEN 58. BO-LAN 59. FEI-PAN 60. JIAN-JING 75. JU-LIAO 76. DA-KUA. 77. XIAO-KUA. 78. DA-ZHUAN 80. YANG-WA. 82. LUE-CAO 83. YANG-LING 84. HOU-SAN-LI 85. QU-CHI 96. SHEN-PENG 100. HUAN-TIAO

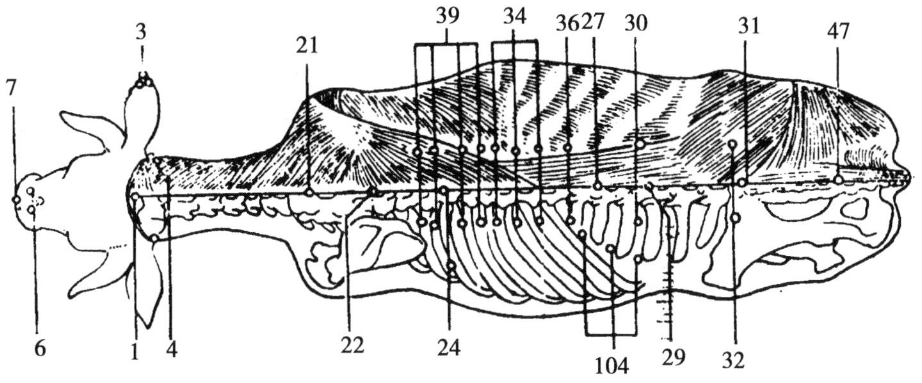

Fig. 4 – 9 Acupoints and the Back of the Ox

1. TIAN-MEN 2. ER-JIAN 4. ER-GEN 6. SHAN-GEN 7. BI-ZHONG 21. DAN-TIAN 22. QI-JIA. 24. SU-QI 27. HOU-DAN-TIAN 29. AN-SHEN 30. YAO-ZHONG 31. BAI-HUI 32. SHEN-SHU 34. LIU-MAI 36. GUAN-YUAN-SHU 39. TONG-QIAO 47. WEI-GEN 104. YAO-PANGSAN-XUE

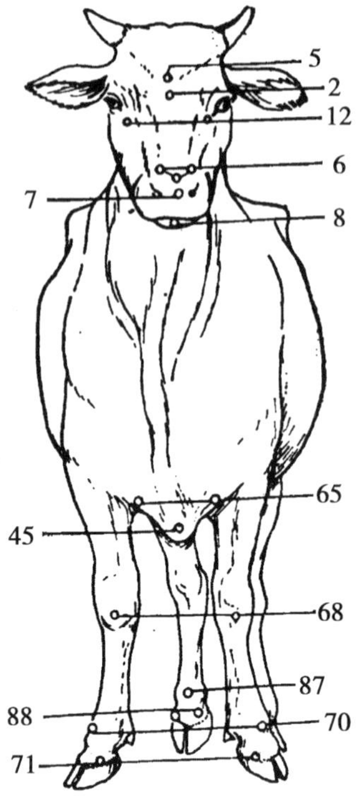

Fig. 4 – 10 Frontal view of the ox showing some Acupoints

2. LONG-HUI 5. TONG-TIAN 6. SHAN-GEN 7. BI-ZHONG 8. CHENG-JIANG 12. SAN-JIANG 45. CHUAN-HUANG 65. XIONG-TANG 70. YONG-QUAN 71. QIAN-TI-TOU 87. DI-SHUI 88. HOU-YI-YOU

8 Acupoints of the Dog

#1. REN-ZHONG

Location (LO): At upper third of Philtrum. One acupoint only.

Anatomy (AN): M. orbicularis oris. N. infraorbitalis of labials superior d N. opthalmlcus.

Method (ME): Straight, about 0.5cm, depth.

Indications (IN): Shock, sunstroke, bronchitis.

#2. SHAN-GEN

LO: At middle of dorsal surface of nose, just at junction of skin d dorsum nasi, one acupoint only.

AN: M. Dorsum nasi, A and V. dorsalis nasi, N, infraorbitalis external nasi.

ME: Bleed, about0.2 ~ 0.5cm depth.

IN: Shock, sunstroke, Sinusitis cold, initial stage of distemper.

#3. SAN-JIANG

LO: Below the modial canthus on the angular vein. Bilateral.

AN: M. levator nasolabialis, masseter, and A and V. Angularis oculi.

ME: Along the blood vessel upwards for 0.2 ~ 0.5cm deep until bleeding.

IN: Constipation, bellyache, conjunctivitis.

#4. CHENG-QI

LO: At middle of infra-orbital Process, Bilateral.

AN: Palpebral conjunctiva and C. bulbis fornicis, depths M. rectus bulbis medialis, M. retractor bulbis, N. cculomotor d. N. abducens.

ME: Straight, Push eyeball up wards. insert about 2 ~ 4cm along the orbit.

IN: Acute and chronic conjunctivitis atrophy of optic nerve, retinitis, cataract.

#5. JING-MING

LO: At the Palpebra superior and inferior of the medial canthus, Bilateral.

AN: A and V. Angularis, N. Infratrochlear.

ME: Straight, Push eyeball externally about 0.2 ~ 0.5cm depth.

IN: Conjunctivitis, keratitis, tumefaction of nictitating membranes.

#6. YI-FENG

LO: Below the ear base and above the mandible articulation. Bilateral.

AN: M. parotid-auricularis, parotid, punch from facial nerve. A and V. Superficial temporal.

ME: Straight, about 1 ~ 3cm depth.

IN: Facial, paralysis, deafness.

#7. SHANG-GUAN

LO: On the outer rim of the M. masseter and in the zygomatic arch. Bilateral.

AN: M. malaris, the front of M. masseter, behind of parotid, deep into M. temporal, N. zygomatic branch of facial n A and V. temporal.

ME: Straight about 3cm depth or moxibustion.

IN: Facial Paralysis, deafness.

#8. XIA-GUAN

LO: In the middle of M. masseter behind, under mandibular articulation. Bilateral.

AN: Parotid, Facial-buccinator nerve and buccacavioris N.

ME: Straight about 3cm depth or moxibustion.

IN: Facial Paralysis, deafness.

#9. ER-JIAN

LO: At apex of convex surface of an auricular, Bilateral.

AN: V. Posterior auricular.

ME: Stop after getting a. little, bloodshed.

IN: Shock, sunstroke, colic and spasm, cold, conjunctivitis.

#10. TIAN-MEN

LO: At middle of Posterior border of occipital bone, one acupoint only.

AN: M. cervico-auriculari superf. M. cervico-aur. Prof major. M. brachio cephalicus, N. ioccipital.

ME: Straight, about 1 ~ 3cm depth, or moxibustion.

IN: Epileptic attack distemper, encephalitis, tetany, convulsion.

#11. JING-MAI

LO: At between the upper and middle 1/3 of the jugular vein. One acupoint only.

AN: Above the M. brachioc-ephalicus, below the M. stermoc-ephalicus.

ME: Puncture needle 0.5 ~ 1cm deep to bleed.

IN: Pneumonia, poisoning, sunstroke.

#12. DA-ZHUI

LO: Between spinous Processes of C_7 and T_1 one Acupoint only.

AN: The back of N. 8th cervicales.

ME: Straight, about 1 ~ 3cm depth, or moxibustion.

IN: Fever, neuralyla. and rheumatism, bronchitis epilepsy. traumatic inflammation, lower the body temperature, strong and healthy prevent diseases.

#13. TAO-DAO

LO: Between spinous processes of T_1 ~ T_2, one acupoint only.

AN: The back of N. 1st thoracic.

ME: Angular, Slightly anterior and downward, about 0.5 ~ 1cm depth or moxibustion.

IN: Neuralgia and sprain of shoulder and fore limb, epilepsy, fever.

#14. SHEN-ZHU

LO: Between spinous processes of T_3 ~ T_4. One acupoint only.

AN: The back of N. 3th thoracic

ME: Angular, anterior and downward about 1 ~ 1.5cm depth or moxibustion.

IN: Pneumonia, bronchitis, distemper, sprain and neuralgia of shoulder.

#15. LING-TAI

LO: Between spinous Processes of T_6 ~ T_7, one acupoint only.

AN: The back of N. 6th thoracic.

ME: Same as SHEN-ZHU

IN: Hepatitis. Pneumonia, bronchitis, stomachache.

#16. ZHONG-SHU

LO: Between spinous Processes of $T_{10} \sim T_{11}$, one acupoint only

AN: The back N. 10th thoracic

ME: Straight, about, 0.5 ~ 1cm depth, or moxibustion.

IN: Gastritis, lack of appetite.

#17. JI-ZHONG

LO: Between spinous processes of $T_{11} \sim T_{12}$, one acupoint only.

AN: The back of N. 11th thoracic

ME: Same as ZHONG-SHU.

IN: Enteritis, lack of appetite, hepatitis.

#18. XUAN-SHU

LO: Between spinous processes of $T_{13} \sim L_1$, one acupoint only.

AN: The back of N. 13th thoracic.

ME: Straight, About 0.5 ~ 1cm depth or moxibustion.

IN: Rheumatism and sprain of loin, indigestion, enteritis, diarrhea.

#19. MING-MEN

LO: Between spinous processes of $L_2 \sim L_3$, one acupoint only.

AN: The back of N. 2nd lumbar.

ME: Same as: XUAN-SHU.

IN: Rheumatism and sprain of loin, chronic enteritis, hormonal imbalance, impotence nephritis, and other urinary disorders, lack of appetite.

#20. YANG-GUAN

LO: Between spinous Processes of $L_4 \sim L_5$, one acupoint only.

AN: The back of N. 4th lumbar

ME: Same as XUAN-SHU.

IN: Hypogonadism, endometritis, metritis, ovaritis, cystic ovary, atrophy of ovary and uterus, Prolonged estrus rheumatism and sprain of loin.

#21. GUAN-HOU

LO: Between spinous Processes of $L_5 \sim L_6$.

AN: The back of N. 5th lumbar

ME: Straight, About 0.5 ~ 1cm depth or moxibustion.

IN: Endometritis, cystic ovary, cystitis, Paralysis of large intestine, constipation.

#22. BAI-HUI

LO: Between spinous Processes of $L_7 \sim K_1$, one acupoint only.

AN: The back of N. 7th lumbar.

ME: Same as XUAN-SHU.

IN: All kinds of nervous disorders, sciatica, posterior paralysis, prolapse of rectum.

#23. WEI-GEN

LO: In the depression in between the processes spinous of the sacral last vertebra and the 1st caudal vertebra. One acupoint only.

AN: The back of N. 3rd sacral

ME: Same as WEI-GAN.

IN: Same as WEI-GAN.

#24. WEI-JIE

LO: Between the spinous processes of $S_2 \sim S_3$, one acupoint only.

AN: The back of N. 2nd sacral.

ME: Same as WEI-GAN.

IN: Same as WEI-GAN.

#25. WEI-GAN

LO: At spinous processes of S1 ~ S2, one acupoint only.

AN: The back of N 1 st sacral.

ME: Straight, about 0.3 ~ 0.5cm depth.

IN: Posterior paralysis, paralysis of tail, prolapse of anus, constipation or diarrhea.

#26. WEI-JIAN

LO: The tip of tail, one acupoint only.

AN: N. spiriales coceygeus.

ME: Straight, needle inserts from the end about 0.5 ~ 1cm, depth.

IN: Shock, sunstroke, gastroenteritis.

#27. ER-YAN

LO: At dorsal sacral foramina 1st and 2nd. Bilateral.

AN: M. tensor fasciae, latae, M glutaeus superficialis d meduus, N. glutaeus cranialis, A. V. anterior gluteal.

ME: Straight, about 0.5 ~ 1cm depth, or moxibustion.

IN: Posterior Paralysis, neuralgia, endometritis.

#28. WEI-BEN

LO: On the blood vessel in the middle of the lower surface of the tail. One acupoint only.

AN: N. cocoygeus. d M. sacro coccygous ventralis, arteria. middle coccygeal.

ME: Puncture needle 0.5 ~ 1cm deep to bleed.

IN: bellyache, Paralysis of tail rheumatism of loin.

#29. JIAO-CAO (HOU-HAI)

LO: Mid point between anus and tail, one acupoint only. At the cross point.

AN: M. Rectococygues, sphincter int and & ext. Posterior haemorrhoidal.

ME: Straight, about 1 ~ 1.5cm depth.

IN: Diarrhea, Prolapse of rectum, Paralysis of sphincter muscles.

#30. FEI-SHU

LO: Between the line of articulatio scapulohumeralis d coxas and in the depression of the back 3rd rib. Bilateral.

AN: M. Latissimus dorsim intercostales ext and int. A and V. Intercostales, N. Thoracalis intercostalis. Thoracalis intercostalis.

ME: Angulsr, about 1 ~ 2cm depth along the intercostal space

IN: Pneumonia, bronchitis, cough.

#32. XIN-SHU

LO: In the depression of the processes anconaeus join the 4 ~ 5 th os costal and cartilagines costales. Bilateral.

AN: M. pectoralis, M. intercostal A and V. intercostal. N intercostal.

Angular. about 1 ~ 2cm depth, along the interocostal space.

ME: Same as FEI-SHU.

IN: Mental stress, heart diseases, epilepsy.

#35. GAN-SHU

LO: At the cross point between the line of articulatio scapulo-humeralis, coxas and in the depreesion of the 9 th rib. Bilateral.

AN: M. obliquus admoninis externus, intercostales ext. & int. A and V, intercostal N. intercostal.

ME: Angular, along the intercostals space about 1 ~ 2cm depth.

IN: Hepatitis, jaundice, eye diseases, neuralgia.

#37. PI-SHU

LO: On the dorsal median line outer surface from the left side about 10cm. down the back of rib 13 th.

AN: M. obliquus abdominis externus & M. obliquus abdominis internus, M. transverses abdominisnn. N. intercostales and A. V. intercostales.

ME: Straight or angular, along posterior border of rib about 1 ~ 2cm depth. or moxibustion.

IN: Indigestion, chronic diarrhea, lack of appetite, vomit, anemia.

#38. WEI-SHU

LO: At between the Articulatio scapulo-humeralis coxas and back in the last but 2nd ~ 3rd intercostal space. Bilateral.

AN: Same as GAN-SHU.

ME: Same as GAN-SHU.

IN: Gastritis, stomach distension, indigestion lack of appetite, enteritis.

#39. SAN-JIAO-SHU

LO: In the musculus iliocostalis groove opposite the last of 1st vertebrae lumbales processus transverse. Bilateral.

AN: Inside M. longissimus dorsi, outside M. ilioscostalis dorsi, int. A. V. 1st lumbar. N. lumbar.

ME: Same as PI-SHU.

IN: Same as PI-SHU.

#40. SHEN-SHU

LO: In the musculus iliocostalis groove opposite the last of 2nd vertebrae lumbales Processus transverse, Bilateral.

AN: Inside M. longissimus dorsi, outside M. ilioscostalis dorsi, int A and V. 2nd lumbar, N, lumbar.

ME: Straight, about 0.5 ~ 1cm. or moxibustion.

IN: Nephritis and other urinary disorders, Polyurea, hypogonadism, and other Sex hormone imbalance, sterility, impotence, rheumatism and sprain of lumbar region.

#41. QI-HAI-SHU

LO: In the M. iliocostalis groove, opposite the last of 3rd vertebrae lumbales processus transverse. Bilateral.

AN: Inside M. longissimus dorsal, cutside M. ilioscostalis, dorsal, int. A and V. 3rd lumbar, N. lumbar.

ME: Same as SHEN-SHU.

IN: Intestinal constipation, inflation of bowel.

#42. DA-CHANG-SHU

LO: In the M. iliocostalis groove, opposite the last of 4th vertebrae lumbales processus transverse. Bilateral.

AN: Inside M. longissmus dorsal, outside M. ilioscostails dorsal, int. A and V. 4th lumbar, N. lumbar.

ME: Same as SHEN-SHU.

IN: Indigestion, enteritis, intestine constipation.

#44. XIAO-CHANG-SHU

LO: Opposite the last of 6th vertebrae lumbales Processus transverse in the musculus iliocostalis groove. Bilateral.

AN: Inside M. Iongissimus dorsi, outside M. ilioscostslis dorsi, int. A and V. 6th lumbar. N. lumbar.

ME: Same as GAN-SHU.

IN: Enteritis, intestinal, lumbago.

#43. GUAN-YUAN-SHU

LO: In the M. iliocostalis groove, opposite the last of 5 th vertebrae lumbales. Processus transverse. Bilateral

AN: Inside M. longissmus dorsi, outside M. ilioscostails dorsi, int. A and V. 5th lumbar, n. lumbar.

ME: Same as SHEN-SHU.

IN: Same as SHEN-SHU.

#45. PANG-GUANG-SHU

LO: About 10cm lateral to the 6 ~ 7th lumbar space. Bilateral.

AN: M. obliquus abdominis externus and M. obliquus abdominis internus, M. transverse abdominis, M. rectus abdominis, A and V. lumbar, N. lumbar, posterior abdominal artery.

ME: Straight, about 1.5 ~ 3cm depth.

IN: Cystitis, hematuria, spasmm of bladder, Urine retention, lumbago.

#46. YI-SHU

LO: About 3cm ventral to acupoint SHEN-SHU. Bilateral.

AN: M. obliquus abdominis externus and M. obliquus abdominis internus, M. Transverses abdominis, A and V. lumbar, N. lumbar.

ME: Steaight, about 0.5 ~ 1cm depth

IN: Pancreatitis, indigestion, chronic diarrhea, diabetes.

#47. LUAN-CAO-SHU

LO: About 3cm. lateral to the 4th lumbar space (4th and 5th transverse Process of lumbar vertebrae). Bilateral

AN: Same as YI-SHU.

ME: Straight, about 1.5 ~ 3cm depth.

IN: Hypogonadism and ovary hormonal insufficienay, hypotrophy of ovary, ovaritis cystic ovary.

#48. ZI-GONG-SHU

LO: About 3cm. lateral to the 5, 6th processus transverse lumbar space. Bilateral.

AN: Same ad YI-SHU.

ME: Straight, about 1.5 ~ 3cm depth.

IN: Cystic uterus, endometritis, metritis,

hypotrophy of uterus, rheumatism of lumbar region.

#51. GAO-HUANG-SHU

LO: On the inner surface of the behind angle of the scapula, Bilateral.

AN: M. latissimus dorsi, M. serratus ventralis, M. subscapularis, N. subscapularis, N. thoracodorsal, Subscapular artery.

ME: Angular, along tuber spine avout 2 ~ 4cm depth, or moxibustion.

IN: Neuralgia, Paralysis and sprain of shoulder, Paralysis of scapular nerve, rheumatism of shoulder. Pneumonia, bronchitis, anaemia, prolonged illness no strong and healthy.

#52. XIONG-TANG

LO: On the vena. cepgalica antebracgii of the thoracic lateral groove near to the between M. triceps brachii d M brachioc Phalicus bilateral.

AN: M. brachiocePgalicus, M. Pectoralis Profundus d. auPerficialis, Plexus brachialis. vena cepgalica antebrachii

ME: Punctur eneedle 0.5 ~ 1cm deep to bleed.

IN: Sunstroke, sprain and neuralgia of shoulder, rheumatism.

#53. JIAN-JING

LO: In the fossa. of the anterior and below border of the shoulders acromion, Bilateral.

AN: M. brachiocephlicus, M. supraspinatus, M. deltoideus, muscul-ocutaneous nerve.

ME: Straight, about 1 ~ 3cm depth.

IN: Neuralgia and paralysis of shoulder and forelimb, sprain of shoulder, paralysis of supraspinatus and brachial nerve.

#54. JIAN-WAI-YONG

LO: In the fossa of the Posterior and below border of the shoulder's acromion, Bilateral.

AN: M. deltold, infraspinatus triceos brachii; N. suprascapularis, radialis.

ME: Straight, about 2 ~ 4cm depth. or moxibustion.

IN: Neuralgia and Paralysis of shoulder and forelimb, sprain of shoulder, Paralysis of srpraspinatus and brachial nerve.

#55. QIANG-FENG

LO: At the point lateral 1/3 between the JIAN-WAI-YONG and ZHOU-SHU, Bilateral.

AN: M. deltoid, tricipitis, N. radialis.

ME: Straight, about 2 ~ 4cm depth.

IN: General anesthesia, nervous disorders of forelimb, sprain.

#56. XI-SHANG

LO: At the point lateral 1/4 between the JIAN-WAI-YONG and ZHOU-SHU, in front and upper of and below ZHOU-SHU. Bilateral.

AN: M. tricipitis, N. radialis.

ME: Same as QIANG-FENG.

IN: Sprain neuralgia and Paralysis of forelimb, Paralysis of brachial and radisl nerve.

#57. ZHOU-SHU

LO: Between epicondyle of brachium and elbow. Bilateral.

AN: M. tricipitis, anconeus, N. radialis.

ME: Same as QIANG-FENG.

IN: Arthritis, neuralgia, Paralysis and sprain of elbow and forelimb.

#58. SI-DU

LO: Between epicondyle of brachium and radius. Bilateral.

AN: M. extensor carpi radialis, M. extensor digitalis, N. radialis.

ME: Straight, about 2 ~ 4cm depth, or moxibustion.

IN: Sprain, neuralgia, and paralysis of forelimb, Paralysis of brachial and radis nerve.

#59. QIAN-SAN-LI

LO: In the fosse of the lateral upper border of the antibrachium 1/4 between m. flexor 5th carpi lateralia and m. extensor digitalis communis. Bilateral.

AN: Below the 5th m. extensor digitalis communis, back the m flexor carpi lateralia, N. radialis.

ME: Same ad SI-DU.

IN: Paralysis of radial and ulna, nerves,

neuralgia and rheumatism of forelimb.

#60. WAI-GUAN

LO: Outside and below 1/4 antibrachium, and between corps radiad and ulnae. Bilateral.

AN: Same as QIAN-SAN-LI.

ME: Straight, about 1 ~2cm depth, or moxibustion.

IN: Paralysis of radial and ulna nerves, neuralgia and rheumatism of forelimb, constipation, lack milk.

#61. NEI-GUAN

LO: Between inside the antibrachium and opposite WAI-GUAN. Bilateral.

AN: Below the corpus radial back the M. flexor carpi radialis and M. deep digital, median nerve and blood vessel.

ME: Same WAI-GUAN.

IN: Neurologic disorders of thoracic limb, stomach and intetinalspasm, colic, heart disease, shock.

#62. YANG-FU

LO: At middle of distal forearm, about 2cm dorsal to the lateral border of YANG-CHI. Bilateral.

AN: V. cephalica. antibrachii, N. Superficial radial.

ME: Straight, about 0.5 ~ 1cm depth, or moxibustion.

IN: Neurologic, disorders of thoracic limb, sprain of carpal tendons, Paralysis of radial nerve.

#63. YANG-CHI

LO: Back of carpal articulation-In the fossa forth radius join os carpi intermedium and ext. distalis radius. Bilateral.

AN: Same as YANG-FU.

ME: Same as YANG-FU.

IN: Sprain of digiti, neuralgia and paralysis of forelimb, cold, arthritis of antibrachiocarpalo.

#64. WAN-GU

LO: At interosseous space of ext distalis ulna and os carpi acces sorium. Bilateral.

AN: N. ulna bone.

ME: Straight, needle inserts from interal side of forearm, about 0.5 ~ 1cm depth, or moxibustion.

IN: Gastritis, wrist, elbow, nail's arthritis.

#65. XI-MAI

LO: Below and inner surface of the os carpale primun and ossa metac-arpalia, 12 th ossa. metacarpalia V. mot volaris superficialis medialis. Bilateral.

AN: V. metacarpea. volaris superficialis medialis.

ME: Puncture needle 0.5 ~ 1cm deep to bleed.

IN: Arthritis of the carpal, fetlook joint, tenositis.

66. YONG-DI (YONG-QUAN and DI-SHUI)

LO: Vena metacarpale (metatarsale) dorsalis medialis between 3 ~ 4th os metacarpale (metatersale) tertium. Bilateral.

AN: V. meta tarsoa (metacarpale) dorsalis medialis, N. 3rd dorsal digital.

ME: Same as Xi-mai

IN: Sprain of digitis, sunstroke abdominal Pain, rheumatism, sinusitis.

#67. LIU-FENG

LO: Ab at caudal to the crumial border of imereligital skinfold; 3 acupoints on each leg, bilateral 6 acupoints.

AN: Dorsal common digitalis, branch of radial (fore-leg): branch of superficial Peroneal (hind-leg)

ME: Angular, about 1 ~2cm depth, or stop after getting a little.

IN: Sprain and Paralysis of digits.

#68. HUAN-TIAO

LO: At trochanter fossa, up and back of the trochanter major of femur. Bilateral.

AN: M. Gluteus superficialis and biceps femors, A. buttock, s.

ME: Straight, about 2 ~4cm depth, or moxibustion.

IN: Posterior paralysis, neuralgia, and Paralysis of Pelvic limb, sciatica, Paralysis of femoral nerve.

#69. XI-SHANG

LO: Ab it 0.5cm. above the paseella and 0.5cm lataral. Bilateral.

AN: M. quadriceps femoris, femoral nerve.

ME: Straight, about 0.5 ~ 1cm depth.

IN: Neurologic disorders of Pelvic limb, arthritis of knee.

#70. XI-AO

LO: Between medial femora and condylus lateralis ttbiae. Bilateral.

AN: M. biceps femoris, M. gastrocnemius, N. Peronacus.

ME: Same as XI-SHANG.

IN: Same as XI-SHANG.

#71. XI-XIA

LO: Between Patella and tuberosity of tibia, in the depression between the lateral and middle batellar ligaments.

AN: The lateral and middle Patellar ligaments, go deep into capsula articularis of knee.

ME: Straight, about 0.5 ~ 1cm depth, or moxibustion.

IN: Sprain, neuralgia, neuralgia, and arthritis of knee.

#72. HOU-SAN-LI

LO: On the upper border 1/4 the Patella, and in the depression in between the fibula. and about 5cm ventral to the head of fibula bilateral.

AN: M. Tibialis anterior, M. long digital extensor (fore). M. deep digital flexor (hind), N. peroneal.

ME: Straight, about 1 ~ 1.5cm depth, or moxibustion.

IN: Posterior Paralysis, neuralgia and Paralysis of Pelvic limb, gastroenteritis, intestinal spasm and colic, arthritis, febrile symptoms, dyspepsia, and Prevent diseases, strong healthy.

#73. JIE-XI

LO: At the middle of tibiae and tarsal joint. Bilateral.

AN: N. superficial Peroneal, go deep into capsula. articularis.

ME: Straight, about 0.5cm depth, or moxibustion.

IN: Sprain, neuralgia, and Paralysis of hind foot.

#74. ZHONG-FU

LO: In the depression of medial calcaneus, 0.5cm to JIE-XI. Bilateral.

AN: N. plantaris.

ME: Same as JIE-XI.

IN: Same as JIE-XI.

#75. HOU-GEN

LO: In the depression above ossa. tarsi and ext distalis of fivule. Bilateral.

AN: N. plantaris.

ME: Same as JIE-XI.

IN: Same as JIE-XI.

#76. SHEN-TANG

LO: On the saphenous vein of the inner surface of the thigh bilateral.

AN: M. gastrocnemius, M. hyoePiglotticus, M. peronacus superficialis, N. saphenous. V. Sphenopalatine.

ME: Same as XIONG-TANG.

IN: Arthritis of hipbone, sprain and neuralgia.

see Fig. 4 - 11, Fig. 4 - 12 and Fig. 4 - 13

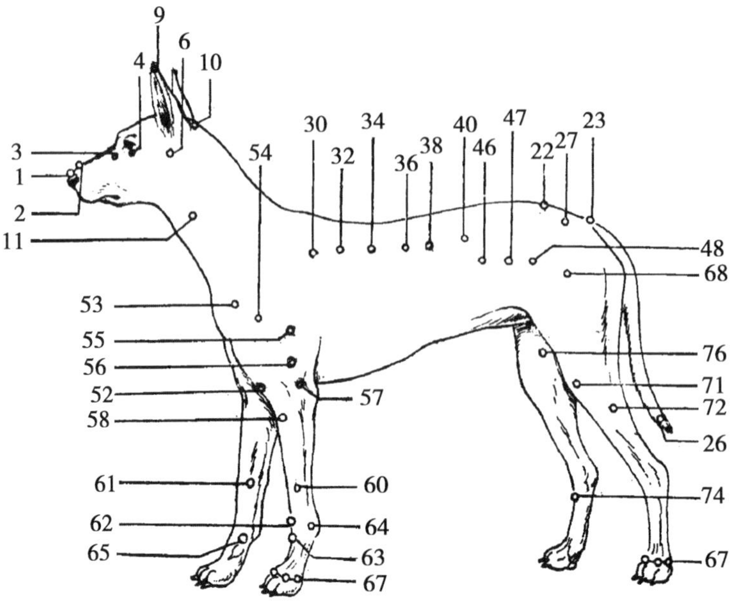

Fig. 4 – 11 Acupoints and body surface of Dogs

1. REN-ZHONG 2. SHAN-GEN 3. SAN-JIANG 4. CHENG-XI 6. YI-FENG 9. ER-JIAN 10. TIAN-MEN 11. JING-MAI 22. BAI-HUI 23. WEI-GEN 26. WEI-JIAN 27. ER-YAN 30. FEI-SHU 32. XIN-SHU 34. GE-SHU 36. DAN-SHU 38. WEI-SHU 40. SHEN-SHU 46. YI-SHU 47. LUAN-CAO-SHU 48. ZI-GONG-SHU 52. XIONG-TANG 53. JIAN-JING 54. JIAN-WAI-YU 55. QIAN-FENG 56. XI-SHANG 57. ZHOU-SHU 58. SI-DU 60. WAI-GUAN 61. NEI-GUAN 62. YANG-FU 63. YANG-CHI 64. WAN-GU 65. XI-MAI 67. LIU-FENG 68. HUAN-TIAO 71. XI-XIA. 72. HOU-SAN-LI 74. ZHONG-FU 76. SHEN-TANG

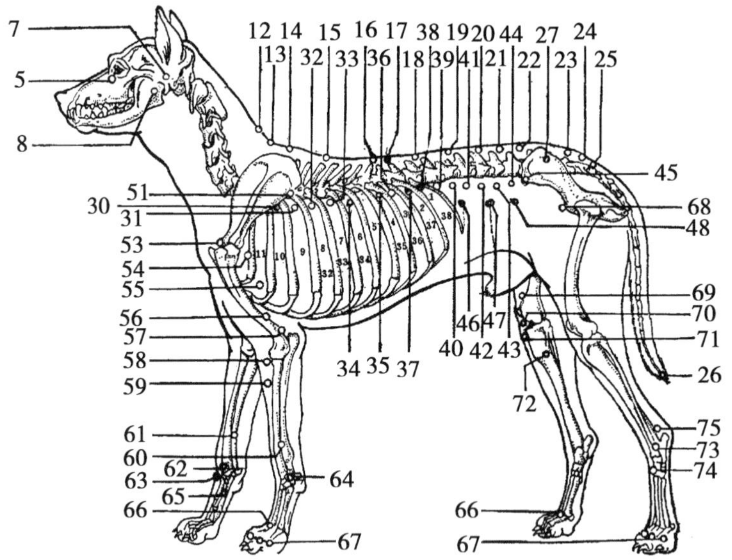

Fig. 4 – 12 Acupoints and Skeleton of Dogs

5. JING-MING 7. SHANG-GUAN 8. XIA-GUAN 12. DA-ZHUI 13. TAO-DAO 14. SHEN-ZHU 15. LING-TAI 16. ZHONG-SHU 17. JI-ZHONG 18. XUAN-SHU 19. MING-MEN 20. YANG-GUAN 21. GUAN-HOU 22. BAI-HUI 23. WEI-GEN 24. WEI-JIE 25. WEI-GAN 26. WEI-JIAN 27. ER-YAN 30. FEI-SHU 31. JUE-YIN-SHU 32. XIN-SHU 33. DU-SHU 35. GAN-SHU 36. DAN-SHU 37. PI-SHU 38. WI-SHU 39. SAN-JIAO-SHU 40. SHEN-SHU 41. QI-HAI-SHU 42. DA-CHANG-SHU 43. GUAN-YUAN-SHU 44. XIAO-CHANG-SHU 45. PANG-GUANG-SHU 46. YI-SHU 47. LUAN-CAO-SHU 48. ZI-GONG-SHU 51. GAO-HUANG-SHU 53. JIAN-JING 54. JIAN-WAI-YU 55. QIANG-FENG 56. XI-SHANG 57. ZHOU-SHU 58. SI-DU 59. QIAN-SAN-LI 60. WAI-GUAN 61. NEI-GUAN 62. YANG-FU 63. YANG-CHI 64. WAN-GU 65. XI-MAI 66. YONG-DI 67. LIU-FENG 68. HUAN-TIAO 69. XI-SHANG 70. XI-AO 71. XI-XIA. 72. HOU-SAN-LI 73. JIE-XI 74. ZHONG-FU 75. HOU-GEN

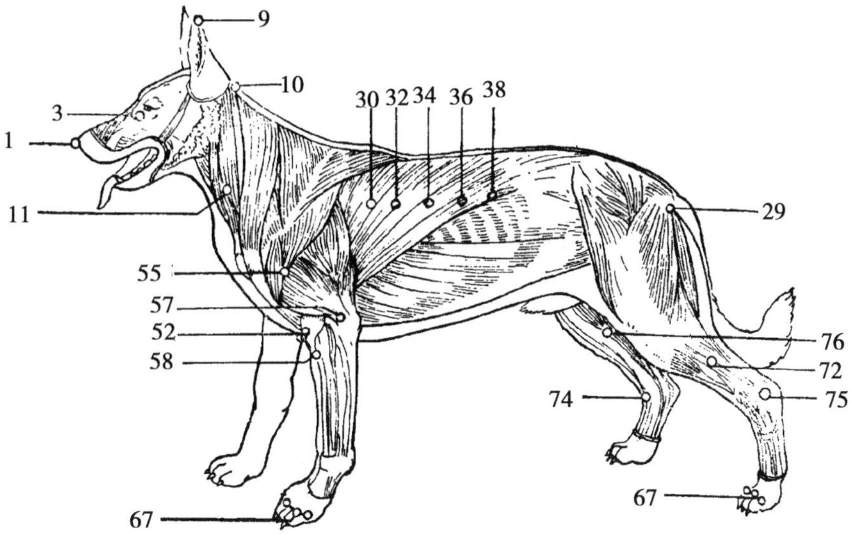

Fig. 4 – 13 Acupoints and Musculature of Dogs

1. REN-ZHONG 3. SAN-JIANG 9. ER-JIAN 10. TIAN-MEN 11. JING-MAI 29. HOU-HAI 30. FEI-SHU 32. XIN-SHU 34. GE-SHU 36. DAN-SHU 38. WEI-SHU 52. XIONG-TANG 55. QIANG-FENG 57. ZHOU-SHU 58. SI-DU 67. LIU-FENG 72. HOU-SAN-LI 74. ZHONG-FU 75. HOUGEN 76. SHEN-TANG

Lecture Five

Internal Medicine of the Traditional Chinese Veterinary Medicine

TCVM internal medicine is a clinical subject dealing with the law of syndrome differentiation and treatment of internal diseases under the guidance of TCVM theory. The internal diseases of TCVM include two categories: exogenous and endogenous diseases. The exogenous diseases refer to febrile diseases that are mainly caused by the invasion of exogenous cold, wind-heat, summer-heat, damp-heat, etc., and can be differentiated and treated according to the theory of six meridians, theory of wei, Qi, Ying and blood, and theory of triple energizer. The endogenous diseases include diseases of internal organs and meridian system that are mainly caused by endogenous pathogenic factors. They can be differentiated and treated according to the theory of zang-fu organs, theory of Qi, blood and body fluid, and meridian theory.

1 Characteristics of Internal Diseases in TCVM

1.1 Endogenous Pathogenic Factors-The Major Causes of Internal Diseases

Most internal diseases are caused by endogenous pathogenic factors. Overstrain may cause edema, diabetes. Improper fodder and drinking may cause stomachache, vomiting, belching and hiccup.

Among the causes of internal diseases, there are some endogenous pathogenic factors secondary to the dys-function of internal organs. These endogenous pathogenic factors may cause disorders with the manifestations similar to those caused by six exogenous pathogenic factors and they are traditionally called five endogenous pathogenic factors, i.e., endogenous wind, endogenous cold, endogenous dampness, endogenous dryness and endogenous fire. For example, endogenous wind may cause syndromes such as numbness and convulsion of limbs. Endogenous cold may cause syndromes such as stomachache and abdominal pain with manifestations of cold limbs, vomiting of clear fluid, and abdominal cold pain. Endogenous dryness may bring about the syndromes such as cough and constipation with the manifestations of dry mouth and throat, cough without sputum, dry skin and constipation with dry stools. Endogenous fire may lead to hemorrhagic syndrome, and wind stroke with manifestations of red eyes, irritability, dry mouth, etc. Endogenous dampness may cause such disorders as diarrhea and epigastric flatulence with manifestations of vomiting, poor appetite, loose stools, etc..

Also, phlegm and blood stasis are two common endogenous pathogenic factors of internal diseases. They are pathological products due to the dysfunction of zang-fu organs and become secondary pathogenic factors to the animal body. Many disorders are associated with these two secondary pathogenic factors. For example, the accumulation of phlegm in the lungs may lead to cough and asthma, and blood stasis may give rise to stomachache and abdominal pain, etc.

1.2 Concurrence of cold, heat, deficiency and excess-the pathogenesis of internal diseases

In most cases, internal diseases persist for a relatively long period. They are usually caused by accumulation of pathogenic factors such as cold and heat and by deficiency of Qi, blood, Yin and Yang in the zang-fu organs as well. Thus the pathogenesis of internal diseases is complicated with the concurrence of cold, heat, deficiency and excess.

Generally the pathogenesis of internal diseases can be classified into two categories: excess and deficiency. Deficiency of Qi, of blood, of Yin and of Yang, or deficiency of both Qi and blood, deficiency of both Yin and Yang are in the category of deficiency, while Qi stagnation, blood stasis, fluid retention, damp-heat and phlegm accumulation are ascribed to the category of excess. To identify whether the case is the heat syndrome or the cold syndrome is to identify the nature of a disease. Generally the cold syndrome and heat syndrome due to internal damage are caused by dysfunction of internal organs and imbalance of Yin and Yang and they are most likely to be complicated by the syndromes of excess or deficiency.

2 The Procedure and Principles of Diasnosis and Treatment of Internal Diseases in TCVM

The procedure of the diagnosis and treatment for internal diseases include the following five steps: to recognize the disease based on the information obtained with the four diagnostic methods, to identify the location and nature of a disease, to infer the cause and judge the pathogenesis of a disease, to design the therapeutic principle and method, and to choose drugs and prescriptionte a prescription.

2.1 To recognize the disease based on the information obtained with the four diagnostic methods

The complete information about a. disease can be collected by four diagnostic methods, i. e., inspection, olfaction and auscultation, interrogation and palpation. It includes the history and manifestations of the disease, and the influence of external environment on the onset and development of the disease. From the information collected, evidence will be provided for differentiating the syndromes, designing the therapeutic principle and organizing a prescription. Therefore the four diagnostic methods are an important procedure for collecting essential information about the disease and a. prerequisite in syndrome differentiation.

Whether the information collected by applying the four diagnostic methods is accurate or not greatly influences the accuracy of syndrome differentiation. So the four diagnostic methods should be performed from different angles to avoid one-sided views and with prominence to the major aspects of the disease.

2.2 To identify the nature and location of a. disease

The fundamental pathogenesis of a disease is the conflict between pathogenic factors and anti-pathogenic Qi, which will lead to the imbalance of Yin and Yang. The excess of Yang will cause heat syndrome, while the excess of Yin will lead to cold syndrome. For this reason, cold and heat indicate the nature of a disease. On the other hand, the conflict between pathogenic factors and antipathogenic Qi will also be reflected as excess or deficiency. Therefore all syndromes can be classified into four categories, i. e., deficiency, excess, cold and heat. And the therapeutic principles such as replenishing deficiency, reducing excess, clearing away heat or warming cold should be applied accordingly. Identifying the nature of a disease is an important step in syndrome differentiation and helps draw an outline of a disease and enables a practitioner to design the general therapeutic principle for a disease.

The location of one disease means the depth, i. e. , exterior and interior; or the site that a. disease affects, i. e. , zang-fu organs, Qi and blood, or meridian. It is important to identify whether the exterior or interior is involved for the differentiation of exogenous diseases, while it is essential to identify whether the zang-fu organs, the meridians or Qi and blood are involved for the differentiation of miscellaneous diseases. Miscellaneous diseases are usually classified into Qi-fen disease and xuefen disease if they are located according to the conditions of Qi and blood. Of all the methods for identifying the location of a. disease, some are simple, and some are comprehensive. Each has a given scope of application, and may be used alone or together with others according to the clinical situations.

2. 3 To infer the cause and judge the pathogenesis of a disease

In order to refine the diagnosis and syndrome differentiation, it is necessary to infer the cause of the disease from the animals' complaints, the results obtained with the four diagnostic methods and the results of the laboratory tests, and from the time, the season, the environmental factors and climate when the disease develops as well, so as to make a synthetical analysis and find out where the crux lies. By so doing, the treatment principle can be decided with much ground.

2. 4 To design the therapeutic principle and method

The therapeutic principle and method are designed according to the result of syndrome differentiation. For example, the therapeutic principle of clearing away liver fire and lowering the adverse rising lung Qi is designed for a cough categorized as syndrome of liver fire invading the lung. The therapeutic principle and method designed must be consistent with the pathogenesis of the disease.

2. 5 To choose drugs and prescriptionte a. prescription

Syndrome differentiation and therapeutic principle designed are the bases for prescriptionting an appropriate prescription. Usually a traditional prescription is organized for a specific syndrome or disease. It is composed of drugs carefully selected in light of their compatibility and indicated for certain cases. To select an adequate prescription for a clinical case requires the knowledge about the ingredients, organizing principle, compatibility of the ingredients, indications and application of the prescription. Traditional prescriptions are the summary of clinical experience over thousands of years. They reflected the treatment strategy and philosophy created by practitioners from generation to generation for certain diseases under the guidance of TCVM theory. And we are encouraged to study, apply and develop these prescriptions in clinical practice. In prescriptionting a prescription for a given disease, simply piecing together the drugs according to a designed therapeutic principle should be avoided and it is better to follow a traditional prescription and modify it as required in actual situation. A traditional prescription is usually chosen based on the pathogenesis of a disease and its reasonable therapeutic principle. Sometimes a prescription may be prescriptionted by selecting drugs simply based on individual conditions, rather than by following any traditional prescription. Nevertheless the therapeutic strategy and philosophy must be embodied in the prescription. Through clinical practice this skill can be gradually gained. Selection of the traditional prescription should be guided by the therapeutic principle and the prescription selected must be consistent with the pathogenesis and the therapeutic principle of the disease.

To choose drugs means to modify a selected

prescription by adding or deleting drugs according to the individual case. As clinical manifestations are complicated and variable, it is hard to get a traditional prescription that completely fits the clinical situation. Modification to a tradi- tional prescription is often required. Usually drugs are added to or deleted from a traditional prescription to make a modified one, suitable for both the major problem and complications of a disease. And for individual cases, modifications should be made flexibly so as to achieve desired results.

In summary, the above five-step procedure is followed in the process of syndrome differentiation and treatment for a disease. Actually it includes the four aspects: Li (theory), Fa (principle), Fang (prescription) and Yao (drug). Li refers to the etiology and pathogenesis of the disease; Fa, the therapeutic principle and method; Fang, the traditional prescription; and Yao, the drugs in a prescription. The consistency in these four aspects for a. specific disease is the premise for good therapeutic effect. Meanwhile this five-step procedure can be divided into two stages, i. e. , stage of syndrome differentiation and stage of treatment. Of the five steps, the first three are at the stage of syndrome differentiation, while the rest two at the stage of treatment. It is for the convenience of learning that the procedure in making syndrome differentiation and treatment is divided into five steps or two stages, but clinically it is not necessary to follow it rigidly step by step. When working on a specific disease, a practitioner may follow the procedure in a flexible way. For example, identifying the location and nature of a disease and inferring the cause and judge the pathogenesis of a disease are being conducted when the four diagnostic methods are performed to collect the information for analyzing the disease.

3 Therapeutic Principles and Methods of Internal Diseases in TCVM

The therapeutic principles are prescriptionted according to the basic theory of TCVM and syndrome differentiation. These principles should be followed throughout the process of designing therapeutic methods, selecting prescriptions, and administering drugs.

3.1 Routine treatment and treatment contrary to the routine

Routine treatment refers to a. therapeutic principle of treating a. disease with the method and drugs contrary to its nature, i. e. , treating heat syndrome with drugs of cold nature, treating cold syndrome with drugs of warm or hot nature, treating deficiency syndrome with replenishing method, and treating excess syndrome with reducing method.

Treatment contrary to the routine refers to a therapeutic principle applied for a. specific condition, in which there exist false phenomena against the essential nature of disease. In this case a treatment aiming at the false phenomena is given, i. e. , treating the false cold phenomena (in a real heat syndrome) with drugs of cold nature, treating the false heat phenomena (in a real cold syndrome) with drugs of warm or hot nature, treating the false phenomena of stagnation (in a. real deficiency syndrome) with replenishing method, and treating diarrhea. (in a. real excessive heat syndrome) with purging method. These treatments seemingly aim at the false phenomena, e. g. , cold method for cold manifestations, and warming method for heat manifestations, but they are actually against the essential pathogenesis of a disease. Clinically Baihu Tang (White Tiger Decoction), which is indicated for excessive heat in Yangming meridian or excessive heat in Qi-fen in a febrile disease, is used to treat the sick animal with false cold phenomena in an excessive heat syndrome; and various Purgative Decoctions are used to treat the sick animal with false phenomena such as diarrhea with accumulated heat in the large intestine.

3.2 Strengthening anti-pathogenic Qi and eliminating pathogenic factors

Strengthening anti-pathogenic Qi is a. therapeutic principle of using replenishing methods for deficiency syndromes; eliminating pathogenic factors is a. therapeutic principle of using reducing or removing methods for excess syndromes. The development of a. disease is the conflict between the anti-pathogenic Qi and pathogenic factors. The disease progresses when pathogenic factors overcome the anti-pathogenic Qi, and it subsides when anti-pathogenic Qi defeats the pathogenic factors. The treatment of strengthening anti-pathogenic Qi and eliminating pathogenic factors will restore the anti-pathogenic Qi to resist pathogenic factors, which is conducive to the recovery of diseases.

The methods for strengthening anti-pathogenic Qi include replenishing Qi, nourishing blood, nourishing Yin and invigorating Yang. The methods for eliminating pathogenic factors include dispelling superficial pathogens, pur gation, eliminating dampness, promoting diuresis, pro moting digestion, and resolving blood stasis. Being mutu ally supportive, strengthening anti-pathogenic Qi will facilitate resisting pathogenic factors, while eliminating pathogenic factors from the body will help protect and restore anti-pathogenic Qi.

Generally methods for strengthening anti-pathogenic Qi are used for deficiency syndromes with indistinct pathogenic factors, while methods for eliminating pathogenic factors are used for excess syndromes with indistinct deficiency of anti-pathogenic Qi. A complicated syndrome with anti-pathogenic Qi deficiency and pathogenic factors accumulation will require a method of both strengthening antipathogenic Qi and eliminating pathogenic factors. In clinical practice it is necessary to differentiate which is dominant, excess or deficiency, in a complicated syndrome. When deficiency is dominant, serious and acute, the complicated syndrome should be treated with replenishing methods supplemented with a mild eliminating method; while a treatment for eliminating pathogenic factors will be adopted together with a. mild replenishing method if the acute and serious excess is the major problem in a. complicated syndrome. In a. deficiency syndrome complicated with excess, when predominant deficiency does not permit an eliminating treatment, or the eliminating treatment may further damage anti-pathogenic Qi, then the eliminating method should be applied only after the antipathogenic Qi is restored. Likewise, in an excess syndrome complicated with mild deficiency or evident deficiency, the replenishing method may foster the pathogenic factors. In that case replenishing method should be used after the pathogenic factors are eliminated, in short, it is important to avoid the retention of pathogenic factors when the method of strengthening anti-pathogenic Qi is applied, and not to damage the anti-pathogenic Qi when the method for eliminating pathogenic factors is used.

3.3 Replenishing and eliminating methods for zang-fu organs

As previously discussed, TCVM holistic view postulates that the animal body is an organic whole, and that zang-fu organs are interrelated physiologically and are affected one another pathologically. Disease of a given organ will affect other organs; and, in turn, disorder of an affected organ may affect the original organ. The replenishing and eliminating methods are given according to the principles defined by the five-element theory among zangfu organs, e. g., inter-generation, inter-restriction, and exterior-interior relation. The principles include replenishing the mother-organ for deficiency syndrome and reducing the child-organ for excess syndrome; replenishing water to inhibit pathogenic Yang and invigorating fire to eliminate pathogenic Yin; and treating the exterior fu-organ to relieve the problem in corresponding interior zangorgan or treating the interior zang-organ to relieve the problem in corresponding exterior fu-organ.

3.4 Different principles for different cases

This means that the principles of treatment should vary with individual cases, climatic conditions, environmental factors, etc.

4 Commonly Used Treatments for Internal Diseases

4.1 Diaphoretic therapy

This is a. treatment for expelling pathogenic factors from the exterior of the body by opening the pores of the skin to induce perspiration. It has the activities of relieving exterior syndrome, subsiding fever and swelling, expelling wind and dampness, and letting out skin eruptions. So it is widely applied for diseases or syndromes with exogenous pathogenic factors on the superficial part of the body, e.g., exogenous exterior syndromes, some acute infectious disease at the early stage, edema and Bi-syndrome with the manifestations of exterior syndrome, and measles at the early stage. In order to avoid improper or excessive diaphoresis the dosage of diaphoretic drugs should be carefully controlled.

4.2 Emetic therapy

This is a. treatment for eliminating harmful substances and pathogenic factors from the body by the application of vomitive drugs, or by physical stimuli that can induce vomiting, including potent emesis, mild emesis, emesis by physical stimuli, and gastrolavage. It is used for disorders with the retention of phlegm fluid, undigested food and poisonous substances in the throat, esophagus and stomach, e.g., manic-depressive syndrome, excess syndrome with coma, food poisoning, and over intake of alcohol, but it is contraindicated in the elderly and debilitated sick animals, critically ill sick animals, sick animals with history of pneumorrhagia. or hemorrhage of upper digestive tract, and pregnant and postpartum animals.

4.3 Purgative therapy

This is a. treatment for relieving constipation, eliminating indigested food, excessive heat and pathogenic fluid by the application of purgatives or lubricants, including purgation with drugs of cold nature, that with drugs of warm nature, that with lubricants, and drastic purgation to eliminate retained fluid. It can be applied to various interior-excess syndromes, for instance, obstruction in the stomach and intestines such as constipation, retention of undigested food, malnutrition due to parasitic infestation, blood stasis, retention of phlegm, etc.; or to the disorders like endogenous heat accumulation, dysentery at the early stage, food or drug poisoning, etc. The timing is very important in using purgation. And the intensity of purgation should be carefully controlled to the appropriate extent.

4.4 Regulating therapy

This is a treatment for eliminating pathogenic factors from ShaoYang meridian, strengthening anti-pathogenic Qi to conduct pathogenic factors out of the body, and coordinating the functions of the viscera. It can be used to coordinate the interior and the exterior, and the Ying-fen and Wei-fen for exogenous febrile disease, but mainly used to coordinate the functions of the liver and spleen, of the gallbladder and stomach, and of the stomach and intestine for endogenous diseases. Though the drugs used in regulating therapy are mild in nature, this therapy should by no means be abused.

4.5 Warming therapy

Warming therapy is a. treatment to invigorate Yangqi and dispel pathogenic cold by the application of drugs of warm or hot nature. It can revive Yang for resuscitation, warm the middle energizer to dispel cold and warm the meridians to activate

Yang. Therefore, it is widely used for interior cold syndromes such as vomiting, diarrhea, cold limbs and intolerance to cold, and spasmodic pain. The therapy should be used with cautions for sick animals with Yin deficiency constitution, or sick animals with a. history of hemorrhage such as haematemesis and epistaxis.

4.6 Heat-clearing therapy

Heat-clearing therapy is a. treatment to clear away pathogenic heat by the application of drugs of cold or cool nature, including clearing away heat from Qi-fen, clearing away heat from Ying-fen and cooling blood, clearing away heat and toxic material, clearing away heat from the viscera, and clearing away heat to conserve Yin and body fluid. It is widely used for disorders due to heat pathogenic factors such as fever, hyperfunction of the viscera, restlessness, bleedings, rashes, and ulceration and pyogenic infection of the skin. Attention should be paid to the differentiation of true or false cold or heat, of deficient fire or excessive fire, and to the differentiation of syndromes and differentiation of individuality of sick animals as well.

4.7 Resolving therapy

Resolving therapy is a. treatment to resolve the accumulation of tangible pathogenic factors by the application of drugs effective for promoting digestion, promoting the flow of Qi and blood circulation, resolving phlegm, and softening hard masses. It is indicated for retention of indigested food, abdominal mass, subcutaneous nodule, scrofula, calculus, ulcer and pyogenic infection of the skin. It should be used with caution for sick animals with Qi and blood deficiency, and sick animals with cold syndrome due to deficiency of kidney Yang and spleen Yang.

4.8 Invigorating therapy

Invigorating therapy is a. treatment to replenish Qi, blood, Yin and Yang, or nourish the viscera. of the animal body by the application of tonics. It is widely used for congenital and postnatal deficiency, and deficiency of Qi, blood, Yin, Yang and body fluid after a. prolonged illness or serious illness. It is also applied to protecting the antipathogenic Qi of the sick animal undergoing purgative or resolving therapies. When this therapy is applied, consideration should be given simultaneously to Qi and blood, Yin and Yang, and the five zang-organs should be invigorated accordingly. Any unconsidered or excessive use of invig

5 Treatment of Diseases

5.1 Colds

General Description

Colds are virus-infected diseases in the upper respiratory tract and classified as common cold and influenza. They are mainly manifested as nasal stuffiness and nasal discharge, sneezing, cough, aversion to cold, fever. Common cold, usually mild, results from rhinovirus infection and is characterized mainly by nasal symptoms. With a. high morbidity and repeated infection tendency, common cold may occur all the year round but mostly in winter and spring. But influenza, usually widespread, is due to influenza virus infec tion and has pronounced systemic toxic symptoms.

Colds fall into the categories of "Shang-feng " (common cold)," Shi-xing-gan-mao" (influenza), etc. in TCVM according to the clinical manifestations. The chief pathogenic factors are pathogenic wind and pestilential toxicity and the main pathogenesis includes pathogenic factors invading the body surface and disharmony between Wei-fen and the exterior. The syndrome mostly seen is exterior excess of the lung and Wei-fen.

Essentials for Diagnosis

(1) Cold contact history, acute onset.

(2) Nasal discharge and obstruction, sneezing, cough, aversion to cold, fever, little or no

sweating.

(3) Influenza should be considered when cold has a wide spreading tendency and sick animals have high fever, which may be accompanied by vomiting, diarrhea, nasal discharge and severe cough.

(4) Laboratory examination shows total leukocyte count normal or slightly reduced, neutrophil decreased and lymphocyte increased. Higher leukocyte count may be seen in sick animals complicated with bacterial infection.

Syndrome Differentiation and Treatment

Syndrome differentiation for this disease mainly focuses on differentiating between the exterior cold and exterior heat. Sick animals with more chills, less fever and no sweating mostly belong to the exterior cold syndrome, whereas those with less chill, more fever, sweating, red and swelling throat mainly pertain to the exterior heat syndrome. Clinically colds have four types of syndromes that are wind-cold, wind-heat, exterior cold and interior heat, and summer-heat and dampness. The therapeutic principle for cold is to relieve exterior syndrome through elimination of pathogenic factors. Wind-cold syndrome is treated by diaphoresis with pungent and warm herbs; wind-heat by clearing heat with pungent and cool herbs; exterior cold and interior heat by expelling cold and clearing heat; and summer-heat and dampness by clearing summer-heat and removing dampness.

5.1.1 *Wind-cold Type*

Chief Manifestations: Severe chill, mild fever, no sweating, stuffy nose, muffled voice, thin nasal discharge, throat itching, cough with clear and thin sputum, white and thin tongue coating, and superficial or superficial and tense pulse.

Therapeutic Method: To relieve exterior syndrome with pungent and warm herbs.

Prescription: Modified Jing Fang Baidu San (Antiphlogistic Powder with Schizonepetae and Saposhnikoviae), composed of Jingjie (Herba Schizonepetae) 30g, Fangfeng (Radix Saposhnikoviae) 30g, Qianhu (Radix Peucedani) 25g, Xingren (Semen Armeniacae Amarae) 30g, Zhike (Fructus Aurantii) 25g, Chuanxiong (Rhizoma Chuanxiong) 21g, Zisuye (Folium Perittae Acutae) 21g, Guizhi (Ramulus Cinnamomi) 21g, Gancao (Radix Glycyrrhizae) 21g and Shengjiang (Rhizoma Zingiberis Recens) 21g.

Modifications: For severe chills and no sweating, add Mahuang (Herba Ephedrae) 21g and Jiegeng (Radix Platycodi) 24g; for wind-cold combined with dampness, add Baizhi (Radix Angelicae Dahuricae) 30g and Qianghuo (Rhizoma. et Radix Notopterygii) 30g; for stuffy and running nose, add Cang'erzi (Fructus Xanthii) 21g and Xinyi (Flos Magnoliae Biondii) 18g; for forceless cough, pale tongue coating and superficial and weak pulse, add Dangshen (Radix Codonopsis Pilosulae) 30g, Fuling (Poria) 30g and Gegen (Radix Puerariae) 30g.

5.1.2 *Wind-heat Type*

Chief Manifestations: Severe fever, slight chill, obstructed sweating, red and swelling throat, thirst with a. desire to drink, stuffy nose with sticky discharge, cough with yellow sputum, red tongue tip, thin and yellow tongue coating, and superficial and rapid pulse.

Therapeutic Method: To relieve exterior syndrome with pungent and cool herbs.

Prescription: Modified Yin Oiao San (Lonicerae and Forsythiae Powder) and Cong Chi Jiegeng Tang (Decoction of Shallot, Sojae Preparatum and Platycodi), composed of JinYinhua. (Flos Lonicerae) 35g, Lianqiao (Fructus Forsythiae) 35g, Dandouchi (Semen Sojae Preparatum) 30g, Bohe (Herba Menthae, to be decocted later) 15g, Niubangzi (Fructus Arctii) 15g, Xingren (Semen Armeniacae Amarae) 30g, Jiegeng (Radix Platycodi) 24g and Gancao (Radix Glycyrrhizae) 15g.

Modifications: For abundant sputum, add Zhebeimu (Bulbus Fritillariae Thunbergii) 21g and Gualoupi (Pericarpium Trichosanthis) 25g; for sore, swelling and red throat, add Xuanshen (Radix Scrophulariae) 20g, Tuniuxi (Radix Achyranthis Bidentatae) 25g and Yizhi Huanghua (Herba Solidaginis Decurentis) 35g; for severe fever and thirst, add Zhimu (Rhizoma. Anemarrhenae) 30g and Tianhuafen (Radix Trichosanthis) 35g; for pronounced manifestations due to toxic heat, add Daqingye (Folium Isatidis) 40g, Banlangen (Radix Isatidis) 40g and Caoheche (Rhizoma Paridis) 30g; for sick animals attacked by autumnal dryness manifested as cough with little sputum, dry throat, nose and lips, dry and thin tongue coating, add Nanshashen (Radix Adenophorae) 33g, Tianhuafen (Radix Trichosanthis) 35g, for sick animals complicated with cough with little sputum, red tongue with little coating and thready and rapid pulse, add Yuzhu (Rhizoma. Polygonati Odorati) 30g and Baiwei (Radix Cynanchi Atrati) 30g.

5.1.3 *Exterior Cold and Interior Heat Type*

Chief Manifestations: Fever, aversion to cold, no sweating, thirst, stuffy nose, muffled voice, cough, rapid breathing, yellow, sticky and thick sputum, brown urine, constipation, red tongue tip and margins with yellow and white coating, and superficial and rapid pulse.

Therapeutic Methods: To disperse wind and release lung Qi, expel cold and clear away heat.

Prescription: Modified Ma Xing Shi Gan Tang (Decoction of Ephedrae, Armeniacae Amarum, Gypsum Fibrosum and Glycyrrhizae), composed of Mahuang (Herba Ephedrae) 35g, Xingren (Semen Armeniacae Amarae) 40g, Fangfeng (Radix Saposhnikoviae) 30g, Jingjie (Herba Schizonepetae) 40g, Huangqin (Radix Scutellariae) 40g, Zhizi (Fructus Gardeniae) 40g, Lianqiao (Fructus Forsythiae) 30g, Shengshigao (Gypsum Fibrosum, to be decocted first) 50g, Jiegeng (Radix Platycodi) 24g, Bohe (Herba Menthae, to be decocted later) 25g and Shenggancao (Radix Glycyrrhizae) 15g.

Modifications: For aversion to cold due to severe external pathogenic cold, add Zisuye (Folium Perillae) 30g and Jiegeng (Radix Platycodi) 28g; for red, swelling and sore throat due to severe endogenous heat, add Banlangen (Radix Isatidis) 40g and Tuniuxi (Radix Achyranthis Bidentatae) 35g; for constipation, add Zhidahuang (Radix et Rhizoma Rhei Preparata) 25 ~ 30g; for persistent high fever, thirst, crimson tongue with yellow and dry coating and rapid pulse due to transformation of wind and cold into the interior, add Shuiniujiao (Cornu Bubali, to be decocted first) 60g, Shengdihuang (Radix Rehmanniae) 45g, Zhuye (Herba Lophatheri) 30g, Lianqiao (Fructus Forsythiae) 35g, JinYinhua (Flos Lonicerae) 35g, Xuanshen (Radix Scrophulariae) 40g, Huanglian (Rhizoma. Coptidis) 30g, Huangqin (Radix Scutellariae) 30g, Shichangpu (Rhizoma. Acori Graminei) 40g and Gancao (Radix Glycyrrhizae) 15g.

5.1.4 *Summer-heat and Dampness Type*

Chief Manifestations: Fever, slight aversion to cold, little sweating, or sweating without relieving fever, cough with sticky sputum, sticky nasal discharge, thirst without much desire to drink, feeling of oppression in the chest, acid regurgitation, scanty dark urine, thin, yellow and greasy tongue coating, and soft-superficial and rapid pulse.

Therapeutic Methods: To clear away summer-heat, resolve dampness and relieve exterior syndrome.

Prescription: Modified Xinjia Xiangru Yin (Newly Modified Mosla Decoction), composed of Xiangru (Herba Moslae) 36g, JinYinhua (Flos Lonicerae) 30g, Biandouhua (Flos Lablab Album, fresh) 30g, Lianqiao (Fructus Forsythiae) 40g, Heye (Folium Nelumbinis, fresh) 40g, Houpo (Cortex Magnoliae Officinalis) 28g, Lugen (Rhizoma Phragmitis) 35g and Shenggancao (Radix Glycyrrhizae) 15g.

Modifications: For predominant summer-heat,

add Huanglian (Rhizoma Coptidis) 25g and Qinghao (Herba Artemisiae Chinghao) 26g; for dampness attacking Wei-fen and the exterior, add Doujuan (Semen Glycines Siccus) 30g and Peilan (Herba. Eupatorii) 30g; for predominant endogenous dampness, add Cangzhu (Rhizorna Atractylodis) 40g, Baikouren (Semen Arnomi Rotundus, to be decocted later) 35g; for scanty and dark urine, add Liuyi San (Six to One Powder, wrapped) 30g and Fuling (Poria) 45g.

5.2 Cough

General Description

Cough is a. main symptom in pulmonary diseases. Generally, cough with a. sound but without sputum is called "Ke", while cough with sputum but without sound is called "Sou". Since a. cough usually with both sound and sputum, they are simply called "Ke-sou" clinically. Cough is a. common symptom of various diseases as well as an independent disease. From the point of etiology, cough can be divided into two types: exogenous and endogenous. The former is caused by the six exogenous pathogenic factors invading the lung while the latter is due to dysfunction of the zang-fu organs leading to disturbance of the lung by endogenous pathogenic factors. Whether the pathogenic factors are exogenous or endogenous, they would impair the purifying and descending function of the lung and induce abnormal rising of lung Qi, thus causing cough. In western medicine, cough is the main symptom in diseases such as upper respiratory tract infection, acute and chronic bronchitis, bronchiectasis, pneumonia and pulmonary tuberculosis.

Essentials for Diagnosis

(1) Cough with a. sound, or accompanied by itching throat and expectoration.

(2) Exogenous cough is characterized by an acute onset, often accompanied by aversion to cold, fever and other exterior symptoms. Endogenous cough has a. long duration, with repeated attacks and complicated with other syndromes resulting from the dysfunction of zang-fu orgalls.

(3) Auscultation reveals hoarse breathing sounds and dry and moist rales.

(4) Laboratory examination reveals a high count of total leukocyte and neutrophil at the acute stage.

(5) Chest X-ray shows normal or increased bronchovascular shadows.

Syndrome Differentiation and Treatment

Cough is clinically classified into two categories exogenous and endogenous. The former mainly includes syndromes such as wind-cold attacking the lung, windheat attacking the lung, warm-dryness impairing the lung and cool-dryness impairing the lung. The latter principally includes syndromes as retention of phlegm-dampness in the lung, accumulation of phlegm-heat in the lung, liver fire attacking the lung, consumption of lung Yin and deficiency of lung Qi. In the treatment, the healthy Qi and the pathogenic factors, and the deficiency and excess should be distinguished first. Exogenous cough should be treated by eliminating the pathogenic factors to benefit the lung, while endogenous cough should be treated by strengthening the body resistance as well as eliminating pathogenic factors, that is, the treatment for the principal and secondary aspects must be taken into consideration simultaneously.

5.2.1 Exogenous Cough

5.2.1.1 Wind-cold Attacking the Lung

Chief Manifestations: Hoarse cough with thin and white sputum, itching throat, nasal obstruction and discharge, or accompanied by aversion to cold, fever, no sweating, thin and white tongue coating, and superficial and tense pulse.

Therapeutic Methods: To disperse wind, expel cold, release lung Qi and stop cough.

Prescription: Modified Zhisou San (Powder for Relieving Cough) and San'ao Tang (Three Crude-Drug Decoction), composed of Mahuang (Herba Ephedrae) 30g, Xingren (Semen Armeniacae Amarae) 40g, Jingjie (Herba Schizonepetae) 40g, Ziwan (Radix Asteris) 30g, Baibu (Radix Stemonae) 30g, Zisuye (Folium Perillae Aculae) 30g, Baiqian (Rhizoma Cynanchi Stauntonii) 30g, Jiegeng (Radix Platycodi) 24g and Gancao (Radix Glycyrrhizae) 15g.

Modifications: For severe exogenous wind-cold, add Fangfeng (Radix Saposhnikoviae) 30g and Qianghuo (Rhizoma et Radix Notopterygii) 30g; for exogenous cold and endogenous heat, remove Baiqian (Rhizoma Cynanchi Stauntonii) and Ziwan (Radix Asteris) and add Shengshigao (Gypsum Fibrosum, to be decocted first) 40g, Sangbaipi (Cortex Mori Radicis) 45g and Huangqin (Radix Scutellariae) 30g; for severe cough, add Jinfeicao (Herba Inulae) 30g.

5.2.1.2 Wind-heat Attacking the Lung

Chief Manifestations: Hoarse cough with white or yellow sputum, or fever, slight aversion to wind and cold, yellow nasal discharge, thirst, red tongue tip, thin and yellow coating, and superficial and rapid pulse.

Therapeutic Methods: To dispel wind, clear away heat, release lung Qi and resolve phlegm.

Prescription: Modified Sang Ju Yin (Mori and Chrysanthemi Decoction), composed of Sangye (Folium Mori) 40g, Juhua. (Flos Chrysanthemi) 40g, Xingren (Semen Armeniacae Amarae) 40g, Jiegeng (Radix Platycodi) 20g, Lugen (Rhizoma. Phragmitis) 45g, Lianqiao (Fructus Forsythiae) 35g, Bohe (Herba Menthae, to be decocted later) 24g and Gancao (Radix Glycyrrhizae) 15g.

Modifications: For severe cough, add Zhebeimu (Bulbus Fritillariae Thunbergii) 30g, Pipaye (Folium Eriobotryae, wrapped) 30g and Qianhu (Radix Peucedani) 30g; for pronounced fever and thirst, add Huangqin (Radix Scutellariae) 30g, Zhimu (Rhizoma. Anemarrhenae) 40g and Gualoupi (Pericarpium Trichosanthis) 30g; for sore throat with hoarse voice, add Shegan (Rhizoma Belamcandae) 30g and Chishaoyao (Radix Paeoniae Rubra) 35g; for dry mouth and red tongue, add Nanshashen (Radix Adenophorae) 40g and Tianhuafen (Radix Trichosanthis) 40g; for summer-heat, add Liuyi San (Six to One Powder, wrapped) 30g and Heye (Folium Nelumbinis, fresh) 40g.

5.2.1.3 Warm-dryness Attacking the Lung

Chief Manifestations: Cough with little sputum which is difficult to expectorate, dry and sore throat, dry nose and mouth, red tongue tip, thin and yellow tongue coating with little moisture, thready and rapid pulse.

Therapeutic Methods: To dispel wind, clear away lung heat, moisten the lung and stop cough.

Prescription: Modified Sang Xing Tang (Mod and Armaniacae Decoction), composed of Sangye (Fotium Mori) 30g, Xingren (Semen Armeniacae Amarae) 30g, Nanshashen (Radix Adenophome) 40g, Zhebeimu (Bulbus Fritillariae Thunbergii) 40g, Qianhu (Radix Peucedani) 30g, Zhizi (Fructus Gardeniae) 35g and Dandouchi (Semen Sojae Praeparatum) 24g.

Modifications: For pronounced dryness-heat, add Shigao (Gypsum Fibrosum, to be decocted first) 60g and Zhimu (Rhizoma. Anemarrhenae) 30g; for severe sore throat, add Xuanshen (Radix Scrophulariae) 40g and Mabo (Lasiosphaera seu Calvatia) 30g; for epistaxis or bloodstained sputum, add Baimaogen (Rhizoma Imperatae) 45g and Shengdihuang (Radix Rehmanniae) 40g; for severe consumption of body fluid, add Mairnendong (Radix Ophiopogonis) 30g and Yuzhu (Rhizoma Polygonati Odorati) 40g.

5.2.1.4 Cool-dryness Attacking the Lung

Chief Manifestations: Cough with little or no spu-tum, itching throat, dry nose and lips, aversion to cold, fever, no sweating, thin, white and dry tongue coating, and superficial and tense pulse.

Therapeutic Methods: To dispel wind and cold, moisten the lung and stop cough.

Prescription: Modified Xingsu San (Powder

of Armeniacae Amarum and Perillae), composed of Xingren (Semen Armeniacae Amarum) 40g, Zisuye (Folium Perillae Acutae) 40g, Jiegeng (Radix Platycodi) 30g, Baiqian (Rhizoma Cyanchi Stauntonii) 40g, Baibu (Radix Stemonae) 40g, Ziwan (Radix Asteris) 30g, Kuandonghua. (Flos Farfarae) 30g, Chenpi (Pericarpium Citri Reticulatae) 35g and Gancao (Radix Glycyrrhizae) 15g.

Modification: For severe aversion to cold, add Jingjie (Herba Schizonepetae) 40g and Fangfeng (Radix Saposhnikoviae) 40g.

5.2.2　Endogenous Cough

5.2.2.1　Retention of Phlegm-dampness in the Lung

Chief Manifestations: Cough with muffled sound, profuse white sticky sputum, or thick sputum in lumps, which is aggravated in the morning, poor appetite, lassitude, white and greasy tongue coating, and soft, superficial and smooth pulse.

Therapeutic Methods: To invigorate the spleen, dry dampness, resolve phlegm and stop cough.

Prescription: Modified Erchen Tang (Erchen Decoction) and Sanzi Yangqin Tang (Decoction of Perillee, Sinapis and Coicis), composed of Chenpi (Pericarpium Citri Reticulatae) 40g, Fabanxia (Rhizoma Pinelliae Preparata) 30g, Cangzhu (Rhizoma Atractylodis) 35g, Houpo (Cortex Magnoliae Officinalis) 35g, Fuling (Poria) 45g, Zisuzi (Fructus Perillae Acutae) 30g, Baijiezi (Semen Sinapis Albae) 30g, Laifuzi (Semen Raphani) 40g and Yiyiren (Semen Coicis) 40g.

Modifications: For severe cold-phlegm, white, sticky and foamy sputum and aversion to cold, add Ganjiang (Rhizoma Zingiberis) 30g and Xixin (Herba Asari cum Radice) 25g; for lassitude due to spleen deficiency in prolonged cases, add Dangshen (Radix Codonopsitis) 40g and Baizhu (Rhizoma Atractylodis Alda) 40g; for yellow sputum due to transformation of phlegm-dampness into heat, add Huangqin (Radix Scutellariae) 40g and Gualoupi (Pericarpium Trichosanthis) 40g.

5.2.2.2　Accumulation of Phlegm-heat in the Lung

Chief Manifestations: Cough with hoarse breath, profuse sputum which is sticky and yellow or hot and stinking, or blood-stained and difficult expectoration, distending sensation in the chest and hypochondrium, chest pain induced by cough, flushed face, fever, dry mouth with a. desire to drink, red tongue with yellow and greasy coating, and smooth and rapid pulse.

Therapeutic Methods: To clear away heat, resolve phlegm, release lung Qi and stop cough.

Prescription: Modified Oingjin Huatan Tang (Decoction for Clearing away Lung Heat and Resolving Phlegm), composed of Sangbaipi (Cortex Mori Radicis) 45g, Huangqin (Radix Scutellariae) 40g, Zhizi (Fructus Gardeniae) 40g, Zhebeimu (Bulbus Fritillariae Thunbergii) 30g, Zhimu (Rhizoma Anemarrhenae) 30g, Jiegeng (Radix Platycodi) 30g, Quangualou (Fructus Trichosanthis) 35g, Juhong (Exocarpium citri Rubrum) 26g and Gancao (Radix Glycyrrhizae) 15g.

Modifications: For purulent or yellow sputum, add Yuxingcao (Herba Houttuyniae) 50g, Jinqiaomai (Rhizoma Fagopyri Cymosi) 60g and Dongguazi (Semen Benincasae) 45g; for the feeling of oppression in the chest, severe cough, profuse sputum and constipation, add Tinglizi (Semen Lepidii seu Descurainiae) 30g and unprepared Dahuang (Radix et Rhizoma Rhei, to be decocted later) 25g; for severe thirst due to body fluid consumption, add Nanshashen (Radix Adenophorae 40g, Maimendong (Radix Ophiopogonis) 30g and Tianhuafen (Radix Trichosanthis) 40g.

5.2.2.3　Liver Fire Attacking the Lung

Chief Manifestations: Paroxysmal irritable cough with pain in hypochondriac region, flushed face, conjuncrival congestion, dry throat, restlessness, bitter mouth, feeling of the throat obstructed by phlegm which is scanty and sticky and difficult to expectorate, thin and yellow tongue coating with little moisture, and smooth and rapid pulse.

Therapeutic Methods: To clear away lung heat, soothe the liver, lower Qi and purge fire.

Prescription: Modified Xiebai San (Powder for Clearing away Lung Heat) and Dai Ge San (Powder of Indigo Naturalis and Concha Meretricis seu Cyclinae), composed of Sangbaipi (Cortex Mori Radicis) 45g, Digupi (Cortex Lycii) 45g, Qingdai (Indigo Naturalis, to be infused separately) 30g, Haigeke (Concha Meretricis seu Cyclinae) 30g, Huangqin (Radix Scutellariae) 30g, Zhimu (Rhizoma. Anemarrhenae) 30g, Tianhuafen (Radix Trichosanthis) 40g and Gancao (Radix Glycyrrhizae) 15g.

Modification: For frequent cough due to predominant heat, add Mudanpi (Cortex Moutan Radicis) 40g and Zhizi (Fructus Gardeniae) 30g.

5.2.2.4 Deficiency of Lung Qi

Chief Manifestations: Long-standing cough with low sound and asthmatic breathing, white and thin sputum, poor appetite, shortness of breath, feeling of oppression in the chest, tiredness, pale and tender tongue with white coating, and thready and weak pulse.

Therapeutic Methods: To replenish lung Qi, resolve phlegm and stop cough.

Prescription: Modified Bufei Tang (Decoction for Replenishing Lung Qi), composed of Huangqi (Radix Astragali) 45g, Dangshen (Radix Codonopsis) 40g, Ziwan (Radix Asteris) 30g, Sangbaipi (Cortex Mori Radicis) 30g, Wuweizi (Fructus Schisandrae) 46g, Fabanxia (Rhizoma Pinelliae Preparata) 30g, Fuling (Poria.) 40g, Baizhu (Rhizoma Atractylodis Alba) 40g, Chenpi (Pericarpium Citri Reticulatae) 35g and Gancao (Radix Glycyrrhizae) 15g.

Modifications: For white and foamy sputum, add Ganjiang (Rhizoma. Zingiberis) 30g, Xixin (Herba Asari cum Radice) 30g and Wuzhuyu (Fructus Evodiae) 35g; for severe cough and shortness of breath, add Hezi (Fructus Chebulae) 35g and Buguzhi (Fructus Psoraleae) 40g; for aversion to cold and cold limbs, add Rougui (Cortex Cinnamomi, to be decocted later) 30g and Zhifuzi (Radix Aconiti Lateralis Preparata.) 30g.

5.2.2.5 Consumption of Lung Yin

Chief Manifestations: General dryness, cough in short bursts with sticky and scanty white sputum, or blood-stained sputum, dry mouth and throat, red tongue with little coating, and thready and rapid pulse.

Therapeutic Methods: To nourish Yin, moisten the lung, stop cough and resolve phlegm.

Prescription: Modified Shashen Maidong Tang (Glehniae and Ophiopogonis Decoction), composed of Nanshashen (Radix Adenophorae) 45g, Maimendong (Radix Ophiopogonis) 42g, Yuzhu (Rhizoma Polygonati Odorati) 42g, Baihe (Bulbus Lilii) 32g, Tianhuafen (Radix Trichosanthis) 42g, Chuanbeimu (Bulbus Fritillariae Cirrhosae) 30g, Xingren (Semen Armeniacae Amarum) 30g, Sangye (Folium Mori) 30g and Gancao (Radix Glycyrrhizae) 15g.

Modifications: For cough with shortness of breath, add Wuweizi (Fructus Schisandrae) 40g and Hezi (Fructus Chebulae) 40g; for profuse night sweating, add Wumei (Fructus Mume) 46g and Fuxiaomai (Fructus Tritici Levis) 45g; for cough with yellow sputum, add Zhimu (Rhizoma Anemarrhenae) 40g and Huangqin (Radix Scutellariae) 40g; for blood-stained sputum, add Mudanpi (Cortex Moutan Radicis) 40g, Zhizi (Fructus Gardeniae) 40g and Oujie (Nodus Nelumbinis Rhizomatis) 40g.

5.3 Constipation

General Description

Constipation refers to a. disorder of prolonged defecator difficult defecation. This disorder is mainly caused by improper fodder, lack of physical exercise, constitutional Yang preponderance and weak constitution after an illness. The pathogenesis of constipation is mainly due to the dysfunction of the large intestine in transmission and also related to the dysfunction of the spleen, stomach, liver and kidney. Pathologically it can be divided into excess type and deficiency type. The excess type includes heat accumulation in the intestine and stomach and Qi stagnation; while the deficiency type includes Yang Qi deficiency and Yin-blood deficiency.

The disorder can be seen in functional constipation, intestinal neurosis and constipation due to the side effect of drugs in western medicine.

Essentials for Diagnosis

(1) Reduced bowel movements, prolonged circle of defecation, or hard stools and difficulty in emptying the bowels.

(2) Accompanied symptoms include abdominal distention, abdominal pain, poor appetite, halitosis, bloody stools, sweating, shortness of breath.

(3) The occurrence is related to invasion of cold and heat, fodder, dysfunction of Zang and Fu organs, lack of physical exercise, aging and weak constitution. The onset and the development are slow.

(4) Fibercoloscope and other laboratory examinations are usually helpful to the diagnosis.

Syndrome Differentiation and Treatment

Constipation may be excessive or deficient. The excess syndromes include heat accumulation in the intestine and stomach, Qi stagnation in the intestine and retention of Yin cold; the deficiency syndromes include spleen Qi deficiency, deficiency of blood and body fluid and deficiency of spleen and kidney Yang. The treatment principle is chiefly the purgation, but clinically purgation should be combined with other treatments according to the differentiation of the syndrome. For excess syndromes, clearing heat and purging fire, regulating Qi and expelling stagnation, expelling cold and warming the middle energizer are adopted; for deficiecy syndromes, tonifying spleen Qi, nourishing blood and moistening the intestine and warming Yang and relieving constipation are often used.

5.3.1 Heat Accumulation in the Intestines and Stomach

Chief Manifestations: Dry stools, abdominal distention, abdominal pain, fever, thirsty, bad breath, scanty deep yellow urine, red tongue with yellow and dry coating, smooth and rapid pulse.

Therapeutic Methods: To clear away heat, relieve stagnation, moisten intestines and promote defecation.

Prescription: Modified Maziren Wan (Cannabis Pill), composed of Huomaren (Semen Cannabis) 35g, Baishaoyao (Radix Paeoniae Alba) 30g, Zhishi (Fructus Aurantii Immaturus) 30g, Houpo (Cortex Magnolias Officinalis) 30g, Dahuang (Radix et Rhizoma Rhei, to be decocted later) 30g, Xingren (Semen Armeniacae Amarae) 30g and white honey 100ml.

Modifications: For damage of body fluid, add Shengdihuang (Radix Rehmanniae) 30g, Xuanshen (Radix Scrophulariae) 40g, Maimendong (Radix Ophiopogonis) 40g; for irritability and red eyes, add Mudanpi (Cortex Moutan Radicis) 21g and Zhizi (Fructus Gardeniae) 30g; for bloody stools caused by heat, add Huaihua (Flos Sophorae) 30g and Diyu (Radix Sanguisorbae) 30g.

5.3.2 Qi stagnation in the intestines

Chief Manifestations: Dry or not very dry stools, urgency to defecate but having difficulty in passing stools, borborygrnus, passing gas, distending pain in the abdomen, fullness in the chest and hypochondriac area, frequent belching, poor appetite, thin and greasy tongue coating, wiry pulse.

Therapeutic Methods: To restore the free flow of Qi and relieve stagnation.

Prescription: Modified Liumo Tang (Limo

Decoction), composed of Binglang (Semen Arecae) 30g, Wuyao (*Lindera strychnifolia*) 35g, Guangmuxiang (Radix Aucklandiae) 30g, Zhishi (Fructus Aurantii Immaturms) 30g, Chenxiangfen (Lignum Aquilariae Resinatum, powdered and to be taken with the decoction) 33g, Dahuang (Radix et Rhizoma Rhei, to be decocted later) 40g and Yuliren (Semen Pruni) 30g.

Modifications: For Qi stagnation turning into fire, add Huangqin (Radix Scutellariae) 30g, Zhizi (Fructus Gardeniae) 30g and Longdancao (Radix Gentianae) 30g; for vomiting with adverseness of Qi, add Zhibanxia. (Rhizoma Pinelliae Preparata) 20g, Xuanfuhuati (Flos Inulae, wrapped) 20g and Daizheshi (Haematitum, to be decocted first) 50g; for constipation after trauma or abdominal surgery with Qi and blood stagnation, add Taoren (Semen Persicae) 40g, Honghua (Flos Carthami) 30g and Chishaoyao (Radix Paeoniae Rubra) 30g.

5.3.3 Retention of Yin cold

Chief Manifestations: Difficult defecation, abdominal pain and distention aggravated by pressure, pain in the hypochondriac area, cold limbs, hiccup and vomiting, white and greasy tongue coating, wiry and tense pulse.

Therapeutic Methods: To warm the interior, expel cold, promote bowel movement and relieve pain.

Prescription: Modified Dahuang Fuzi Tang (Rhei and Aconiti Lateralis Decoction), composed of Shufuzi (Radix Aconiti Lateralis Preparata) 30g, Dahuang (Radix et Rhizoma. Rhei, to be decocted later) 30g, Xixin (Herba Asari cum Radice) 20g, Zhishi (Fructus Aurantii Immaturus) 30g, Houpo (Cortex Magnoliae Officinalis) 30g, Muxiang (Radix Aucklandiae) 30g, Ganjiang (Rhizoma. Zingiberis) 25g, and Xiaohuixiang (Fructus Foeniculi) 20g.

Modification: For cold pain in the abdomen, add Rougui (Cortex Cinnamomi) 30g and Wuyao (*Lindernia strychnifolia*) 25g.

5.3.4 Spleen Qi Deficiency

Chief Manifestations: Neither dry nor hard stools, urgency to defecate with difficulty in passing stools, sweating, shortness of breath, fatigue after bowl movemere, pale complexion and lassitude, tiredness of the limbs, dislike of speaking, pale tongue with white coatiag, weak pulse.

Therapeutic Methods: To tonify Qi and lubricate the intestines.

Prescription: Modified Huangqi Tang (Astragali Decotion), composed of Zhihuangqi (Radix Astragali, rosated) 45g, Dangshen (Radix Codopsitis Pilosulae) 40g, Chenpi (Pericarpium Citri Reticulatae) 30g, Huomaren (Semen Cannabis) 40g, Danggui (Radix Angelicae Sihensis) 30g and white honey 100ml.

Modifications: For severe Qi deficiency, add Dangshen (Radix Codonopsitis Pilosulae) 40g and Baizhu (Rhizoma Atmetylodis Macrocephalae) 40g; for prolapse of the anus caused by sinking of Qi, add Shengma (Rhizoma Cimicifugae) 30g, Chaihu (Radix Bupleuri) 30g and Jiegeng (Radix Platycodi) 25g; for dry stools and difficult defecation, add Yuliren (Semen Pruni) 40g, Xingren (Semen Armeniacae Amarae) 40g and Roucongrong (Herba Cistanches) 30g; for pale finger nails, add Shengheshouwu (Radix Polygoni Multifeori) 30g, Shengdihuang (Radix Rehmanniae) 40g; for shortness of breath, add Gejiefen (Gecko, powdered and to be taken with the decoction) 20g; for poor appetite, add Chaomaiya (Fructus Hordei Germinatus, stir-baked) 45g.

5.3.5 Deficiency of Blood and Body Fluid

Chief Manifestations: Dry stools, lusterless hair, palpitation, shortness of breath, insomnia, dream-disturbed sleeping, poor memory, pale lips, pale tongue with white coating, thready pulse.

Therapeutic Methods: To nourish blood and moisten the dryness.

Prescription: Modified Zunsheng Runchang Wan

(Zunsheng Pill for Lubricating Intestine), composed of Zhidahuang (Radix et Rhizoma Rhei Preparata.) 40g, Danggui (Radix Angelicae Sinensis) 42g, Shengdihuang (Radix Rehmanniae) 30g, Huomaren (Semen Cannabis) 50g, Taoren (Semen Persicae) 40g, Zhike (Fructus Aurantii) 30g, Shengheshouwu (Radix Polygoni Multiflori) 40g and Baiziren (Semen Ptatycladi) 40g.

Modifications: For blood deficiency and internal heat, add Zhimu (Rhizoma Anemarrhenae) 30g and Huhuanglian (Rhizoma Copitdis) 35g; for recovery of the blood and fluid and dry stools, add Yuliren (Semen Pruni) 40g and Songziren (Semen Pini Koraiensis) 30g.

5.3.6 Deficiency of Spleen Yang and Kidney Yang

Chief Manifestations: Dry or not dry stools, difficult bowl movement, large quantity of clear urine, cold limbs, cold pain in the abdomen relieved by warmth, cold pain in the waist and knees, pale tongue with white coating, deep and slow pulse.

Therapeutic Methods: To warm Yang and promote bowel movement.

Prescription: Modified Jichuan Jian (Jichuan Decoction), composed of Danggui (Radix Angeelicae Sinerusis) 42g, Huainiuxi (Radix Achyranthis Bidentatae) 30g, Rousongrong (Herba Cistanches) 45g, Shengma (Rhizoma Cimicifugae) 30g, Zhike (Fructus Aurantii) 30g, Ganjiang (Rhizoma Zingiberis) 35g, Zhifupian (Radix Aconiti Lateralis Prelmrata, to be decocted first) 30g and Rougui (Cortex Cinnamomi) 30g.

Modifications: For nocturnal polyuria, add Jinyingzi (Fructus Rosae Laevigatae) 35g, Wuyao (*Lindera Strychnifolia*) 30g and Shanyao (Rhizoma. Dioscoreae) 45g; for severe abdominal pain, add Muxiang (Radix Aucklandiae) 30g and Yanhusuo (Rhizorna Corydalis) 30g.

5.4 Urinary Infection

General description

Urinary infection is one of the common infective diseases, referring to the inflammation caused by the reproduction of the pathogens in the urine involving the mucous membrane or tissue of the urinary tract. It is clinically categorized into upper urinary tract infection (ureteritis and pyelitis) and lower urinary tract infection (cystitis and urethritis). Lower urinary tract infection may be present alone, while the upper urinary tract infection is often complicated with the inflammatory symptoms of the lower urinary tract, which makes it difficult to differentiate between the two clinically. Pyelitis can be further classified into the chronic stage and the acute stage which are mostly attributive to the infection of lower urinary tract. Chronic pyelitisis is one of the major causes of chronic renal dysfunction.

In TCVM, urinary infection pertains to the categories of "Lin Zheng" (stranguria), "Long Bi" (uroschesis), "Yao Tong" (lumbago). It is mainly caused by inversion of dirty pathogenic factors from the urethra leading to retention of heat in the bladder, which is then transmitted from Zang to Fu organs; improper fodder with excessive intake of greasy, sweet or pungent food leading to dysfunction of the spleen and production of damp heat which moves downward to the lower energizer; weakness of the elderly people with deficiency of kidney Qi leading to failure of the urinary bladder to control urination. The diseased parts are the kidneys and bladder, while the pathogenesis is the accumulation of damp heat in the lower energizer and obstruction of Qi activities in the urinary bladder. In prolonged cases, healthy Qi is consumed by damp heat, resulting in deficiency of the kidneys and spleen.

Essentials for Diagnosis

(1) Urinary infection, at the acute stage, is primarily manifested as the general symptoms like frequent, urgent and difficult urination, often accompanied by chills, fever, loss of appetite, nausea, etc. At the chronic stage, there may appear the same symptoms as those at the acute stage, but

at a. sudden onset of chronic urinary infection, the symptoms can be just as severe at the acute stage.

(2) There may be tenderness on costovertebral point, and positive percussion pain in the renal regions.

(3) Pyuria can be found in routine urine test. Under the high power lens, white cell count is often more than 5 per field, and leukocyte cast can also be found. Midstream urine reveals culture colony positive.

Syndrome Differentiation and Treatment

Syndrome differentiation of this disease mainly concerns with whether the syndromes are excess or deficiency ones. Excess syndromes result from the accumulation of damp heat in the lower energizer as well as obstruction of Qi activities in the urinary bladder. The disease does not persist very long, often with difficult and painful urination, red tongue with yellow tongue coating, and rapid, full pulse. Deficiency syndromes are caused by deficiency of both the spleen and kidneys, and obstruction of Qi activities in the urinary bladder, usually with a. long course, frequent and urgent, but not very difficult and painful urination, pale tongue with thin coating, and rapid, thready pulse. For excess syndromes, the therapeutic method of clearing away heat is employed; for deficiency syndromes, that of nourishing the kidney and spleen is applied.

5.4.1 Damp Heat in the Urinary Bladder

Chief Manifestations: Frequent, dripping, and urgent urination, and discharge of dark yellow urine, vomiting, constipation, thin, yellow tongue coating, and soft-superficial, rapid pulse.

Therapeutic Methods: To induce diuresis, treat stranguria, clear away heat and remove toxic substances.

Prescription: Modified Bazheng San (Eight Health Restoring Powder), composed of Cheqianzi (Semen Plantaginis, wrapped) 42g, Qumai (Herba Dianthi) 32g, Mutong (Caulis Akebiae) 30g, Huashi (Talcum, wrapped) 45g, Bianxu (Herba Polygoni Avicularis) 32g, Shengdahuang (Radix et Rhizoma. Rhei, to be decocted later) 30g, Gancaoshao (Apex Radix Glycyrrhizae) 25g and Dengxincao (Medulla Junci) 30g.

Modifications: In cases of urinary stones, add Shiwei (Folium Pyrrosiae) 35g, Jineijin (Endothelium Corneum Gigeriae Galli) 30g, Jinqiancao (Herba Lysimachiae) 60g and Yujin (Radix Curcumae) 40g; in cases of high fever, add JinYinhua (Flos Lonicera) 45g, Huangqin (Radix Scutellariae) 42g and Huangbai (Cortex Phellodendri) 40g; for thick, greasy tongue coating and retention of damp heat, add Yiyiren (Semen Coicis) 40g and Zhizi (Fructus Gardeniae) 40g.

5.4.2 Impairment of Blood Vessels by Heat

Chief Manifestations: Frequent, urgent and difficult urination, discharge of scanty, dark red urine, sometimes with blood clot or urethral spasm, lumbago with tenderness, high fever with aversion to cold, red tongue with thick, dry and yellow coating, smooth and rapid pulse.

Therapeutic Methods: To clear away heat, induce diuresis, cool blood to stop bleeding.

Prescription: Modified Xiaoji Yinzi (Cirsii Decoction), composed of Shengdihuang (Radix Rehmanniae) 45g, Xiaoji (Herba Cephalanoplosis segeti) 35g, Huashi (Talcum, wrapped) 42g, Mutong (Caulis Akebiae) 30g, Puhuang (Pollen Typhae, wrapped) 10g, Danzhuye (Herba Lophatheri) 6g, Danggui (Radix Angelicae Sinensis) 10g, Zhizi (Fructus Gardeniae) 30g and Zhigancao (Radix Glycyrrhizae, roasted) 21g.

Modification: For predominant heat, high fever and aversion to cold, add JinYinhua. (Flos Lonicerae) 40g and Lianqiao (Fructus Forsythiae) 45g; for conspitation, add Shengdahuang (Radix et Rhizoma Rhei, to be decocted later) 40g; for impaired Yin by heat, hyperactivity of fire due to Yin deficiency, add Zhimu (Rhizoma. Anemarrhenae) 40g, Huangbai (Cortex Phellodendri) 40g, Nüzhenzi (Fructus Ligustri Lucidi)

42g and Hanliancao (Herba Eeliptae) 42g.

5.4.3 Deficiency of Both the Spleen and Kidneys

Chief Manifestations: Prolonged course with alternation in severity, aggravated by overwork and cold, difficult urination lessened with discharge of dark yellow urine, persistent dripping urination, nocturia, pale tongue, and feeble and rapid pulse.

Therapeutic Methods: To strengthen the spleen and nourish the kidney.

Prescription: Modified Wubi Shanyao Wan (Powerful Dioscoreae Pill), composed of Shanyao (Rhizoma. Dioscoreae) 40g, Fuling (Poria) 42g, Zexie (Rhizoma Alismatis) 32g, Shudihuang (Radix Rehmanniae Preparata) 45g, Shanzhuyu (Fructus Corni) 42g, Bajitian (Radix Morindae Officinalis) 42g, Tusizi (Semen Cuscutae) 42g, Duzhong (Cortex Eucommiae) 32g, Niuxi (Radix Achyranthis Bidentatae) 32g, Wu-weizi (Fructus Schisandrae) 45g and Roucongrong (Herba Cistanches) 30g.

Modifications: For Yin deficiency and retention of damp heat, manifested as dry throat and lips, frequent and scanty urine and painful, difficult urination, add Zhimu (Rhizoma. Anemarrhenae) 40g and Huangbai (Cortex Phellodendri) 40g; for retention of damp heat, add Bianxu (Herbe Polygoni Avicutaris) 35g and Oumai (Herba Dianthi) 35g.

5.5 Acute Catarrhal Conjunctivitis

General description

Acute catarrhal conjunctivitis is a common ocular disease as a result of bacterial infection. Clinically it manifests mainly the symptoms of obvious congestion of the conjunctiva, purulent and mucous secretion and a tendency of natural cure. The disease is very infective and often epidemic in the warm seasons of a year.

The most common pathogens of the disease are KochWeek's bacillus, Diplococcus pneumoniae, staphylococci and bacillus infiuenzae.

From the clinical manifestations, the disease should pertain to "Baofeng KeRe" (pseudomembranous conjunctivitis) in traditional Chinese veterinary medicine. Externally, it mainly arises as the sudden invasion of pathogenic wind and heat from outside, and internally due to the interior pathogenic heat and excessive yang. The chief pathogensis of the disease is that the pathogenic wind and heat combine with each other externally and internally, which invades the eye and causes the sudden onset of the disease.

Essentials for Diagnosis

Clinical Manifestations

(1) The symptoms in mild cases include itching and uncomfortable feeling of the affected eye as if caused by a. foreign body. In severe cases, there is photophobia with a burning fever and heavy sensation of the eyelid. Sometimes, because of hypersecretion, the vision becomes blurring but can restore when the secretion is cleared.

(2) Congestion occurs in the palpebral conjunctiva and fornix in mildcases and; in severe cases, the congestion of the bulbar conjunctiva is obvious, even with chemosis, swelling of the lid and large amount of mucous and purulent secretion in the conjunctival sac. In some cases, there may be petechial and patchy hemorrhage.

(3) The disease is usually of bilateral typesimutaneous binocular onset or one eye after another. Usually, a. mild start comes to its utmost in 3 to 4 days and relieves in 8 to 14 days. Cases due to Diplococcus pneumoniae arrive to the worst in 8 to 10 days and then turn back to recovery while cases owing to Koch-Week's bacillus, which are serious, need two to four weeks and then back to recovery.

Laboratory Examination

Koch-Week's bacillus, Diplococcus pneumoniae, staphylococci or bacillus infiuenzae can often be found in the secretion culture of the conjunctival sac. These bacteria multiples quickly during the onset of the first 3 to 4 days and can not easily be found at the later stage or after the administration of medicine.

Syndrome Differentiation and Treatment

Generally, the disease is mostly caused by the pathogenic wind and heat in the lung meridian. In clinic, it is chssified into the syndromes of wind exceeding heat, heat excccding wind and excess of both wind and heat. In principle, the treatment should focus on diffusing lung-Qi to dispel wind and removing pathogenic heat and fire. Also it should be clearly differentiated whether the key pathogenic factor is wind or heat or both together. If wind is the main factor, the treatment should be chiefly dispelling the wind, if heat is the main factor, the treatment should be removing the heat, and if both wind and heat are the main pathogenic factors, the treatment should be dispelling the wind and removing the heat.

5.5.1 Syndrome of Pathogenic Wind Exceeding Pathogenic Heat

Chief Manifestations: Slight swelling of the lid, mild redness of the bulbar conjunctiva with itching and unsmooth feeling, photophobia and delacrimation, fever and aversion to wind; thin and whitish or light yellowish tongue coating and floating pulse.

Therapeutic Method Expelling wind with diaphoresis plus heat-removing drugs.

Prescription The modified Qianghuo Shengfeng Decoction: Qianghuo (Rhizoma et Radix Notopterygii) 40g, Fangfeng (Radix kedebouriellae) 40g, Duhuo (Radix Angelicae Pubescentis) 40g, Jingjie (Herba Schizonepetae) 30g, Chaihu (Radix Bupleuri) 30g, Baizhu (Rhizoma. Atractylodis Alba) 40g, Bohe (Herba Menthae, to be decocted later) 30g, Baizhi (Radix Angelicae Dahuricae) 30g, Chuanxiong (Rhizoma. Ligustici Chuanxiong) 30g, Zhike (Fructus Aurantii) 30g, Huangqin (Radix Scutellariae) 40g, Jiegeng (Radix Platycodi) 25g, Qianhu (Radix Peucedani) 25g, and Gancao (Radix Glycyrrhizae) 21g.

Modification In the cases of less pathegenic wind, Qianghuo and Duhuo are removed while in the cases of obvious pathogenic heat, 40g of Jinyinhua (Flos Lonicerae), Lianqiao (Fructus Forsythiae) and Sangbaipi (Cortex Mori Radicis) and 30g of Juhua (Flos Chrysanthemi) are added respectively.

5.5.2 Syndrome of Pathogenic Heat Exceeding Pathogenic Wind

Chief Manifestations: Serious redness and swelling of the bulbar conjunctiva with hot tears and mucopurulent secretion, palpebral swelling, restlessness and thirst, yellowish urine and constipation; red tongue with yellowish coating and rapid strong pulse.

Therapeutic Method Removing pathogenic heat and fire accompanying dispelling wind.

Prescription The modified Xiefei Beverage. Shigao (Gypsum Fribrosum, to be decocted first) 40g, Huangqin (Radix Scutellariae) 20g, Sangbaipi (Cortex Mari Radicis) 20g, Lianqiao (Fructus Forsythiae) 20g, Shanzhizi (Fructus Gardeniae) 20g, Chishaoyao (Radix Paeoniae Rubra) 20g, Zhike (Fructus Aurantii) 20g, Mutong (Caulis Akebiae) 20g, Jingjie (Herba Schizonepetae) 30g, Fangfeng (Radix Ledebouriellae) 30g, Baizhi (Radix Angelicae Dahuricae) 30g, Qianghuo (Rhizoma et Radix Notopterygii) 20g, and Gancao (Radix Glycyrrhizae) 15g.

Modification In the cases of excessive toxic heat, Qianghuo and Baizhi are removed, and30g of Jinyinhua (Flos Lonicerae) and 35g of Pugongying (Herba Taraxaci) and 35g of Yuxingcao (Herba Houttuyniae) are added, in the cases of serious constipation, 35g of Dahuang (Radix et Rhizorna Rhei) and 36g of Mangxiao (Natrii Sulphas) are added.

5.5.3 Syndrome of Severe Pathogenic Wind and Heat

Chief Manifestations: Redness and swelling of the lid and bulbar conjunctiva, alternate pain and itching, photophobia with a burning sensation, hot tears and mucopurulent secretion, fever and aversion to cold, constipation and yellowish urine; red tongue with yellowish coating and rapid strong pulse.

Therapeutic Methods Dispelling pathogenic wind externally and removing pathogenic fire-heat internally.

Prescription The modified Fangfeng Tongshengsan Decoction. Fangfeng (Radix Ledebouriellae) 30g, Jingjie (Herba Sehizonepetae) 30g, Jiegeng (Radix Platycodi) 26g, Bohe (Herba. Menthae, to be decocted later) 26g, Lianqiao (Fructus Forsythiae) 30g, Shanzhizi (Fructus Gardeniae) 30g, Huangqin (Radix Scutellariae) 30g, Huashi (Talcum) 30g, Chuanxiong (Rhizoma. Ligustici Chuanoxiong) 30g, Danggui (Radix Angelicae Sinensis) 30g, Baishaoyao (Radix Paeoniae Alba) 30g, Dahuang (Radix et Rhizoma Rhei) 26g, and Gancao (Radix Glycyrrhizae) 15g.

External Therapies

(1) Eye Drops a. Eye Drops of Coptis and Watermelon Frost. The eye is dropped 3 to 4 times a. day, which is applicable to the mild cases of the disease. b. Qianli Guang Eye Drops. The eye is dropped 3 to 4 times a. day, which is applicable to the mild condition of the disease, c. Xiongdan Eye Drops. The eye is dropped 3 to 4 times a. day, which is applicable to all syndromes of the disease.

(2) Fumigation-washing Method 20g of fresh Yejuhua (Mos Chrysanthemi Indici) and respective 30 g of Pugongying (Herba Taraxaci), Cheqiancao (Herba Plantaginis), Zihuadiding (Herba Violae), all the ingredients are decocted with water for the fumigation and washing of the eye once or twice every day. It is applicale to the severe cases of the disease.

5.6 Intestinal Obstruction

General description

Intestinal obstruction refers to a. condition in whih the contents in the intestinal cavity are unable to pass through the intestine caused by torsion, spasm and paralysis of the intestine or adhesion of cord, clinically characterized by abdominal pain, abdominal distension, vomiting, no flatus and no defecation, and by complicated causative reasons, changing condition, swift development and multiple complications. This disease belongs to the categories of "Obstruction and Rejection", "Intestinal Accumulation" and "Abdominal Distension", mostly related to improper fodder intake, dysfunction of the spleen in transportation, internal accumulation of dampness and heat, internal retention of stagnant blood, or stagnation of pathogenic cold, internal accumulation of dry stool, retention of roundworm, leading to failure of the purgative and descending function due to inability of Qi activity, stagnation of Qi and blood.

Essentials for Diagnosis

(1) It is characterized by abdominal pain, abdominal distension, vomiting, no flatus and defecation.

(2) Abdominal pain related to mechanical intestinal obstruction is mostly paroxysmal colic pain. If it develops to be a. strangulatied intestinal obstruction, the pain is characterized by progressive colic pain. If the position of the intestinal obstruction is higher, vomiting is frequent and the vomiting substances are fodders or gastric juice. If the position of the intestinal obstruction is lower, abdominal distension is obvious and vomiting appears later and less frequently, with feces-like vomiting substances. In the late stage of the intestinal obstruction, there can be shock symptoms of dry lips, dry tongue, sunken eyes, loss of skin elasticity, scanty or no urine, pale complexion, extremely cold limbs.

(3) In the abdominal examination in simple intestinal obstruction, tenderness or wandering mass can be present. But, no rebound pain and muscular tension present usually. When the condition develops further, tenderness in the abdomen

can be progressively aggravated, accompanied by fixed mass, rebound pain and muscular tension, and also visible peristalsis, hyperactive bowel sound in auscultation, with water or metal sound. If paralytic intestinal obstruction appears, bowel sound decreases or disappears in auscultation.

(4) In the laboratory examination, the total count of peripheral leukocyte and neutrophil ratio are obviously enhanced in the patients. When obstruction lasts for a. comparatively long time and is complicated with dehydration, hemoglobin and packed cell volume can be remarkably enhanced. In complication of acidosis, carbon dioxide combining power can decrease remarkably in blood biochemical examination. Radibgraphy is a. reliable method to judge intestinal obstruction. Accumulation of air in the small intestine can be seen in abdominal radiography in the early stage of the pathological condition. Four to six hours after intestinal obstruction, several stepladder-like airfluid levels can be found in the position above the intestinal obstruction in the abdominal plain film (standing and lying position), and obvious accumulation of air can be seen in the whole intestinal cavity in the paralytic intestinal obstruction.

Syndrome differentiation and treatment

you can accord the "General description" analysing fodder, heat, blood, dry stool, roundwom, etc..

5.6.1 Syndrome of Qi Stagnation in Intestines

Chief manifestations It is similar to early intestinal obstruction and adhesive intestinal obstruction in Western medicine, manifested by paroxysmal distending pain in the abdomen, with distending sensation worse than pain, intermittent appearance of mass in the abdomen, drum-like abdomen in percussion, slight tenderness, accompanied by nausea, vomiting, no flatus and defecation, slightly red tongue, thin and whitish coating, and wiry pulse.

Therapeutic Methods Promote flow of qi, disperse accumulation, dredge the interior for purgation.

Prescription Modified "Da. Cheng Qi Decoction" with raw Dahuang (Radix et rhizoma Rhei, decoet later) 36g, Houpo (Cortex Magnoliae Officinalis) 27g, Zhishi (Fructus Aurantii Immaturus) 30g, Huangqin (Radix Scutellariae) 36g, Wuyao (Radix kingerae) 30g, Qingpi (Pericarpium Citri Reticulatae Viride) 27g, Muxiang (Radix Aucklandiae) 30g, Laifuzi (Semen Raphani) 36g, Mangxiao (Natrii Sulfas) (take after infused with water) 30g, and raw Gancao (Radix Glyeyrrhizae) 18g.

Modification For serious abdomiml pain, add Yanhusuo (Rhixoma Corydalis) 36g and Chuanlianzi (Fructus Meliae Toosendan) 30g. For severe nausea and vomiting, add prepared Banxia (Rhizoma Pinelliae Praeparata) 24g and ginger-prepared Zhuru (Caulis Bambusae in Taeniam) 30g.

5.6.2 Intestinal Stagnation and Accumulation Syndrome

Chief manifestations It is similar to various types of intestinal obstruction with disturbance of blood circulation of different degrees or tumorous intestinal obstruction, manifested by serious abdominal pain, with pain worse than distending sensation, fixed location of pain, aggravated by pressure, coffee-like vomiting substance sometimes, accompanied by dark red tongue, or with purple spots, yellowish and greasy coating, and hesitant pulse.

Therapeutic Methods Disperse stagnation, promote flow of Qi and dredge the interior for purgation.

Prescription Modified "Taohe Cheng Qi Decoction" with Taoren (Semen Persicae) 30g, Danggui (Radix Angelicae Sinensis) 36g, Chishaoyao (Radix Paeoniae Rubra) 30g, Danshen (Radix Salviae Miltiorrhizae) 45g, Honghua. (Flos Carthami) 18g, raw Dahuang (Radix et Rhizorna. Rhei) (decoct later) 36g, Houpo (Cortex Magnoliae Officinalis) 30g, Zhishi

(Fructus Aurantii Immaturus) 30g, Qingpi (Pericarpium Citri Reticulatae Viride) 27g, Muxiang (Radix Aucklandiae) 30g, Laifuzi (Semen Raphani) 36g, and Mangxiao (Natrii Sulfas) (take after infused with water) 36g.

Modification For severe abdominal distension, add Xiangfu (Rhizoma Cyperi) 36g and Chuanlianzi (Fructus Meliae Toosendan) 30g. For those who are accompanied by obvious mass in hard nature, add baked Ruxiang (Resina Olibani) 18g, baked Moyao (Myrrha) 18g and Sanleng (Rhizoma Sparganii) 30g.

5.6.3 Syndrome of Obstruction due to accumulation of Worms

Chief manifestations This syndrome has often a. history of roundworm, manifested by paroxysmal abdominal pain, with constant attacks, palpable roundworm mass in the abdomen, intermittent flatus, pale tongue, thin and whitish coating, and wiry pulse.

Therapeutic Methods Expel worm, circulate Qi and dredge Fu organs for purgation.

Prescription Modified "Qu Hui Cheng Qi Decoction" with Binglang (Semen Arecae) 30g, Wumei (Fructus Mume) 36g, Chuanlianzi (Fructus Meliae Toosendan) 30g, raw Dahuang (Radix et Rhizoma Rhei) (decoct later) 30g, Houpo (Cortex Magnoliae Officinalis) 27g, Zhishi (Fructus Aurantii Immaturus) 30g, Qingpi (Pericarpium Citri Reticulatae Viride) 18g, Muxiang (Radix Aucklandiae) 30g, Laifuzi (Semen Raphani) 30g, and Mangxiao (Natrii Sulfas) (take after infused with water) 30g.

Modification For retention of fodder, add Shenqu (Massa Fermentata Medicinalis) 36g and fried Maiya (Fruoctus Hordei Germinatus) 36g. For vomiting, add ginger-prepared Zhuru (Caulis Bambusae in Taeniam) 30g.

5.6.4 Synodrome of Dryness in

Chief manifestations This syndrome is often seen in the old and feeble patients with a. history of habitual constipation, manifested by difficult defecation and constipated stool, progressive abdominal distension, abdominal pain, palpable feces mass in the left lower abdomen, accompanied by poor appetite, dry tongue with scanty fluid, thick and greasy coating, thready and rapid pulse.

Therapeutic Methods Nourish yin, moisten the intestine and dredge the interior for purgation.

Prescription Modified "Zeng Ye Cheng Qi Decoction" with raw Dihuang (Radix Rehmanniae) 45g, Xuanshen (Radix Scrophulariae) 36g, Maimendong (Radix Ophiopogonis) 45g, raw Dahuang (Radix et Rhizoma. Rhei) (decoct later) 30g, Houpo (Cortex Magnoliae Officinalis) 27g, Zhishi (Fructus Aurantii Immaturus) 30g, Qingpi (Pericarpium Citri Reticulatae Viride) 18g, Muxiang (Radix Aucklandiae) 30g, Laifuzi (Semen Raphani) 36g, Mangxiao (Natrii Sulfas) (take after infused with water) 30g, and raw Gancao (Radix Glycyrrhizae) 18g.

Modification For old age and weak constitution, add Dangshen (Radix Codonopsis Pilosulae) 30g and Danggui (Radix Angeticae Sinensis) 30g. For thirst with preference for drinks, add Shihu (Herba Dendrobii) 45g and Shashen (Radix Adenophorae) 30g.

5.6.5 Syndrome of Retention of Water and Obstruction of Dampness

Chief manifestations This syndrome is similar to various types of intestinal obstruction complicated with accumulation of fluid in the intestines, manifested by progressively aggravated abdominal pain, constant borborygmus, abdominal distension aggravated by pressure, accompanied by nausea, vomiting, thirst without preference for drinks, no flatus and no defecation, scanty urine, reddened tongue, whitish and greasy coating, slippery and rapid pulse.

Therapeutic Methods Circulate qi, expel water and dredge Fu organs for purgation.

Prescription Modified "Gansui Tong Jie Decoction" with powder of Gansui (Radix Euphorbiae Kansui) (take after infused with water) 3g,

Taoren (Semen Persicae) 30g, fried Zhike (Fructus Aurantii) 30g, Chishaoyao (Radix Paeoniae Rubra) 30g, Houpo (Corter Magnoliae Officinalis) 30g, raw Dahuang (Radix et Rhizoma Rhei) (decoct later) 30g, Muxiang (Radix Aucklandiae) 30g, and Niuxi (Radix Achyranthis Bidentatae) 30 g.

Modification For severe abdominal distension, add Xiangfu (Rhizoma Cyperi) 30g and Chuanlianzi (Fructus Meliae Toosendan) 10g. For obvious vomiting, add ginger-prepared Banxia (Rhizoma Pinelliae) 30g.

5.7 Acute Mastiffs

General description

Acute mastitis refers to an acute suppurative disease caused by invasion of bacteria into the breasts. The pathogenic bacteria are staphylococcus aureus or streptococcus. This disease often happens in the animals less than one month after delivery. Its clinical characteristics are high fever, aversion to cold, mass in the breasts with local redness, swelling, heat and painful sensation, unsmooth lactation, and even pus discharge after rupture. It is believed in Chinese veterinary medicine that this disease belongs to the category of "Breast Carbuncle". The problem occurring during lactation is termed "external breast abscess" and the problem happening during pregnancy is termed "internal breast abscess". This disease is mostly caused by accumulation of heat due to improper fodder intake after delivery, leading to obstruction of the collaterals in the breasts and suppuration by heat resulted from accumulation of milk. Besides, this disease can also be caused by deformity of the puerpera's breasts, hesitant lactation, or by suppuration related to invasion of pathogenic factors due to rupture of the nipples.

Essentials for Diagnosis

(1) This disease often happens in the animals during lactation 3 - 4 weeks after delivery.

(2) In accordance with the typical clinical manifestations, this disease can be divided into three stages.

Initial Stage Swelling and pain in the breasts, no red or slightly red color in the local skin, unsmooth lactation, with or without mass in the breast.

Middle Stage Palpable mass in different size in the breast, obvious tenderness, then progressively enlarged mass in the breast, with swelling in red color and burning pain, throbbing pain in the local area, and fluctuating sensation in palpation.

After Rupture Cheesy pus after rupture of breast carbuncle, with heat relieved after pus discharge, fresh granulation, quick healing of the wound. Lingering purulent fluid only in a. few cases with unfresh granulation and lingering wound, even sack pus.

(3) In the initial and middle stages, this disease can be accompanied by obvious general symptoms, such as aversion to cold, fever, aching pain in the head and body, stuffy chest, nausea, vomiting, yellow and brown urine, constipation. In those with weak anti-pathogenic ability after rupture, the symptoms of lustreless complexion, lassitude, low spirit and reluctance in speaking may occur.

(4) In the initial and middle stages of this disease, the total count of peripheral leukOcyte and neutrophil ratio are obviously enhanced.

Syndrome differentiation and treatment

Auording to the zotuafion of description, analysing foddor intake, heat, deformity or nathogenic factors, etc..

5.7.1 Accumulated Milk and Exterior Heat Syndrome

Chief manifestations Distending pain in the breast, with the skin in neither red or slightly red color, unsmooth lactation, palpable mass in the breast, accompanied by aversion to cold, fever, constipation, slightly red tongue, thin and whitish coating, superficial and rapid pulse.

Therapeutic Methods Expel wind, clear away

heat, promote lactation and disperse accumulation.

Prescription Modified "Gualou Niubangzi Decoction" with Quangualou (Fructus Trichosanthis) (smashed) 45g, Niubangzi (Fructus Arctii) 45g, Jinyinhua (Flos Lonicerae) 45g, Lianqiao (Fructus Forsythiae) 45g, Chaihu (Radix Bupleuri) 30g, Chenpi (Pericarpium Citri Reticulatae) 20g, Huangqin (Radix Scutellariae) 30g, Chishaoyao (Radix Paeoniae Rubra) 30g, Lulutong (Fructus Liquidambaris) 30g, Wangbuliuxing (Semen Vaccariae) 30g, Jingjie (Herba Schizonepetae) 30g, Fangfeng (Radix Ledebouriellae) 30, and raw Gancao (Radix Glycyrrhizae) 15g.

Modification For retention of milk during lactation, add Lujiaoshuang (Cornu Cervi Degelatinatum) 27g and Loulu (Radix Rhapontici seu Eehinopsis) 30g. For obvious mass in the breast, add Dangguiwei (Radix Angelicae Sinensis) 30g, and Juhe (Semen Citri Reticulatae) 60g. For Qi stagnation, add Chuanlianzi (Fructus Melliae Toosendan) 30g and Zhike (Fructus Aurantii) 45g. For lingering postpartum lochia, add Chuanxiong (Rhizoma Ligustici Chuanxiong) 30g and Yimucao (Herba Leonuri) 60g.

5.7.2 Syndrome Accumulation of Heat in the Liver and Stomach

Chief manifestations Redness, swelling and burning pain in the breast, burning sensation in the skin of the mass with obvious tenderness, even softened mass with fluctuant sensation when touched by fingers, accompanied by strong heat sensation and restlessness, excessive thirst with preference for drinks, yellow and brown urine, constipation, reddened tongue, yellowish coating, slippery and rapid pulse.

Therapeutic Methods Clear away heat, dissolve toxin, diminish tumefaction and disperse accumulation.

Prescription Modified "Xian Fang Huo Ming Drink" with Jinyinhua (Flos Lonicerae) 30g, Lianqiao (Fructus Forsythiae) 45g, Zaojiaoci (Spina Gleditsiae) 60g, processed Chuanshanjia (Squama Manitis) (decoct first) 30g, Danggui (Radix Angelicae Sinensis) 36g, Mudanpi (Cortex Moutan Radicis) 30g, Shanzhizi (Fructus Gardeniae) 45g, Chishaoyao (Radix Paeoniae Rubra) 30g, Pugongying (Herba Taxi) 90g, Quangualou (Fructus Trichosanthis) (smashed) 45g, and raw Gancao (Radix Glycyrrhizae) 15g.

Modification For excessive heat and higher body temperature, add raw Dihuang (Radix Rehmanniae) 30g and raw Shigao (Gypsum Fibrosum) (Decoct first) 60g. For serious pain, add Chuanlianzi (Fructus Meliae Toosendan) 30g and prepared Ruxiang (Resina Olibani) 18g and prepared Moyao (Myrrha) 18g. For retention of milk after delactation, add raw Maiya (Fructus Hordei Germinatus) 90g and raw Shanzha (Fructus Crataegi) 90g.

5.7.3 Heat Accumulation and Stagnation Syndrome

Chief manifestations It is mostly seen in those with constant mass formed by application of large dosage of antibiotics m the treatment of acute mastitis, manifested by stiff and hard mass in the breast, unsmooth lactation, insidious pain, purple dark tongue, thin and whitish coating, and wiry pulse.

Therapeutic Methods Disperse stagnation, circulate qi, promote lactation and dissipate accumulation.

Prescription Modified "Tao Hong Si Wu Decoction" with Taoren (Semen Persicae) 30g, Honghua (Flos Carthami) 30g, Danggui (Radix Angelicae Sinensis) 30g, Chishaoyao (Radix Paeoniae Rubra.) 30g, Danshen (Radix Salviae Miltiorrhizae) 30g, Chuanxiong (Rhizoma. Ligustici Chuanxiong) 30g, Ruxiang (Resina Olibani) 18g, Moyao (Myrrka) 18g, Zaojiaoci (Spina. Gleditsiae) 20 g, Chenpi (Pericarpium Citri Reticuiatae) 18g, and Lulutong (Fructus Liquidambaris) 30g.

Modification For hard mass in the breast, add Sanleng (Rhizoma Sparganii) 30g, Ezhu (Rhizoma Zedoariae) 30g and Juhe (Semen Citri Reticulatae) 60g. For insufficiency or obstruction of lactation, add Tongcao (Medulla Tetrapanacis) 20g and Wangbuliuxing (Semen Vaccariae) 45g.

5.7.4 Qi and Blood Deficiency Syndrome

Chief manifestations Thin pus after rupture, relieved tumefaction and pain, slightly red granulation in the wound and difficult to be healed, accompanied by lustreless complexion, low spirit and lassitude, pale tongue, thin coating, and thready pulse.

Therapeutic Mathods Replenish Qi and blood, promote granulation and heal the wound.

Prescription Modified "Shi Quan Da. Bu Decoction" with Danggui (Radix angelicae Sinensis) 30g, raw Huangqi (Radix Astragali) 60g, Dangshen (Radix Codonopsis Pilosulae) 30g, Baizhu (Rhizoma. Atractylodis Alda) 30g, Fuling (Potia) 30g, prepared Dihuang (Radix Rehmanniae Praeparata) 30g, Taizishen (Radix Pseudostellariae) 30g, Maimendong (Radix Ophiopogonis) 30g, Honghua. (Flos Carthami) 20g, Chishaoyao (Radix Paeoniae Rubra) 30g, Chuanxiong (Rhizoma Ligustici Chuanxiong) 30g, Danshen (Radix Salviae Miltiorrhizae) 30g, and baked Gancao (Radix Glycyrrhizae) 20g.

Modification For lingering pathogenic factors, add Jinyinhua (Flos Lonicerae) 30g and Lianqiao (Fructus Forsythiae) 45g. For greasy tongue coating, add Chenpi (Pericarpium Citri Reticuiatae) 20g and Sharen (Fructus Amomi) (Decoct later) 20g. For poor appetite, add charred Shanzha (Fructus Crataegi) 30g and Shenqu (Massa Fermentata Medicinalis) 60g.

5.8 Sterility

Sterility is a. common disease. In the animal reproduction procedure, a. number of factors may affect pregnancy and lead to sterility.

5.8.1 Primary sterility

The kidney governs reproduction, stores essence and functions as the prenatal base of life. Only when kidney Qi is superabundant, essence and blood is sufficient, the thoroughfare and conception vessels are full can sexual intercourse result in pregnancy. Any factors that affect any links in this procedure may cause sterility. So sterility is closely related to the kidney. Since the kidney is the dominant one among the fiver zang and six fu organs, asthenia. of the kidney may lead to dysfunction of the liver, spleen and heart. The dysfunction of these viscera further leads to insufficiency of the liver and kidney, asthenia of both the spleen and kidney as well as disharmony between the heart and kidney that result in disturbance of Qi and blood in the thoroughfare and conception vessels and failure of the uterus to conceive fetus.

Essentials for Diagnosis

(1) Examination concentrates on the type of build and the secondary sex characteristic, etc. Gynecological examination is done to see whether there are organic changes.

(2) Assay of ovarian function. Smear examination of the exfoliated cells in the vagina: If the smear examination indicates simple action of estrin or hypofunction of estrin, it suggests no ovulation. Examination of cervical mucus, cervical marks-estimation is for monitoring of ovulation.

(3) Hormone level test is done to analyze whether ovary, pituitary gland and hypothalamus have affected ovulation.

(4) Type B ultrasonic examination is helpful for detecting development of ovum and understanding whether there are organic changes of the uterus and ovary.

(5) Hysterosalpingography is helpful for detecting the pathological changes of the uterus and the conditions of the oviduct.

(6) Abdominoscopy and uterioscopy are helpful for direct observation of the pelvis and the uterus.

Syndrome differentiation and treatment

The root cause of primary sterility lies in the kidney, but the liver, spleen and Qi and blood must be taken into consideration.

5.8.1.1 Syndrome of kidney Qi asthenia

Chief manifestations: No pregnancy long after repeated amphimixis, clear and profuse urine, loose stool, small uterus, light-colored tongue with whitish fur, deep and thin pulse or deep and slow pulse.

Therapeutic methods: Warming the kidney and replenishing essence, nourishing the thoroughfare and conception vessels.

Prescription: Modified Yulinzhu composed of 45g of Dangshen (Radix Codonopsis Pilosulae) 10g Fuling (Potia), 45g of Shanyao (Rhizoma Dioscoreae), log of Shudihuang (Rhizoma Rehmanniae Praeparata), 30g of Shanzhuyu (Fructus Corni), 10g of Danggui (Radix Angelicae Sinensis) 30g of Chuanxiong (Rhizoma Chuanxiong), 30g of Xuduan (Radix Dipsaci), 30g of Tusizi (Semen Cuscutae), 30g of Duzhong (Cortex Eucommi) 45g of Lujiaojiao (Colla Cornus Cervi) and 30g of Ziheche (Placenta Hominis).

Modification: For lower abdominal cold, 20g of Xiaohuixiang (Fructus Foeniculi), 30g of Zishiying (Fluoritum) and 30g Yinyanghuo (Herba Epimedii) are added; for loose stool, 20g of Paojiang (baked Rhizoma Zingiberis), 20g of toasted Muxiang (Radix Aucklandiae) and 30g of stir-baked Baibiandou (Semen Lablab Album) are added; for frequent urination, 30g of Yizhiren (Fructus Alpiniae Oxyphyllae) and 30g of Sangpiaoxiao (Oötheca Mantidis) are added, for emaciation, 30g of Mudanpi (Cortex Moutan Radicis), 30g of Guiban (Plastrum Testudinis) (to be decocted first), 20g of Baiwei (Radix Cynanchi Atrati) and 30g of Zhimu (Rhizoma. Anemarrhenae) are added.

5.8.1.2 Syndrome of liver and kidney deficiency

Chief manifestations: No pregnancy long after repeated amphimixis, red tongue with scanty fur, taut pulse or thin and taut pulse.

Therapeutic methods: Nourishing the liver and kidney, replenishing essence and nourishing yin.

Prescription: Modified Erzhi Dihuang Tang composed of 45g of Nüzhenzi (Fructus Ligustri Lucidi), 45g of Hanliancao (Herba Ecliptae), 45g of Shanyao (Rhizoma Dioscoreae), 30g of Shudihuang (Rhizoma Rehmanniae Praeparata), 30g of Shanzhuyu (Fructus Corni), 30g of Danggui (Radix Angelicae Sinensis), 30g of Baishaoyao (Radix Paeoniae Alba), 30g of Fuling (Potia), 30g of Xuduan (Radix Dipsaci), 30g of Gouqizi (Fructus Lycii), 30g of Ziheche (Placenta Hominis) and 15g of Gancao (Radix Glycyrrhizae).

5.8.1.3 Syndrome of Qi and blood asthenia

Chief manifestations: No pregnancy long after repeated amphimixis, lusterless skin, physical weakness, accompanied by maldevelopment of the uterus, light-colored tongue with white fur, thin and weak pulse.

Therapeutic methods: Nourishing Qi and blood, strengthening and invigorating the uterine collaterals.

Prescriptions: Modified Bazhen Tang composed of 45g of Huangqi (Radix Astragali), 36g of Dangshen (Radix Codonopsis Pilosulae), 30g of Baizhu (Rhizoma. Atractylodis Alba), 30g of Fuling (Poria), 60g of Shanyao (Rhizoma Dioscoreae), 30g of Danggui (Radix Angelicae Sinensis), 20g of Chuanxiong (Rhizoma Chuanxiong), 30g of Baishaoyao (Radix Paeoniae Alba), 30g of Shudihuang (Rhizoma Rehmanniae Praeparata), 36g of Nüzhenzi (Fructus Ligustri Lucidi), 36g of Ejiao (Colla CoriiAsini) and 30g of Xiangfu (Rhizoma Cyperi).

Modification: For aversion to cold, 8g of Fuzi (Radix Aconiti Praeparata) and 10g of Buguzhi (Fructus Psoraleae) are added.

5.8.1.4 Syndrome of liver and heart Qi stagnation

Chief manifestations: No pregnancy long after repeated amphimixis, deep-red tongue with thin and white fur, taut pulse or thin pulse.

Therapeutic methods: Soothing the liver to relieve depression and regulating Qi and blood.

Prescription: Modified Kaiyu Zhongyu Tang composed of 36g of Danggui (Radix Angelicae Sinensis), 36g of Chishaoyao (Radix Paeoniae Rubra), 36g of Baishaoyao (Radix Paeoniae Alba), 30g of Baizhu (Rhizoma Atractytodis Alba), 30g of Fuling (Poria), 30g of Xiangfu (Rhizoma Cyperi), 20g of Qingpi (Pericarpium Citri Reticulatae Viride), 20g of Chaihu (Radix Bupleuri), 30g of Yujin (Radix Curcumae), 20g of Chuanlianzi (Fructus Toosendan), 30g of Yanhusuo (Rhizoma Corydalis), 30g of Danshen (Radix Salviae Miltiorrhizae) and 30g of Niuxi (Radix Achyranthis Bidentatae) are added.

Modification: For distending pain and nodules in the breasts, 30g of Juye (Folium Citri Tangerinae), 30g of Juhe (Semen citri Reticulatae), 36g of Quangualou (Fructus Triehosanthis) and 30g of Lulutong (Fructus Liquidambaris) are added.

5.8.2 Secondary sterility

Failure to be pregnant after last pregnancy (including delivery and abortion) is called secondary sterility. Secondary sterility is usually due to improper care after delivery and invasion of pathogenic factors into the uterus that lead to downward migration of blood stasis or damp-heat to block the uterine collaterals as well as the thoroughfare and conception vessels; or due to stagnation of liver qi, disharmony between Qi and blood as well as obstruction of the thoroughfare and conception vessels; or due to obesity lead to endogenous phlegm-dampness obstructing the thoroughfare and conception vessels. The conception of sterility in TCM is similar to sterility caused by salpingitis, endometriosis and immune disorder in Western medicine.

Essentials for Diagnosis

(1) Failure to be pregnant after last pregnancy accompanied by mass in the abdomen, occasional low fever.

(2) Hysterosalpingography is helpful for detecting morbid changes in the uterus and the conditions of oviduct; abdominoscopy and uteroscopy are helpful for direct observation of the organs in the pelvis and uterus, such as inflammation and mass in the reproductive system.

(3) Diagnostic curettage and histological examination of endometrium is helpful for understanding both the morbid changes in the cervix and uterus and the functions of yellow body of ovary. Hyperplasic endo metrium indicates no ovulation while secretory endometrium suggests ovulation; early secretory endometrium or insufficiency of secretion suggests ovulation with hypofunction of yellow body of ovary.

(4) Type B ultrasonic examination is helpful for detecting development of ovum and understanding whether there are organic changes of the uterus and ovary.

(5) Hormone level test is done to analyze whether ovary, pituitary gland and hypothalamus have affected ovulation.

Syndrome differentiation and treatment

The examination of secondary sterility may be done in light of pregnancy and abortion or history of other disease that may cause sterility, usually marked by asthenia of the root aspect and sthenia. of the branch aspect or mixture of asthenia and sthenia. Clinically various factors have to be taken into consideration in differentiating syndromes in light of the differentiation of diseases.

5.8.2.1 Syndrome of Qi stagnation and blood stasis

Chief manifestations: Experience of pregnancy, sterility due to abortion, premature delivery and gynecological operation, unpressable pain in the lower abdomen, purplish and blackish tongue or with ecchymoses, taut and unsmooth pulse.

Therapeutic methods: Activating blood, resolving stasis and dredging uterine collaterals.

Prescription: Modified Shaofu Zhuyu Tang composed of 36g of Danggui (Radix Angelicae

Sinensis), 30g of Chishaoyao (Radix Paeoniae Rubra), 20g of Chuanxiong (Rhizoma Chuanxiong), 30g of Taoren (Semen Persicae), 30g of Honghua (Flos Carthami), 30g of Chuanniuxi (Radix Cyathulae), 36g of Wulingzhi (Faeces Trogopterori), 30g of Xiangfu (Rhizoma. Cyperi), 30g of Wuyao (Radix Linderae), 30g of Zhike (Fructus Aurantii), 45g of Danshen (Radix Salviae Miltiorrhizae) and 36g of Yanhusuo (Rhizorna Corydalis).

Modification: For severe abdominal pain, 20g of Moyao (Myrrha), 20g of prepared Ruxiang (Olibanum) and 27g of Gegen (Radix Puerariae) are added; for lower abdominal cold pain, 30g of Aiye (Folium Artemisiae Argyi), 20g of Rougui (Cortex Cinnamomi) and 30g of Xiaohuixiang (Fructus Foeniculi) are added; for abdominal pain with low fever, 45g of Baijiangcao (Herba Patriniae), 30g of Mudanpi (Cortex Moutan Radicis), 30g of Pugongying (Herba Taraxaci) and 30g of Digupi (Cortex Lycii Radicis) are added; for serious blood stasis with strong physique, Poxiao Dangbao Tang can be used.

5.8.2.2 Syndrome of interior retention of phlegm-dampness

Chief manifestations: Sterility for years due to abortion, obesity, lassitude and fatigue, light-colored bulgy tongue with whitish greasy fur and slippery pulse.

Therapeutic methods: Drying dampness and resolving phlegm, strengthening the spleen and regulating qi.

Prescription. Modified Qigong Wan composed of 30g of Banxia (Rhizoma Pinelliae), 30g of Cangzhu (Rhizoma Atractylodis), 30g of Baizhu (Rhizoma Atractyloidis Alba), 30g of Fuling (Poria), 30g of Chenpi (Pericarpium Citri Reticulatae), 30g of Shenqu (Massa Medicata Fermentata), 27g of Shichangpu (Rhizoma Acori Graminei) 18g of Houpo (Cortex Magnoliae Officinalis), 30g of Xiangfu (Rhizoma Cyperi), 18g of Chuanxiong (Rhizoma Chuanxiong), 30g of Yuanzhi (Radix Polygalae) and 45g of Haizao (Sargassum).

Modification: For anorexia, 45g of Yiyiren (Semen Coicis) and 30g of Peilan (Herba. Eupatorii) are added.

5.8.2.3 Syndrome of downward migration of dampheat

Chief manifestations: Sterility for years due to gynecological operation, red tongue with yellow and greasy fur as well as taut and rapid pulse.

Therapeutic methods: Clearing away heat and eliminating dampness and regulating the thoroughfare and conception vessels.

Prescription: Modified Simiao Wan combined with Hongteng Baijiang San composed of 30g of Huangbai (Cortex Phellodendri), 30g of Cangzhu (Rhizoma Atractylodis), 30g of Niuxi (Radix Achyranthis), 45g of Yiyiren (Semen Coicis), 45g of Hongteng (Caulis Sargentodoxae), 36g of Baijiangcao (Herba Patriniae), 30g of Fuling (Poria), 30g of Xiangfu (Rhizoma Cyperi), 30g of Yanhusuo (Rhizoma Corydalis), 36g of Lulutong (Fructus Liquidambaris), 45g of Tianxianteng (Herba Aristolochiae) and 40g of Zhike (Fructus Aurantii).

Modification: For abdominal distending pain, 20g of Moyao (Myrrha) and 20g of prepared Ruxiang (Olibanum) are added; for yellow and greasy tongue fur, 20g of Danxing (Arisaema cum Bile) and 30g of Gualoupi (Pericarpium Trichosanthis) are added; for predominant heat in damp-heat, Longdan Xiegan Tang can be used.

5.9 Oophoritic cyst

Oophoritic cyst is a. commonly encountered tumor in animal obstetrics, either benign or malignant. The following discussion may focus on benign type. Oophoritic cyst is usually caused by dysfunction of the viscera, disharmony between Qi and blood, invasion of wind and cold fight after labor or during menstruation, retention of blood stasis during menstruation and labor that inactivate kidney yang; or by interior impairment due to stagnation of Qi and inteflor retention of phlegm, fluid and stasis.

Essentials for Diagnosis

(1) Mass in the lower abdomen, or accompanied by abdominal distension, pressure symptoms and pain.

(2) Obstetrics examination shows mass beside the uterus with evident margin or mobility.

(3) Cytological examination, puncture with thin needle for biopsy, type B ultrasonic examination, radiological examination, abdominoscopy and tumor signifiers can be used as supplementary examinations of benign and malignant tumor.

(4) Benign ovarian tumor should be differentiated from oncological changes of ovary, oviduct and ovary cyst, hysteromyoma, gravid uterus and ascites; malignant ovarian tumor should be differentiated from endometriosis, inflammation of pelvic connective tissue, tuberculous peritonitis, tumor outside birth canal and metastatic ovarian tumor.

Syndrome differentiation and treatment

Clinically oophoritic cyst is classified into syndrome of Qi stagnation and blood stasis, syndrome of phlegm and dampness retention and syndrome of stagnation of virulent dampness and heat. The therapeutic principle is softening hardness and eliminating mass. The therapeutic methods used are promoting Qi flow, resolving phlegm and clearing away heat and draining dampness.

5.9.1 Syndrome of Qi stagnation and blood stasis

Chief manifestations: Cystic mass in the lower abdomen, abdominal distension and pain, dry mouth without desire to drink water, dry lips, unsmooth urination and defecation, purplish tongue, taut and thin pulse.

Therapeutic methods: Promoting Qi flow and activating blood, softening hardness and eliminating mass.

Prescription: Modified Qizhi Xiangfu Wan combined with Xuefu Zhuyu Tang composed of 36g of Cangzhu (Rhizoma Atractylodis), 36g of Baizhu (Rhizoma Atractylodis Alba), 36g of Danggui (Radix Angelicae Sinensis), 30g of Chishaoyao (Radix Paeoniae Rubra), 30g of Taoren (Semen Persicae), 1.5g of Hupo powder (Succinum) (to be taken orally), 20g of Muxiang (Radix Aucklaneliae), 30g of Shanzha. (Fructus Crataegi), 20g of crude Jineijin (Emdothelium Corneum Gigeriae Galli) and 15g of roasted Zhike (Fructus Aurantii).

Modification: For constipation, 18g of Dahuang (Radix et Rhizoma Rhei) (to be decocted later) is added; for red tongue with scanty fur, 27g of dry Dihuang (Radix Rehmanniae) and 45g of roasted Guiban (Plastrum Testudinis) are added.

5.9.2 Syndrome of phlegm and dampness coagulation

Chief manifestations: Obesity, white and greasy tongue fur, taut and slippery pulse.

Therapeutic methods: Resolving phlegm and promoting Qi flow, softening hardness and eliminating symptoms.

Prescription: Modified Haizao Yuhu Tang composed of 36g of Haizao (Sargassum), 36g of Kunbu (Thallus Laminariae), 36g of Xiakucao (Spica Prunellae), 27g of Shichangpu (Rhizoma Acori Graminei), 27g of Danxing (Arisaema cum Bile), 90g of crude Muli (Cato Ostreae) (to be decocted first), 27g of Cangzhu (Rhizoma Atractylodis), 27g of Ezhu (Rhizoma Curcumae Zedoariae), 27g of Sanleng (Rhizoma Sparganii), 30g of Taoren (Semen Persicae), Chishaoyao (Radix Paeoniae Rubra), 30g of crispy Shanzha (Fructus Crataegi), 30g of roasted Liuqu (Massa Medicata Fermentata) and 15g of Rougui (Cortex Cinnamomi) (to be decocted later).

Modification: For relative predominance of cold, 27g of prepared sliced Fuzi (Radix Aconiti Carmichaeli Praeparata) and 27g of Baijiezi (Semen Sinapis) are added and the quantity of Rougui (Cortex Cinnamomi) is added to 15g.

5.9.3 Syndrome of stagnation of virulent dampness and heat

Chief manifestations: Lower abdominal mass, abdominal distension or pain or fullness, or irregular vaginal bleeding, even accompanied by ascites, dry stool, yellow urine, burning sensation in urination, dry mouth and no desire to drink water, deep-red tongue with thick and greasy fur, taut and slippery or rapid and slippery pulse.

Therapeutic methods: Clearing away heat and draining dampness, eliminating toxin and dissipating mass.

Prescription: Modified Qingre Lishi Jiedu Tang composed of 90g of Banzhilian (Herba Scutellariae Barbatae), 90g of Longkui (Herba Solani Nigri), 90g of Baihuasheshecao (Herba Hedyotis Diffusae), 30g of Chuanlianzi (Fructus Toosendan), 90g of Cheqiancao (Herba Plantaginis), 90g of Tufuling (Rhizoma Smilacis Glabrae), 45 Qumai (Herba Dianthi), 90g of Baijiangcao (Herba Patriniae), 90g of Biejia (Carapax Amydae) and 30g of Dafupi (Pericarpium Arecae).

Modification: For predominant virulent heat, 45g of Longdancao (Radix Gentianae), 45g of Kushen (Radix Sophorae Flavescentis) and 45g of Pugongying (Herba Taraxaci) are added; for severe ascites, 30g of Shuihonghuazi (Fructus Polygoni Orientalis) and 30g of Youhulou (*Gryllus testaceus* Walker) are added.

Lecture Six

Modern Studies on TCVM

1 Sterility

Sterility refers to a. disease which lead domestic animals not to oestrous and hybridization or pregnant. For example in cattle especially diary cow, endometritis and ovary function barrier (mainly including ovary static, persistent corpus luteum, cystic ovaries etc.) are the most common.

1.1 Endometritis

Endometritis pertains to "belt syndrome" which secretion of vagina increase in traditional Chinese veterinary medicine. Persistently or interrupted discharges white, yellow-white pus, either mucous with a. little pus, either thin muddy mucous from the vulva, that is also leucorrhea. Cow manifest not oestrous or oestrous cycle is unusual, but less of them are normal.

Pathogen

In traditional Chinese veterinary medicine, sterility is thought that it is result from germ infection. Staphylococcus, streptococcus, corynebateria pyogenes, E. coli etc.

In traditional Chinese veterinary medicine, it is regarded as casing Qi stagnation and blood stasis, downward migration of damp-heat.

Therapeutics

Therapeutic principle is antibacterial, sterilization, and promotion womb contraction. Antibiotics are mainly used in clinical veterinary medicine. Traditional Chinese veterinary medicine take principle that invigorates blood and resolves stasis, alleviate fever dry or wet, contracts womb and discharges pus fluid, removes stasis and make newly. For example, "Qinggongye" series medicine are typical medicaments.

Qinggongye

Composition: Yimucao (Herba Leonuri Sibirici), Taoren (Semen Persicae), Puhuang (Pollen Typhae), Yujin (Radix Curcumae wenyujin), Lianqiao (Fructus Forsythiae), Baijiangcao (Herba Patriniae Scabiosaefoliae) and few antibacterial medicines.

Indication: Treats each kind of type endometritis.

Character: Brown liquid, smell fragrant.

Usage: The womb pours into, each time 100ml, 4 are a. treatment course every other day.

Curative effect: Cure rate above 85%.

Pharmacological action: The modern Pharmacology research indicated that "Qinggongye" has bacteriostasis sterilization, anti-inflammation, promotion womb contraction, acceleration damage tissue's restoration and adjustment ovary function.

Qinggongye Ⅱ

Composition: Dansheng (Radix Salviae Miltiorrhizae), Lianqiao (Fructus Forsythiae), Rendongteng (Caulis Lonicerae), Honghua. (Flos Carthami) etc.

Character: Sorrel suspension.

Indication: Treats all kind of endometritis.

Usage: The womb pours into, full jolting or

jarring when using it, 100ml each time, 4 is a. treatment course every other day.

Curative effect: Cure rate above 85%.

Pharmacological action: The modern Pharmacology research indicated that "Qinggongye Ⅱ" has bacteriostasis, anti-inflammation, promotion womb contraction, acceleration restoration of damage tissue.

Qinggongye Ⅲ

Character: The purple-red oil suspension.

Indication: Treats each kind of endometritis, suitably uses in the chronic endometritis treatment.

Usage: The womb pours into, full jolting or jarring when using it, once every 4 days, each time 25ml, 3 times is a. treatment course.

Curative effect: Cure rate above 83%.

Pharmacological action: The modern Pharmacology research indicated that "Qinggongye Ⅲ" has function of bacteriostasis, anti-inflammation, promotion womb contraction, acceleration restoration of damage tissues.

1.2 Ovaries function barrier

Ovaries function barrier is that normal period of ovaries is lost, cow don't manifest oestrous in clinical veterinary medicine.

Pathogen

Modern veterinary medicine thought that it is improper management lead to endocrine disorders.

Veterinary medicine thought that it is caused by inappropriate management, then result in cow Qi and blood asthenia, asthenia of both the spleen and kidney lead to sterility.

Therapeutics

Hormone therapy was widely used in modern veterinary medicine, but accurate diagnosis and appropriate dosage is needed. However, serious side effect can appear frequently because diagnosis is inaccurate or accurate the hormone dosage was difficult in the practical application process.

In traditional Chinese veterinary medicine, methods which are supplement Qi activate blood, supplement kidneys and strengthen positive, promotion oestrous and help pregnant were used and no side effect. Typical medicament is "Cuiqingzhuyunye", single medicine is Yanghongshan.

Cuiqingzhuyunye

Composition: Danggui (Radix Angelicae Sinensis), Huangqi (Radix Astragali), Yimucao (Herba Leonuri Sibirici), Tusizi (Semen Cuscutae), Yinyanghuo (Herba Epimedii), Yangqishi (Tremolitum)

Character: Brown red transparent liquid.

Indication: cow is not able to oestrous.

Usage: Pours into the womb, each time 100ml, 4 times are 1 treatment course every other day.

Curative effect: Urges the oestrous rate 87.23%.

Yanghongshan nutrition and activate blood, supplement Qi strengthen spleen, supplement kidney and strengthen positive. The usage, powder of the whole grass of Yang Hong Shan 500 ~ 600g, the boiling water burns dilute starch shape, when temperature or mixes in the fodder feeds, previous day 1 medicinal preparation, 5 medicinal preparations were 1 treatment course. Good effect may be obtained.

2 Mastitis

Mastitis is called Breast Carbuncle in traditional Chinese veterinary medicine.

Pathogen

Modern veterinary medicine thought that it is caused by germ infection and result in inflammatory of nipple.

Traditional Chinese veterinary medicine thought it cased by the external injuries influences or internal poisonous pent up, Channels and collaterals of breast impeded, Qi and blood stagnates.

Therapeutics

In the modern veterinary medicine the antibiotic therapy was used, it may dose by breast injection, intramuscular injection, or intravenous injection. It can enhance the curative effect through the medicine sensitive experiment and choice sensitive antibiotic, but antibiotic residue can occur in the treatment period breast.

Traditional Chinese veterinary medicine take to alleviate fever and disintoxicating, disperse stasis to lessen the swollen, dispel Qi and impede Channels and collaterals as mainly principle to choose the medicine treatment. The traditional way to dose is mainly that takes orally, coordinating the partial external use when it is necessity; the new Chinese native medicine preparation also may use the breast injection, intramuscular injection and intravenous injection. 2 kind of effective preparations are introduced in the following:

Ruyuankang

Composition: Gansu dansheng (Gansu Radix salviae Miltiorrhizae), Lianqiao (Fructus Forsythiae), Huanglian (Rhizoma Coptidis Chinens), Jinyinhua (Flos Lonicerae).

Character: Red transparent liquid.

Indication: Cow mastitis.

Usage: The breast pours into, each regio mammalis 50ml a time, once a day, long-term usage 3 ~ 5 day.

Rukang 2 mixtures

Composition: Pugongying (Herba Taraxiaci), Huangqin (Radix Scutellariae), Difuzi (Fructus Kochiae Scopariae), Jinyinhua (Flos Lonicerae), Wangbuliuxing (Semen Vaccariae).

Character: deep brown liquid, taste micro painstaking.

Indication: Cow clinical mastitis.

Usage: Takes orally, 1.5ml/kg. BW, once a day, serious or chronic mastitis cow take 2 times a day, 5 days are 1 treatment course.

3 Retention of the Afterbirth

Retention of the afterbirth defines as cows don't still discharged placenta after parturitating 12 hours.

Pathogen

The modern veterinary medicine reveals that difficult childbirth, abortion, embryo oversized and so many other reasons causes the cow womb contraction incapably, and then result in retention of the afterbirth.

Traditional Chinese veterinary medicine reveals that improper feeding and management cause impede of Qi and blood movement in cow, uterus moving force weaken, cannot cause the afterbirth normally to discharge.

Treatment

Modern veterinary medicine use excited womb medicine. Commonly used medicines are oxytocin, Ergometrine, estrogen and Neostigmine etc. The treatment result is too bad.

Traditional veterinary medicine take methods that invigorate blood and resolve blood stasis, dispel Qi and drainage as mainly treatment principle, moreover, basis dialectical, alleviates fever, either warms channels and collaterals or makes up Qi.

"Cow Medical classics Warning" carries that prescription for the retention of the afterbirth Huashi (Talcum), Guizhi (Ramulus Cinnamoni), Taoren (Semen Persicae), Xuanmingfen (Natrii Sulfas Exsicatus), Mutong (Caulis Akebiae Quinatae), Sanling (Rhizoma. Sparganii) Guiwei (Chinese Angelica), Daihuang (Radix et Rhizoma Rhei), Houpo (Cortex Magnoliae Officinalis), Ezhu (Rhizoma. Curcumae Zedoariae) fries with the long continuous water, takes with sesame oil". This treatment principle of prescription is that invigorates blood and resolves blood stasis;, move Qi and promote water discharge.

On clinical veterinary medicine shenghuatang were widely used: "Danggui (Radix Angelicae Sinensis), Chuanxiong (Rhizoma Chuanxiong), Taoren (Semen Persicae), Paojiang (Rhizoma Zingiberis Preparata), Gancao (Radix Glycyrrhizae Uralensis)", which coordinate each other, warms Channels and collaterals and dissipates cold, invigorates blood and resloves blood stasis. In the clinical practice, medicines are added or subtracted according to sickness, for example, adding the Dangsheng (Radix Codonopsis Pilosulae), Huangqi (Radix Astragali Mongholici), Baizhu (Rhizoma Atractylodis Alba) and so on can supply its lacking in vital energy; adding the Wulingzhi (Faeces Tropterori) and Xiangfu (Rhizoma) and so on can make Qi moved; adding Honghua (Flos Carthami), Yimucao (Herba Leonuri Sibirici) and fresh Puhuang (Pollen Typhae), Dansheng (Radix Salviae Miltiorrhizae) and so on can strengthens action on invigorate blood and resolve blood stasis; adding Rougui (Cortex Cinnamomi), Aiye (Folium Artemisiae Argyi) can strengthen action on warm channels; adding the Daji (Radix Euphorbiae pekinensis).

Shanzha (Fructus Crataegi Pinnatifidae) and brown sugar may prevent the cow.

Yimucao (Herba Leonuri Sibirici), the brown sugar may prevent the cow afterbirth not under.

4 Experimental Studies on the Fowl Cholera. Disease Prevention of Chinese Medicine

Two tests attacked with strong virus and protected with Chinese medicine were carried out in this study, and now, the methods and results of the test were reported executively in brief as follow

The 1st Test: Test attacked with strong virus No. 1

Materials and methods:

(1). Chickens: Ross species layers, which had been raised for 463 days at Lanzhou Chenchang breeding and processing, plant. After clinical and dissective examination, these Layers were not found any unusual. Except prevented by Newcastle disease vaccines, any other vaccines had not been used to the layers.

(2) Virus: 84. 6. 4 frozen strong virus C48-1, produced by Chinese veterinary medicine control Institute, were removed into the culture solution of brain & heart and cultured for 24 hrs. 0. lml sample were got from this cultured bacterial solution injected intramuscularly at the chest of the chickens to make the bacteria's stronger. After the chickens died (18 hrs), we autopsied and found white pinpoint sized necrosis spot on the Liver surface. The blood of the liver and spleen were cultured on flat utensils to separate the different bacterial. After 24hrs' culture, kept the bacteria's on an inclined surface as the attacking virus. The bacteria's were cultured again in culture solution of brain & heart got this cultured solution and diluted to count. By this, the best temperature to preserve culture medium and examined bacterial solution could be determined after three times examination. After a. night kept at 8'13, the dying ratio of the bacterial solution sample, which had examined and counted was 16. 6, 20, 30 percent respectively.

(3) Chinese medicine and its preparing aeth0dstchinese medicine, named health fowls, was nursed in water for 30mins, and boiled with strong heating, then kept decoction for 20mins. After filtration, added an appropriate water and continued to decoct for 25 mins. Filtrated it again and put the two decoctions together. Fixed the medicine solution to a. certain volume that per 10ml contained the raw drugs 1g and kept at -4℃ in refrigerator for use. The medicine could also be crushed into powders and added to the fodder and let the chickens take freely.

(4) Bacterial vaccine: 731 and B26 bacterial vaccine solation was provided by Chinese veterinary medicine supervisory institute whose activity was killed by culture. The white oil produced by Chinese veterinary supervisory institute was used as adjuvant. 240ml of this adjuvant was added to per 120ml bacteria solution. After centrifugal agitation for 5 mins, at 8000 rpm, 120ml white oil adjuvant

eves added, and centrifuged for 5 mins again. The selected sample was cultured and reacted negative. After being injected intramuscularly at the chest, the layers showed unconscious and ate tess, but returned to normal at the later day.

(5) Test group: The chickens were divided into five groups:

① The vaccine group: 0.6ml white oil vaccine was injected intramuscularly at the chest of nine chickens 24 days ago before virus attacked.

② The powder group: There were 10 chickens in this group. 1g powder of the crushed Chinese medicine was added to the fodder to feed each checked 22 days ago tilt virus attacking.

③ The decoction group A. 10 chickens, the decoction of Chinese herbal medicine, 15mg/each chicken, was orally taken 10 days ago before virus attacked, it did not stop for medicine feeding for 3 days after virus attacking.

④ The decoction 9ronp B: Ten chickens. 15ml decoction was orally taken days ago before virus attacking. It would be kept 3 times till virus attacking.

⑤ Blake control group. Ten chickens. This group was only used as a. contrast to the strong virus attacked group and not treated by anything.

(6) Breeding administrations. The layers were bred at the sane level in the chicken box and picked the fodder freely. The fodder bought from Lanzhou feed company.

(7) The toxicity of virus analysis: The toxicity was analyzed by 10^{-8} count results (20.5 bacteria's per ml) of virus injectoin as Tab. 6 – 1 showed.

Tab. 6 –1 The toxicity of virus analysis

Injected venom quantity (ml)	0.2	0.4	0.6	0.8
The contained bacteria amount by the lst examination	4	8	12	16
The contained bacterial amount by re-examination	3.2	6.4	9.6	12.8
Injected chickens	4	4	4	2
The proportion of dying chickens (%)	4/4	4/4	4/4	2/2

The minimum lethal dose is 3.2 bacteria's.

The chickens analyzed the toxicity began to die orderly in the later morning according to the injected bacterial amount. 4 chickens of them died from 20 to 24 hrs after injection (0.6, 0.8, 0.8, 0.6), and other 10 chickens had all died in 25 ~ 28 hrs.

(8) Strong virus attacking: At 10 ~ 11 o'clock of September 13, 0.1ml venom examined yesterday (32 bacteria/ml) was intramuscularly injected at the chest to each layer. The actual bacterial amount of attacking virus was re-examined on a. brain & heart agar flat utensil。The results of the next day re-examination decided the actual bacterial amount that attacked each layer was 2.12 bacteria's contained in 0.1ml.

Results:

(1) Clinical characteristics: The singular reaction had not been found in 10 hrs after virus attacked on the chickens. In 11 ~ 12 hrs, the powder and control group in succession showed unconscious, poor appetite, eyes closed and inactive. The powder group showed more distinctive than the contrast; and then the chickens began to die

(2) Dying time. Four chickens of powder group and three of the contrast were died in 20 ~ 28 hrs. In 28 ~ 36 hrs the powder group died 2 chickens, the contrast died 4 and the vaccine died 2. In 36 ~ 48 hrs, the control group died I chicken and the vaccine, also died 1. In 48 ~ 72 hrs the control group died 2 chickens and the vaccine died 1 In 72 ~ 108 hrs. The vaccine group died 1 chicken.

(3) The preventing affects See Tab. 6 – 2.

Tab. 6 – 2 The preventing affects

Group \ Type	N	Dying chickens	Dying ratio	Living chickens	Preserving ratio	Others
Powder group	10			4	40%	The medicine was taken 22 days ago. And one day before virus attacking the medicine was not given.
Vaccine group	10		55.6%		44.4%	The vaccine was injected 24 days ago before virus attacking
Control group	10	10	100%	0	0	No treatment
Decoction group A	10		0	10	100%	The medicine was taken 10 days ago. 3 days before virus attacking, the medicine was not given
Decoction group B	10			10	100%	The medicine was taken 2 days ago till virus attacking

(4) Conclusions:

①The effect of the decoction group in this test is the best. Its preserving rate is 100%. But this can only suggests that the decoction of Chinese traditional medicine has an immunity cure effect. It cannot determine the immunity period.

②The medicine powder has a certain anti-cholera affect The dissected liver and spleen are swelling bigger, it stores up and has a poisoning proclivity.

③The decoction of Chinese herbal medicine is good for anti-cholera. Its taking times can be decreased to three properties.

④The time that make the chicken die The shortest time that make the chicken die when virus attacked is 23 hrs (powder group), and the longest is 96 hrs (The chickens in the decoction group A take the medicine continuously).

⑤The surviving chickens have solid immunities. After 15 days observation, they are found unusual.

The 2nd Test; Strong Virus Test Ⅱ:

Materials and Methods:

(1) Chickens: Isa. Brown species layers from Lanzhou Wu-xing-ping Chicken plant, No poultry cholera vaccine had ever injected.

(2) Virus: Strong Cirus C_{48-1} form Chinese Veterinary medicine Supervisory institute.

(3) Chinese medicine and its preparing methods: It was the same to the above.

(4) Vaccine: Provided by Chinese Veterinary Medicine Supervisory Institute.

(5) Group division and virus attacking time:

① Propound No. 1 Group: There were six chickens in this group. Each chicken took 5g Chinese medicines everyday from May 22nd to 26th. In 26th, the medicine was stop to be taken in 29th; the chickens were attacked with virus.

Compound No. 2 Group: Five chickens. The time of medicine being taken and attacking with virus were the same to the above.

② The Period of Prevention and Immunity:

One-month group (compound No 1). 11 chickens From May 4th to 7 th, 5g medicine was fed to each chicken everyday.

Two months group (compound No 2): 11 chickens From April 7 ~ 10 th, 5g medicine was fed to each chicken everyday.

Three months group (componnd No 2): 11 chickens. From March 5 ~ 8 th, 5g medicine was fed to each chicken everyday.

③ Vaccine control Group: 10 chickens. The chickens were given a. subcutaneous injection with 0.1ml of the vaccine (it was 25 days before virus attacking)

④ Strong Virus control Group: 10 chickens. No treatment was given to these chickens; it was only contrased to virus attacking

(6) Breeding Administrations: Same to the and test group

(7) Toxicity Analyzed: 10^{-8} venom , was put into refrigerator, its reviving rate was 90% (95.48 + q8.44 + 76% ~ 90%) The examined venom was injected intramuscularly at the chest of the chickens with 30 bacteria. per ml The injection was taking with 3 group division.

0.1ml, two chickens, the dying chickens were 0/2

0.2ml, two chickens, the dying chickens were 1/2 (36 ~ 40 hrs.).

0.3ml, two chickens, the dying chickens were 2/2 (24 ~ 36 hrs.).

Re-examining the venom, the result was 8 bacteria's per ml. Therefore, the bacteria's contained in the attacking virus were 8 × 0.3 ~ 2.4. Thus. The lethal dose of it was 2.4 bacteria's.

(8) Attacked with strong virus:

8 bacteria's per ml of 10^{-8} examined virus solution counted by 90% was injected intramuscularly, at the chest of the chickens al 10: 30 to 11: 35 of May 29 The bacteria's contained in the virus was 8/ml × 0.4 × 90% = 2.88. Test again with the virus, the actual bacteria's injected was 2.8.

Results See Tab. 6 – 3.

Tab. 6 – 3 The effects of preventing cure

Group \ Effects		N	Dying chickens	The ratio of relative prevention (%)	The ratio of absolute prevention (%)
Control group		10	9	10	10
Preventing cure	opound No. 1 group	6	2	66.6	56.6
	Compound No. 2 group	5	0	100	90
Preventing	One month group (compound No. 1)	11	3	72.3	62.3
	Two months group (Compound No. 2)	11	0	100	90.9
	Three months Group (compound No. 2)	11	1	90.9	80.9
Vaccine group		10	4	60	50

5 Extraction, Isolation and Anti-acarid Effect of Chinese Herbal Medicines-Stemona tuberosa. Louse, etc.

Studies on active compounds extraction, isolation and structural identification of four Chinese herbal medicines were carried out. Active compounds were extracted from Stemona tuberosa Lour, Citrus tangerina, Mentha haplocalyx Briq, and Macleaya codata, and the former three of them were isolated, as well as the structure of them were identified separately. Activity of crude extracts from four Chinese herbs and oxymatrine were investigated and selected primarily. The antimite effect of Stemona, tuberosa Lour, Citrus tangerina, Mentha haplocalyx Briq, and Macleaya codata were also studided. The anti- acarid experiment on Stemona tuberosa Lour and Macleaya codata insecticidal for bee-acarid were taken in vitro.

Active compounds from Stemona tuberosa Lour and Macleaya codata were extracted by dipping in ethanol, Active compounds from Stemona tuberosa Lour, Citrus tangerina, Mentha haplocalyx Briq, and Macleaya codata were isolated by VLC (volume layer chromatography), TLC (thin layer chromatography) and recrystallization. Structure of three compounds were identified by determination mp. and various spectral analyses (UV, 1HMNR, MS), which were tuberostemonine, menthol and thymol。

The results of the primary experiment showed that different concentration of Stemona, tuberosa. Lour (0.1%, 0.1%, 0.2%) and aerial parts of Macleaya. Codata (1%, 2%, 3%) and Citrus tangerina had potent activity to kill Varroa jacobsoni as well as Tropilaelaps clareae. However, Sophora, flavescense Ait which has been used to kill psoroptic can't kill Varroa, jacobsoni and Tropilaelaps clareae significantly. In addition, 25% Citrus tangerina, were indicated to be active to kill Varroa, jacobsoni and Tropilaelaps clareae, but 50% Citrus tangerina, hurt honeybees.

The results of effect experiment showed that three Chinese herbs can kill Varroa jacobsoni and Tropilaelaps. Effect of 25% Mentha haplocalyx Briq were significantly and it can be as optimum concentration. Three Chinese herbs is safety for bees ($P > 0.05$). It is the first time that extracts from Stemona tuberosa Lour and Macleaya codata were used to kill Varroa jacobsoni and Tropilaelaps clareae, so their optimum concentration need to be studied further.

The results of anti-mite experiment showed that three concentration of Stemona tuberosa Lour and Macleaya codata had potent activity to kill Varroa jacobsoni and Tropilaelaps clareae. Effect of 0.05 percent potent activity of extracts from Stemona tuberosa Lour were indicated to be equal to 15% lactic acid, while potent activity to kill Varroa. jacobsoni the former is prior to the latter.

6 Extraction and Isolation of Active Constituents from *Salvia miltiorrhiza* Bunge and the Pharmacology and Clinical Application of Ruyuankang in Cow

The lipid soluble chemical components of *Salvia. miltiorrhiza. Bunge* were extracted by dipping in cold dichloromethane, isolated and purified by TLC, LSC and recrystalization. Four compounds were isolated from the roots of Danshen. On the basis of UV, IR, ^{1}HNMR spectra and m. p., and also by comparison of their spectral data with those reported in literatures, the structures were identified as compound 1 (tansheninoneII_A), 2 (dansheninoneI), 3 (cryptotanshinone), 4 (dihydrotanshinone). The purity is above 97%.

Ruyuankang injection has been made by dansheninones which were eatacted and the other Chinese herbal medicine chemical constitunents. Confirmed by a. series of facts, the injection is steady. The antibacterial test results shows that the injection can kill and restrain the mastitis pathogenic bacterial. The rabbit test shows that the injection has no local stimulability. The acute toxicity reaction showed that the LD50 of Ruyuankang by ig. is36.80g/kgB. W, the 95% reliability is 32.12 ~ 41.47g/kgB. W.; the LD50 of Ruyuankang by ip. is 13.85g/kgB. W., the 95% reliability is 12.76 ~ 14.94g/kg B. W. These facts indicate that Ruyuankang Injection has no acute toxicity. Assessing anti-inflammatory by dimethylbenzene-inuced ear oedema in mice, the result indicated that Ruyuankang may eliminate inflammatory ($P < 0.05$). The effect of anti-inflammatory of Ruyuankang is same as hydrogen cortisone ($P > 0.05$).

The mastitis of 96 cows' 122quarters is treated with injecting Ruyuankang, penicillin and streptomycin, ciprofloxacin into mammae, and the total effective rate is 94.7%, 93.7%, 93.2%, the total heal rate is 81.7%, 70.8%, 82.2%. The clinical results shows that Ruyuankang injection is safty and high heal rate in dairy cattles mastitis. Especially it can reduce the temperature and lessening of quarter. This experiment was firstly shown that Chinense herbal medicine, which is made with different chemical constituents exracted from herbs, treat mastits in cows.

7 The Specifications of Liuqiansu

Liuqiansu, one of the active substance in

Rubia. cordifolia. L. , has been proved by clinical that has good therapy effect on infection via. Respiratory tract, Alimentary tract, Urinary tract and wound, especially for bacteroidal diarrhea. and dysentery.

From 1991, we have synthesized Liuqiansu artificially and done the test of pharmacology and pharmacodynamics. The test was proved that Liuqiansu was a. kind of broad-spectrum anti-bacteria. drugs with lower toxin (giving drug to little mouse by intramuscular: LD_{50}: 230. 73mg/kg B. W; by intravenous: LD_{50}: 227mg/kg B. W). It can strongly restrain Salmonella, Poultry Choleraic Pasteurella, Escherichia coli, Staphylococcus aureus, Streptococcus agalactiae, etc. , as does to Epiphyte. As a. natural herbal component, Liuqiansu has more merits than others by its low toxicity, no bearing-drugs, and a. superior applied value in veterinary clinic.

[character] White crystalline powder

[function and therapy] Having broad-spectrum anti-bacteria drugs and the merits of low toxicity, no bearing-drugs, and good therapy effect, mainly used for cure chicklin diarrhea resulted by Salmonella, milch cow mastitis, calf diarrhoea and piglet diarrhea resulted by Escherichia coli.

[use method] For the chicklin diarrhea resulted by Salmonella, it is taken 0. 5 ~ 1g and mixed with 1kg fodder to feed chicklin 5 ~ 7day. For the milch cow mastitis, it is injected at part cow udder with a. dosage 400mg per day, The therapy period is 3days. And for the piglet diarrhea resulted by Escherichia coli, it is taken orally or injected 7days successively with a dosage 80 ~ 120mg per day, 2 ~ 3times per day, The therapy period is 3days.

[Storage] Conserved at the airproof and dry condition.

8 Instruction of Qiancaosu Injection

Based on successful development of Liuqiansu, Qiancaosu is invented with QSAR study, which carried through its antibacterial activation by using Topiss, Free-Wilson models and Hansch analytical method. By combined with the three motheds preferably, a compound which has better antibacterial activation is filtrated A lot of clinical test indicate that the compound, Qiancaosu, have better clicinal effect than Liuqiansu to cure cow mastitis、uterine innermembranous inflammation、hydropsy of piglets.

Qiancaosu have the characters of good curative effect、lower toxicity、no residua、detumescence and quick return milk when it is used to treat cow mastitis and uterine innermembranous inflammation. Futhermore, it also can avoid matter of medcine-resistence of bacterium individual caused by abusing antibiotics. It is recognized and favored by numerous culturist deeply. At present many regions of china. extensively use it and obtain favorable therapeutic effect. Qiancaosu can effective control hydropsy of piglets when use it to cure the disease. Moreover, it heal one ill-pig only spend about ten to fourteen yuan, and is generalized easily. Qiancaosu become preferred medicine to cure hydropsy of piglets in many provinces, citys and regions.

[character] White crystalline powder.

[function and therapy] Qiancaosu mainly treat cow mastitis、uterine innermembranous inflammation, hydropsy of piglets, and have the character of good antibacterial

curative effect, lower toxicity, no medcine-resistence, no residua、detumescence and

quick retune milk and vile price.

[direction and dosage]

①Cow mastitis: it is melted in sterile water and led to mammary-pool using milk-pipe, each time 200 ~ 400mg per mammary-section, when getting more serious, it can add to 600 ~ 800mg each time , twice every day. A. period of treatment is three days.

②Uterine innermembranous inflammation: it is melted in sterile water then lead to womb directly, 800 ~ 1 200mg each time, when muscle injection use same dosage, once every day. A. period of thertment is eight days.

③Hydropsy of piglets: it is melted in sterile water and given by intramuscle (i. m.). Dosage is 100 ~ 200mg each time (4 ~ 8mg/kg), 2 ~ 3times per day.

[specs] 400mg each bottle, 50 bottles each box, 20 boxes each tank.

[storage] Air proof, conservation in dry places.

9 The Specifications of Cuirusu

Cuirusu, been made up of several herbal medicines which mainly contains Radix, has obviously biologic activity on animal in vivo. It can accelerate animal latex excretion, improve propagate capability, increase immunity and has some other merits involving little dosage, quick effect and low toxicity. It has been improved that Cuirusu can improve latex excretion by 5-10 percent on lower milk excreted milch cow. and has nothing influence on milch cow body and its production capability.

[main component] Several herbal medicines mainly containing Radix.

[main therapy] It is mainly used to increase milk excretion for the lower milk excreted in milch cow bodies, and cure the disease of lacking milk. It also has function of dredging main and collateral channe, invigorating the circulation of blood, and adjusting the two opposing principles in nature.

[use method] For the cow and horse, it is taken 50g and mixed with fodder to feed every day, as for the sow and sheep, it is taken 30g. The therapy period is 3 ~ 5days

[Storage] Three years.

[packaging] One bag contains 100g or 300g.

10 The Specifications of Hypericin

Hypericin, one of the most active substances in Hypericum, has been studied focusing on treating depression and virosis diseases (including HIV) in oversea institutes recently. The oversea anti-virus experiments indicate that Hypericin has a. strong activity on restraining HIV and other retrovivus (such as HSV-1, HSV-2, Parainfluenza. virus and Cowpox virus) in vitro and vivo.

New Veterinary Pharmaceutic Project Laboratory, Lanzhou Institute of Animal & Veterinary Pharmaceutice scince, CAAS, has extracted and separated Hypericin from Hypericum successfully. Furthermore we have finished identifing the structure and doing the work such as elementary pharamacological and toxicological experiments. We also have carried out systemic anti-virus tests in vitro and vivo. The results indicated that Hypericin can kill H_5N_1& H_9N_1 AIV and FMDV effectively and heal ill animals obviously.

Now, we have combined Hypericin with functional protein and manufactured the Hypericin's compound which can improve Hypericin's stability and be used in clinical.

Description: Browned-black powder and slightly bitter

Action: To cure the disease derived from Influenzavirus A (AIV), Foot-and-Mouth Disease Virus (FMDV) and other retrovivus's disease.

Dosage: when cure disease, Hypericin's compound 25g is added to one kilogram fodder and feed five days continuously; while prevent disease, Hypericin's compound 0. 5g is added to one kilogram fodder and feed continuly.

Storage: Preserve in tightly closed containers, low temperature and dark condition

Address: 335, Little west Lake Jiangouyan, Lanzhou, Gansu

Linkman: Liang Jian-ping

Postcode: 730050

11 Chinese Herbal Medicine

Chinese herbal medicine refers to Chinese traditional medicine and herbal medicine. Usually, Chinese traditional medicine refers to medicine that the traditional Chinese veterinarian commonly used; the herbal medicine refers to the medicines that are applied in the folk. However, with the unceasing excavate, research and the promoted application of herbal medicine, some good curative effect herbal medicines have been gradually applied by traditional Chinese veterinarian. Thus, the discrimination the Chinese traditional medicine from the herbal medicine even more and more difficult, therefore, we will call the Chinese traditional medicine and the herbal medicine as the Chinese herbal medicine.

According to the natural, Chinese herbal medicine are possible to divide into:

Plant medicine: for example, Radix Angelicae Sinensis, Radix Astragali Mongholici, Radix et Rhizoma. Rhei, Flos Lonicerae etc;

Animal medicine: for example, Hairy Antler, Fossilia. Ossis Mastodi, and Hippocampus etc;

Mineral medicine: for example, gypsum, alum, borax etc.

According to the literature reported, there are 12, 694 kinds of medicinal plant, 11, 020 kinds of plant medicine, 1, 590 kinds of animal medicine and 84 kinds of mineral medicine in our country. Actually, traditional Chinese medicines, which have been used for commodity, are only 1,000 kinds.

11.1 Research content of Chinese herbal medicine

11.1.1 Origin and classification

According to the Origin: Wilding and artificial cultivation (or raising);

According to the action of medicine: for example, antidote, antipyretic, medicine for regulating the flow of vital energy, the medicine for invigorating blood and resolving blood stasis etc.;

According to partial for medicine: for example, root class, leaf class, flower class, skin class, kind of subclass, rhizome class and entire grass class;

According to the active components: for example, Chinese herbal medicine that contains alkaloid, or flavones, or volatile oil or glycosides etc.

11.1.2 gathering, processing and concocts

Gathering: Chinese herbal medicine containing components are the material bases which the medicine prevents and treats disease. But quality and quantity of the active components are extremely related to gathering season and time as well as the method of the Chinese herbal medicine. It is important for Chinese traditional medicine to gather at the right moment and is one of medicine quality guarantee condition; it is also important factor to affect the performance and curative effect of medicine quality.

In the different growth developmental stage, accumulation of chemical composition in the plant is different; meanwhile, due to influence of climate, habitat, soil and so many other factors, they are very different. Firstly, ages of plant growth are closely related to quality of the chemical composition. For example, the glycyrrhizinate is one of the principal components from Radix Glycyrrhizae, if growing three or four years, content of glycyrrhizinate is more than a. time one year's. Also, the total Ginsenoside content in the ginseng is highest by 6 ~ 7 year recovery. Secondly, in the plant growth process, along with the change of month, content of active component is also various. For example, in the different growth developmental stage of Radix Glycyrrhizae, carrying on determination content of the glycyrrhizin, results showed that glycyrrhizin content is 10% in the preliminary blossoms and it's the highest, therefore the earlier period in blossom recovery Radix Glycyrrhizae is suitable.

Processing: Chinese herbal medicine processing is mainly elimination the impurity and prompts dry in order to avoid rotting and to be advantageous for the preservation and the application.

Concocts: Chinese herbal medicine concoct are the further processing crafts which needs to be according to its theory on clinical using and the medicine configuration. This is closely related with the drug efficacy. The practice proved that, through concocting can eliminate or reduce the toxicity or the side effect of the medicine, change property or enhances the curative effect of the medicine, is advantageous for the smashing to process and to store and so on. For example, the Rhizoma Corydalis, whose function is invigorates the blood, regulates the flow of vital energy and stop pain, but all the function of the vinegar concocting Rhizoma. Corydalis enhance distinctively.

A. Commonly used concocting method:

Fries: Fries without additional ingredients, fries with earth and fries with the bran etc to eliminate the moisture content, to cause tissue to be loose and be benefit for leaching and smashing components, in addition, it make the enzyme in the plant deactivated to be advantageous for store.

Roasts: Medicine and the liquid supplementary material fries altogether, which causes the supplementary material permeating medicine. Commonly used methods are honey roasts, salt roasts, vinegar roasts, liquor roasts, oil roasts and so on.

Calcines: Medicine high-temperature treatment, which causes medicine crisp and be pure as well as advantageous for smashing and leaching active components.

Steams: The method, which heats up, by the steam. Commonly used methods are steams, vinegar steams, liquor steams etc.

Boils: Medicine and the supplementary material boils altogether. Commonly used method is the vinegar boils.

Waters: Methods are including rinse, soak, rotten and "Shuifei" which grinds repeatedly together with water etc.

11.1.3 Identification

There are 6 kinds commonly used methods: Identification original plant (based on original plant), character identification, micro identification, physics and chemistry identification, bioassay, quality analysis.

11.1.4 Chemical compositions

The compositions of Chinese herbal medicine are very complex, usually including sugar, amino acid, protein, fat, organic acid, volatile oil, alkaloid, glucosides, tannin, inorganic salt, pigment, Vitamin and so on. Each kind of Chinese medicine possibly includes many kinds of components, some of them has obvious biological activity and medical function, it is often called the active component, such as alkaloid, glucosides, volatile oil, amino acid etc. For example, berberine of Rhizoma, Coptidis Chinens, ephedrine of Ephedra Herb and Radix Scutellariae etc. But others once revealed the invalid components such as polysaccharide, protein, tannin etc became active components because of discovering of biological activity. For example, pachmaram, Radix Angelicae Sinensis polysaccharides, Wolfberry polysaccharides etc.

11.1.5 Pharmacological actions

Under Chinese medicine theory instruction, our country general medical persons, union modern science theory, method and method, close union clinical, has carried out the Chinese traditional medicine pharmacology research thoroughly systematically.

At present, to prescribe medicine pharmacological action and clinical practice, researching of Chinese traditional medicine compound prescription and simple come to be main factor; On the other hand, application modern pharmacology method to extract and separate biological activity components from the Chinese traditional medicine, to explain action mechanism of Chinese traditional medicine and identify structure. Finally, synthesis and innovation, develops the new drugs. The ant- malarial artemisinin development is a. very good example.

Pharmacology research of each kind of Chinese traditional medicine

Medicine for invigorating blood and resolving blood stasis: Radix Angelicae Sinensis, donkey-hide gelatin, Tuber Fleeceflower etc..

Medicine for alleviating fever and antidote: Fructus Forsythiae, Indigowoad Root, Herba. houttuynaie etc..

Medicine for alleviating fever and cold blood: Radix arnebiae, China. Pulsatilla. etc..

Medicine for alleviating fever and make dry or wet: Sophora. japonica., Rhizoma. Coptidis Chinens, Radix Scutellariae etc..

Supplement benefits medicines: Radix Codonopsis Pilosulae, Radix Astragali Mongholici, medlar etc..

11.1.6 Clinical practices

Chinese medicine performance It refers to every medicine inherent nature and the function of itself.

"Four QI and five senses" Four Qi refers to cold, hot, warm, cool of the traditional Chinese medicine, which are four kinds of different nature. The five senses are the acid, painstaking, sweet, pungent, and salty.

Fluctuations and vicissitude It refers to the medicine function in the organism four kind of trends.

Guijing It mainly refers to selective act on some internal organs or the channels and collaterals. In other words a. medicine can cure illness, this medicine is Guijing of the medicine.

Medicines reciprocities In ancient times, each kind of medicine reciprocity was summarized: coordinate, assistance, elimination, counterbalance, expiration, opposite

Medicines usage Generally, Usage of Chinese herbal medicine margin for safety is quite big, but certain virulent medicines still had to be paid extremely attention to and could not excessive.

11.2 Chinese traditional medicine and compound prescription Chinese traditional medicine

Chinese traditional medicine and Chinese traditional medicine atlas Here we emphasis several medicines.

Chinese traditional medicines compound prescriptions preparation In recent years, Chinese herbal medicine preventing and controlling poultry disease had demonstrated the extremely good prospect, arouses the domestic and foreign colleagues' strong interest. For example, the cow mastitis, cow reproduction disease, cow intestinal tract disease, pig dysentery, pig respiratory tract disease as well as birds and beasts' all kinds of diseases treat with the Chinese traditional medicine compound prescription preparation and obtain the very good effect. The Chinese Institute of Agricultural Sciences Institute of Traditional Chinese Veterinary Medicine develops the Chinese traditional medicine compound prescription preparation- "Diqiuling" which are mainly used in prevention and treatment of coccidiosis. By promotion and application the medicine in 20 provinces and cities in our country and millions feathers chickens, the statistical data demonstrated that prevention effect above 98%, cure rate 93% ~98%; Next, through clinical practice compound prescription preparation "Qingwenwang", the results showed that it is has the good effect on not only the birds and beasts (chicken, duck, goose) cholera, chicken dysenteria. alba. and bacterial illness so on but also the IBDV, ND and other epidemic virus disease. Moreover, some Chinese traditional medicine compound prescription preparations such as "Qinggongye" can treat cow endometritis. By ten thousands of clinical examples treatment experiment, effectiveness 94.5%, cure rate 87.84%, after the latter three oestrus conception rate is 84.97%. However, there are many compound prescriptions, every does not state in detail.

12 Study on Detection of Acoustic Emission Signals (AES) Propagated along 14 Meridians in Sheep

Abstract: In order to explore the objective index of Propagated Sensation along Meridians (PSM) and diagrams of acupoints along meridians in animals, Acoustic Emission Signals (AES) at 87 acupoints and 43 control points along 14 meridians in 223 sheep (a. total of 22467 acupoints or point times) were detected. The results showed that AES propagates bidirectionally along meridians, is about 6.67cm/sec, and that AES propagated along meridians in Sheep is a. biophysical signal following biological phenomena, and that the property of AES are associated with all tissues, chiefly with muscles and their nerves.

Key Words: Acoustic Emission Signals; 14 Meridians; Acupoint; Sheep

Introduction

Jingluo, or the Meridian is one of the fundamental theories of the Traditional Chinese Veterinary Medicine (TCVM). It has been used to guide the clinical practices of TCVM and Chinese Veterinary Acupuncture (CVA) for several thousand years. The Meridian phenomenon has been objectively demonstrated through an investigation on Propagated Sensation along Meridians (PSM) in hundreds of thousands of people. In order to explore the objective index of PSM and diagrams of acupoints along meridians in animals, after achieving success in clinical studies on CVA and acupuncture anesthesia. (AA), especially in AA with electro-rotated needles instead of electro-acupuncture in dogs and sheep, we have done the experimental study on the detection of AES propagated along 14 meridians in 223 sheep.

Materials and methods

Experimental Animals: 223 local sheep (108 female, 115 male) from Lanzhou, Xi' an and Hefei were tested. They were all in good health, with no clinical signs, and ranged in age from 0.5 ~ 8 years, in weight from 15 ~ 60 kg, in body temperature from 38.0 ~ 40.0℃, in heart rate from 70 ~ 80 t/min, in respiratory rate from 12 ~ 23 t/min. They were stalled for a. week for healthy observations before the experiment. Hay and water were available, ad libitum during the experiment.

Experimental Instruments: Physiological recorder and monitor, a computer data analysis system, the model JSF4-1 detection on AES along meridians, the model PZT-5 transducer for AES reception, the stabilization-rack for animal with netting (SRAN), the nylon-button belts, scalpels, etc..

Reagents: chloralos puress, Ethyl carbamate C.P., Acequindox, etc..

Acupoints (AP) and Control Points (CP): 87APs were detected. Their location was based on our clinical experiences of CVA. and transpositioned the acupoints along 14 meridians in human, 43CPs were selected and located the spots 2 ~ 5cm left to the AP at the same horizontal line. See Tab. 6 – 4.

AES Detection:

(1) The sheep were stabilized on SRAN at lateral position, and the wool on AP and CP areas was cut, and the skin on the areas was cleared with the cotton balls with anhydrous alcohol.

(2) The transducers, connected with the physiological recorder and monitor or the model JSF4-1 detection on AES along meridians and computer, and daubed with a. little of vaseline, were fixed on the APs and the CPs with nylon-button belts.

(3) An acupoint, chosen on the same meridian with the APs, was pressed and stimulated for 10 seconds with a spring stick with a rubber end and a pressure gauge at one time.

(4) A record was divided into 3 sections. The first section was recorded for 10seconds before the acupoint was pressed and stimulated, and the second section was recorded for 10 seconds during the acupoint was pressed and stimulated, and the third section was recorded for 10 seconds after stopping pressing and stimulating the acupoint.

(5) the signals recorded at the second section were looked upon as the detected signals (DS) and were used to compare with the first section and the third section signals. The first and third signals

were looked upon as the basic signals (BS). AES was determined to be remarkable if DS/BS was larger than 4 : 1. The mild AES was determined if DS/BS was smaller than 4 : 1 but wasn't 1 : 1. no AES was determined if DS/BS was 1 : 1, The AES were inhibited if DS/BS was smaller than 1 : 1. No AES was counted as a. negative reaction (NR), and the others were counted as a. positive reaction (PR).

Tab. 6 – 4 Acupoints and Control Points of the Experiment

Meridian Name	Acupoints	Control Points
The lung meridian of forelimb *Taiyin* (Lu)	4	3
The large intestine meridian of forelimb *Yangming* (LI)	7	3
The stomach meridian of hind limb *Yangming* (St)	13	13
The spleen meridian of hind limb *Taiyin* (Sp)	9	7
The heart meridian of forelimb *Shaoyin* (H)	3	2
The small intestine meridian of forelimb *Taiyang* (SI)	6	3
The pericardium meridian of forelimb *Jueyin* (P)	4	2
The triple-warmer meridian of forelimb *Shaoyang* (TW)	6	2
The gallbladder meridian of hind limb *Shaoyang* (GB)	10	5
The liver meridian of hind limb *Jueyin* (Liv)	6	3
The urinary bladder meridian of hind limb *Taiyang* (UB)	6	*
The kidney meridian of hind limb *Shaoyin* (K)	5	*
The *Du* (the dorsal midline) meridian (Du)	4	*
The *Ren* (the vental midline) meridian (Ren)	4	*
total	87	43

* Because of time etc, no CPs were selected along the latter 4 meridians, and their APs were compared with the average value of CPs along the other meridians.

Analysis on Frequency and Amplitude Ranges of AES: The frequency and amplitude were separately analyzed with the computer data. analysis system or the instrument for analysis of frequency and amplitude range, and the frequency and the amplitude were compared with electromyogram (EMG's), electroencephalogram (EEG's) and electrogastrogram (EGG's).

Analysis on Correlation between AES and Tissues: In 11 anesthesia. sheep, Acupoint *Li-dui*, Acupoint *Shang-ju-xu* and Acupoint *Zu-san-li* were chosen for the experiment. AES of Acupoint *Li-dui* was recorded as Acupoint *Zu-san-li* was stimulated with needle separately before the skin was cut (BSC), as the skin was transected (ST), the skin was cut in a. ring (SCR), the muscles were transected (MT), the muscles were cut in a. ring (MCR) and the bone was transected (BT) from Acupoint *Shang-ju-xu* of one hind limb.

Analysis on Biological Characters of AES: After 11 sheep were put to death with anesthetic and the other hind limb was cut off from hip joint, AES of Acupoint *Li-dui* were recorded as Acupoint *Zu-san-li* was stimulated with needle every 10 min and didn't finish the experiment until there was no AES for detection.

Calculation on Velocity of AES Propagated along Meridians: AES of Acupoint *Zhong-feng* , Acupoint *Xing-jian* at Lu and a. CP were simultaneously recorded during Acupoint *Tai-chong* at the same meridian was stimulated. The velocity of AES propagated along meridians was calculated by V =

D/T (D is the difference of the distance between Acupoint *Zhong-feng* and Acupoint *Xing-jian* and the distance between Acupoint *Tai-chong* and Acupoint *Zhong-feng*; T is the difference of the time between AES appearance of Acupoint *Xing-jian* and Acupoint *Zhong-feng*).

Results

A. Detection of AES along 14 Meridians: See Tab. 6 – 5.

Tab. 6 – 5 Comparison Between AES PR Rates of Aps and CPs

		Cases in all	PR cases	PR rates (%)	Comparison
Lu	AP	1 086	896	82.96	$P<0.01$
	CP	243	75	30.96	
LI	AP	843	698	82.80	$P<0.01$
	CP	281	75	26.69	
St	AP	4 289	3 545	82.65	$P<0.01$
	CP	1 822	540	29.63	
Sp	AP	1 549	1 395	90.05	$P<0.01$
	CP	521	92	17.65	
H	AP	939	794	84.56	$P<0.01$
	CP	274	72	26.28	
SI	AP	1 009	844	83.65	$P<0.01$
	CP	267	82	30.71	
P	AP	160	128	80.00	$P<0.01$
	CP	40	12	30.00	
TW	AP	244	194	79.51	$P<0.01$
	CP	40	11	29.95	
GB	AP	2 163	1 898	87.75	$P<0.01$
	CP	940	280	29.78	
Liv	AP	1 986	1 793	90.28	$P<0.01$
	CP	873	243	27.83	
UB	AP	860	756	87.91	$P<0.01$
K	AP	700	584	83.43	$P<0.01$
Du	AP	336	280	83.33	$P<0.01$
Ren	AP	336	272	80.95	$P<0.01$
Total	AP	16 494	14 077	85.35	$P<0.01$
	CP	5 301	1 482	27.96	

B. Frequency and Amplitude of AES: See Tab. 6 – 6.

Tab. 6 – 6 The Comparison of Frequency and Amplitude of AES with EMG's, EEG's and EGG's

	Frequency (c/s)	Amplitude	Pattern
AES	2 ~ 30.4	0.33 ~ 1.5mv	Impulse
EMG	1 ~ 100	0.15 ~ 20mv	Impulse
EEG	0.5 ~ 35	20 ~ 500μv	Impulse
EGG	4.54 ± 0.35	11974 ± 352μv	Sine wave

C. Velocity of AES Propagated along Meridians: The distance between Acupoint *Tai-chong* and Acupoint *Zhong-feng* was 9cm, and the distance between Acupoint *Tai-chong* and Acupoint *Xing-jian* was 5cm, and their difference was 4cm. The time difference between AES appearances of Acupoint *Zhong-feng* and Acupoint *Xing-jian* was 0.8 seconds, so the velocity of AES propagated along meridians was 6.67cm/sec (V = 4/0.8). See Fig. 6 – 1.

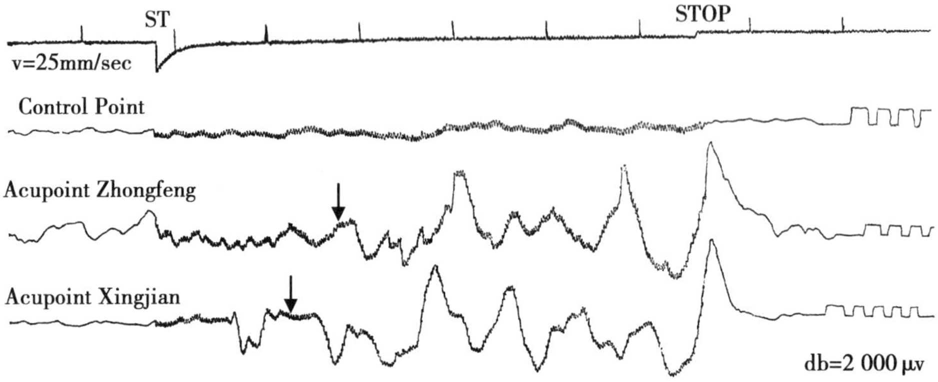

Fig. 6 – 1 AES as Stimulating Acupoint Taichong

No: 02 2, ♂ 16, March 1990

D. Correlation between AES and Tissues: AES could all be affected by cutting different tissues, but chiefly affected by transecting the muscles. See Tab. 4 and Fig. 2.

E. Sustained Time of AES appearance: After the limb was cut off at hip joint, AES were detected at Acupoint *Li-dui* while a. stimulation was given at Acupoint *Zu-san-li*. Sustained Time of AES appearance was 240 (200 ~ 280) mins. See Tab. 6 – 7 and Tab. 6 – 8.

Tab. 6 – 7 AES Changed as Different Tissues Were Cut

	BSC	ST	SCR	MT	MCR	BT
Cases in all	66	75	60	65	70	63
PR cases	54	57	45	11	7	0
PR rate (%)	81.8	76.0	75.0	16.9	10.0	0.0
Reduce of PR rate (%)		5.8	1.0	58.1	6.9	10.0
Amplitude (μv)	520	430	350	130	75	0
Reduce of Amplitude (μv)		90	80	220	55	75

Tab. 6 – 8 Sustained Time of AES appearance

No. Sheep	1	2	3	4	5	6	7	8	9	10	11	average
Times	29	21	23	28	21	29	23	29	21	23	25.9	24.8
Time (min)	280	200	220	270	200	280	270	220	280	200	220	240

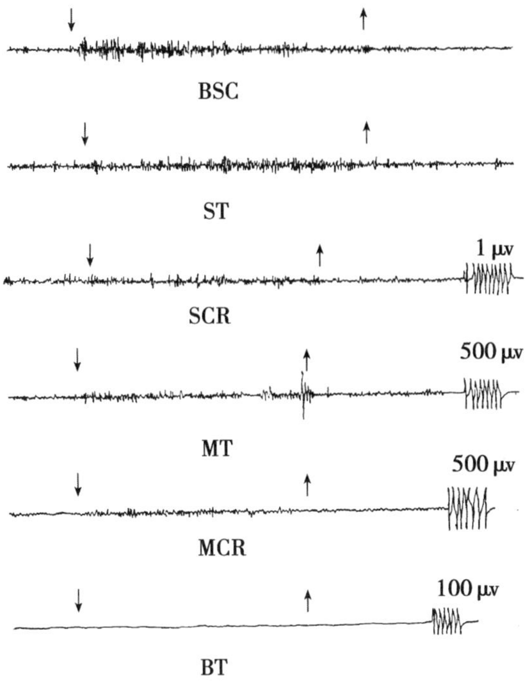

Fig. 6 – 2 AES after Cutting Different Tissues of Sheep Hind limb

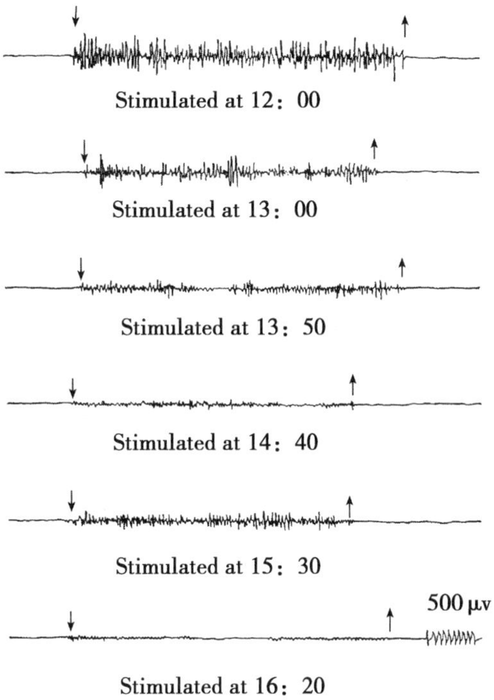

Fig. 6 – 3 AES at Different Time after the Hind limb was Cut off

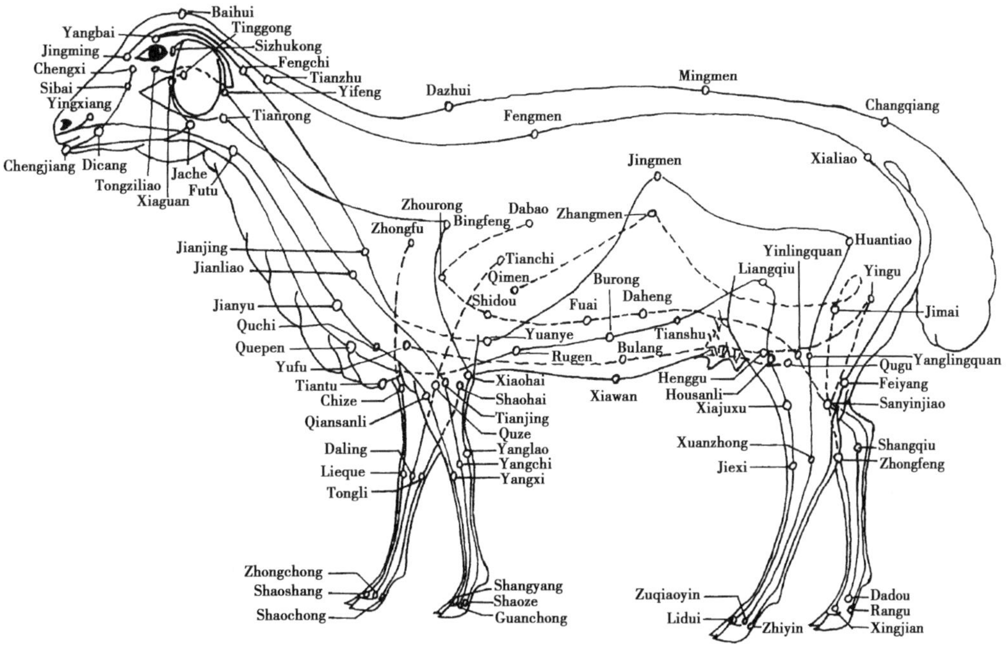

Fig. 6 – 4 Illustration of AES Acupoints of Sheep

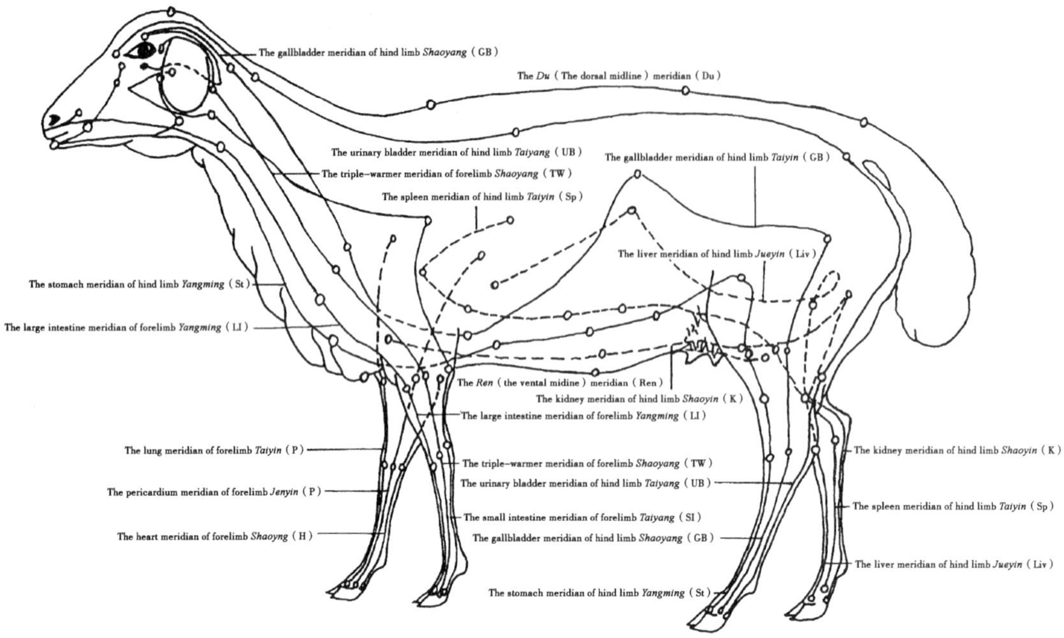

Fig. 6 –5 Illustration of AES Meridians of Sheep

Discussion and brief summary

a. Characters That AES Propagates Bidirectionally along Meridians: The results show that AES propagates bidirectionally along meridians. AES PR rates of APs on the same meridians with the stimulated acupoints, especially at the proximal and the distall ends of the stimulated acupoints on the same meridians, were remarkably different from their CPs.

b. Characters that AES Propagates slowly: The velocity of AES propagated along meridians is 6. 67cm/sec (V = 4/0. 8). It showed no remarkable difference from that the velocity of human AES propagates along meridians is 4. 16cm/sec and that human PSM is usually slower than 10cm/sec, but much slower than nervous velocity (1 ~ 100 m/sec).[1,2]

c. Biological Characters of AES: It showed that AES propagated along meridians in Sheep is a. biophysical signal following biological phenomena. that AES was different from EMG, EEG and EGG in frequency and amplitude, and that its sustained time for detection was no different from the time demanded to rehabilitate the blood circulation of wounded limbs in 4 ~ 6h on re-plantation of severed limbs.[3]

d. It showed that the properties of AES associated with all tissues, chiefly with muscles and their nerves that PR rates and amplitude of AES were reduced separately by 71. 6% (58. 1%/81. 8%) and 42. 3% (220μv /520μv) by muscles transecting.

References

[1] Sun Pingsheng, Zhao Yuzhuo, Li Yulan, Yan Qinglian, Liu Hong. Detection of Acoustic Emission Signals (AES) of Meridians in human. *Liaoning Journal of Traditional Chinese Medicine*, 1987; (12): 33. (in Chinese)

[2] Meng Zhaowei. The third Equilibrium System——Channel and Collateral System. *Chinese Acupuncture & Moxibustiun*, 1983; 3 (1): 25. (in Chinese)

[3] Huang Jiaju, Wu Jieping. *Surgery* (the latter volume): 643, People's Medical Publishing House, October, 1979. (in Chinese)

13 Study On the Electrogastrogram of Spleen-asthenia. Condition in Cows and Goats

Abstract: To look for scientific indices for the spleen-asthenia. condition in cows and goats. *Methods*: The frequency, amplitude range, activity index, reactive region and phase position in electrogastrogram from goats and cows with spleen-asthenia, changed remarkably before experiment and after treatment. *Result*: The symptom with weakness/asthenia in stomach and spleen (*Pi Wei Xu Ruo*, Spleen/Stomach deficiency) showed an EGG at lower frequency and amplitude index. The spleen asthenia pattern in cattle and goat is a. kind of integrated pathological conditions of vegetative nervous system. The pathological change in cellular metabolism and from spleen asthenia animal, is morphological features of disorders in mitochondrion is probably material base of symptoms of spleen asthenia animals.

Key Words: Electrogastrogram (EGG); Spleen- asthenia. Condition; Cows and Goats

Introduction

see Fig. 6 - 1, 6 - 3, 6 - 4 and Fig. 6 - 5.

The spleen, according to Traditional Chinese Veterinary Medicine (TCVM), is essential for postnatal growth, and has a. very important effect in the physiology, the pathology, the prevention and treatment of diseases in animals. A series of degenerated changes of spleen-asthenia condition is a syndrome with a complex mechanism and varieties. Many modern scientific indices have been used for its study, but there were a. few with the syndrome peculiarity, and it was difficult either to be used in clinic or to guide the Determination of Treatment Based on the Differentiation of Symptoms/Patterns. Electrogastrogram (EGG) used successfully in oxen and goats would open up a. new train of thought. [1] With financial support of the China. National Natural Science Fund, we have studied EGG of spleen-asthenia condition in oxen and goats.

Materials and methods

Animals: 150 cows being in good health without clinical signs of spleen-asthenia. were used as the study of health animal; 16 goats as the models of spleen-asthenia condition; 201 cows with the clinical spleen-asthenia condition as the disease.

Materials: The model EGG-2B Electrogastrogragh, 3 pairs of ZC-1 Ag-Agcl plate electrode, anhydrous ethanol, physiological saline, adhesive plastered.

Methods: As the animals were standing at ease, the wool on the projected positions of four stomachs was cut of and the skin was degreased with anhydrous ethanol. The electrodes and the cotton balls with physiological saline were put and fixed on the skin with nylon buckles, and the control electrodes were put and fixed on the external skin Of the upper forearm. Then all electrodes were connected with the model EGG-2B Electrogratrogragh. After the animals were quite, each lead was recorded for 10 min, and 5 min data chosen from every lead was counted.

Experimental Indices:

The frequency (F) and amplitude (A).

The difference of gastric phases (Dp): The postponed phase of the slow gastric waves.

Activate indices: The product of the frequency times amplitude in 3 min (AI).

Reactive area: The area. under the curve of each EGG slows waves or pressure waves and classified waves by their duration into two bins (RA).

All experimental datum are statistically analyzed with T-Test.

Experimental results

A. EGG of Health Cows

The Waves of EGG: According to the frequency & amplitude features, EGG wave of 150 cows with a. good health, can be divided into the single-phase waves with low amplitude (29.5%), the high frequency waves with high amplitude (9.7%) and the complex waves (60.8%).

Frequency and Amplitude: A. difference of the frequencies between rumen and abomasums and that between omasums and abomasums was very significant ($P < 0.01$). The amplitude difference between rumen and reticulum and the difference between omasum and abomasums were remarkable ($P < 0.05$). The changes of frequencies and the amplitudes were similar ($r = 0.7258$, $r > 0.05$, $P < 0.05$), so the changes of frequencies were correlated with the amplitudes.

Difference of gastric phase: The differences of gastric phase between rumen and reticulum, omasum, abomasum separately were (1.13 ± 4.6) sec, (0.16 ± 3.3) sec, and (0.2 ± 2.6) sec. The differences of gastric phase between reticulum and omasum, abomasum were (0.02 ± 4.1) sec, (0.41 ± 3.4) see. The difference between omasum and abomasum was (0.66 ± 4.3) sec. The statistical results showed that it was remarkably different to compare the differences of gastric phase between rumen and reticulum with the different between rumen and omasum, the those between rumen and abomasum, and the difference between reticulum and omasum ($P < 0.01$), and it was different or remarkably different to compare the difference of gastric phase between omasum and abomasum with the those between rumen and omasum, and the those between reticulum and abomasum, and the others were not statistically different.

Activity Indices (AI) and Reactive Area. (RA): AI of rumen and abomasum were higher than AI of reticulum and omasum. The differences of AI between rumen and reticulum, omasum and abomasum were very remarkable, and it was separately remarkable too ($P < 0.01$) to compare abomasum with reticulum and omasum. The change of RA was similar to that of AI. RA of rumen and reticulum were not remarkably different from those of abomasum and omasum separately, but RA. of rumen and abomasum were remarkably different from those of reticulum and omasum ($P < 0.01$). See Tab. 6 – 9.

Tab. 6 – 9 Frequency, Amplitude AI and RA. in Healthy Cows (M ± SD)

	F (c/min)	A. (μv)	AI (C × μv /3min)	RA (mm^2/3min)
Rumen	2.75 ± 0.22	844.4 ± 219	63 246 ± 6 083	1 489 ± 521
Reticulum	2.67 ± 0.38	633.2 ± 262	42 753 ± 2 427	1 129 ± 552
Omasum	2.73 ± 0.21	626.9 ± 258	41 491 ± 2 065	1 134 ± 125
Abomasum	2.79 ± 0.21	795.7 ± 265	52 021 ± 2 214	1 391 ± 564

B. EGG of Spleen-asthenia. Model in Goats:

Three goats were randomly chosen from 16 healthy goats and assigned in a. healthy control group (HCG), and the others were made to be spleen-asthenia. models with the Traditional Chinese Medicine[2]. 5 goats, randomly chosen from the latter, were assigned in a spleen-asthenia control group (SCG), and another 5 goats were assigned in the treatment group (TMG), and treated using *Si-jun-Z*i Decoction (Four gentlemen herbal soup) until they recovered. Other 3 goats with the TCVM Spleen-asthenia model were selected, assigned in the natural recovery group (NRG), given no treatment, and monitored for the spontaneous recovery. EGG, h*emoglobin* (*Hb*), Color-index, *red blood cell* (*RBC*) *and acetylcholine enzyme* (*AchE*)[3], *Rumen Homeostasis*[4], *the activity of ptyalin at eating* [5], *va-*

nillyl mandelic acid (*VMA*)[6], *the absorbent rate of xylose*[7], glucagons and gastrin, cyclic adenosine monophosphate (*cAMP*) *and* cyclic *guanosine* monophosphate (*cGMP*)[8] *etc were separately examined and recorded* in HCG, SCG, TMG and NRG.. Then all goats were euthanized and their stomachs, spleen, intestine and liver tissues were collected and were observed for ultrastructure under electronic microscope of transmission Japonic of The JEM—1200EX/S and the trace elements in the ultrastructures were analyzed with the model 9100-EDAX Energy Table separately[9] in HCG, SCG and TMG.

Biochemistry Indices: The Hb, RBC, VMA, the absorbent rate of xylose was lower in SCG than in HCG.. Those parameters were higher in TMG and NRG than in SCG.. The activity of cholinesterase and glucagon was higher in SCG than in HCG, and was lower after the treatment. The activity of ptyalin (AP), stimulated with citric acid, was much higher in HCG and was much lower in SCG than that in the time period previous to stimulation. See Tab. 6 – 10.

Tab. 6 – 10 The Changes of Biochemistry (M ± SD)

		HCG	SCG	TMG	NRG
	Hb (g)	7.3 ±0.6	5.2 ±0.8	6.3 ±0.5	5.8 ±0.8
	RBC (m/mm^3)	8.48 ±0.213	6.2 ±0.16	8.23 ± 0.196	7.25 ±0.392
	Color-index	1.07 ±0.04	0.81 ±0.07	1.03 ±0.06	0.91 ±0, 01
	ARX (%)	1.76 ±0.9	0.42 ±0.2	0.94 ±0.2	0.48 ±0.2
	VMA. (μg)	88.5 ±17.3	20.5 ±11.8	100 ±36.5	40 ±20
	AchE (μ)	10.1 ±3.3	30.4 ±7.3	21.8 ±6.5	26.8 ±8
	Glucagon (pg/ml)	129.8 ±35.9	319.4 ±16.6	161.2 ±76.8	231.2 ±35.1
	AP BCAS	98.8 ±27.7	41.6 ±20.6	56.5 ±67.9	20.9 ±11.1
	ACAS	191.6 ±62.3	15.4 ±14.7	90.4 ±60.6	36.9 ±4.6
	cAMP (pg/ml)	3.58 ±1.3	2.68 ±0.84	3.68 ±0.92	2.33 ±0.15
	cGMP (pg/ml)	0.9 ±0.52	1.16 ±0.75	0.96 ±0.48	0.89 ±0.45
	Gastrin (pg/ml)	1 045 ±91.5	914.2 ±60.6	1 136.9 ±39.23	1 109.9 ±91.81
	Rumen pH	7.0 ±0.02	6.0 ±0.1	6.9 ±0.2	6.5 ±0.3
	Ciliates alive (%)	66.8 ±2.3	28.7 ±9.5	65 ±4.2	46 ±4.0
	ciliates amount/ (m/mm^3)	139.5 ±47	27 ±13	160 ±19.5	56.3 ±25
ciliate sort	big ciliates (%)	9.7	1.8	6.7	4.9
	Middle ciliates (%)	21.5	3.3	17.5	5.4
	Small ciliates (%)	68.8	94.9	75.8	89.7

The Changes of Ultrastructure: The gastric parietal cells were irregular with spaces to adjacent cells when an animal is in a. spleen-asthenia condition. The space in the nuclear membrane is wider than the healthy condition; parts of the nucleus disappear with their membranes to be dissolved. Cytoplasm of the cells falls apart, and a. great number of crippled tubercle disappeared. A great deal of pieces of destroyed mitochondria is in full field of vision. The enlarged endoplasmic reticulum appear in bubbles, to which many nucleoproteins adhere. There is not any secretory into the center of the cells, the membrane rolls inside the cells and forms a. few small tubes that spread into the center of the cells, in which there are many micro-villi at different length. Gastric glandular cells are smaller than gastric parietal cells, space between nuclear and free looks like empty bubbles. There are a great number of red cells among the gastric glandular cells.

Arteries and veins in the liver are dilated, and there are more red cells in there. There are a great number of acidophilus among cells in the tissue. Hepatic cells appear polygon with blunt nucleus whose membrane is dissolved. There are many fatty particles in cytoplasm. The double-layer membrane of mitochondria is difficult to identify, some crippled membrane of tubular are in the vision. Vesicle of tubular capsule hardly is found. The nuclei of splenolymphocytes are bigger than normal, chromatin of the cells is concentrated around the nuclei, and space of double-layer membrane of nuclei is increasing. There are thinner cytoplasm's, less organelle, fewer mitochondria whose double-layer membrane can be identified in the shape of twist with dissolved membrane of tubercle. Big lymphocytes appear bigger with more cytoplasm. The shape of reticulocyte is irregular with an increase in wide of membrane of nuclei, and enlarged endopiasmic reticulum ribosome and phagosome can be found in cytoplasm.

Changes of Trace Elements: The relative weight per cent (WT%) of Na, Mg, Al, Si, P, S, K, Ca, Zn, Ti, Ni and Cr etc were examined in stomach, spleen, intestine and liver tissues of goats. Those of Cu, Fe and Zn among the 20 trace elements changed remarkable. WT% of Cu, Fe, and Zn in ultrastructure units of spleen, stomach and intestine in experimental goats that are in spleen-asthenia. are lower than those in the animals of the control groups. The WT% of the Cu, Fe, and Zn in liver of goats in state of spleen-asthenia. is higher than control group. The index became normal level after treatment with the Chinese herbal formula. See Tab. 6 – 11.

Tab. 6 – 11 The Trace Elements Ultrustructrues of Experimental Goats (M ± SD)

Tissues	Elements	Normal	Spleen- asthenia	*Treatment*
Stomach	*Fe*	2.8 ± 0.9	0.7 ± 0.2	2.5 ± 0.2
	Cu	3.0 ± 0.7	2.2 ± 0.1	3.5 ± 0.6
	Zn	3.4 ± 0.5	2.0 ± 0.1	2.7 ± 0.6
Intestine	*Fe*	1.7 ± 0.6	1.6 ± 0.3	2.1 ± 0.1
	Cu	4.4 ± 0.6	1.8 ± 0.4	3.1 ± 0.9
	Zn	3.0 ± 0.9	1.4 ± 0.2	2.1 ± 0.6
Spleen	*Fe*	6.7 ± 0.3	2.9 ± 0.6	6.7 ± 0.2
	Cu	2.7 ± 0.2	1.7 ± 0.7	1.9 ± 0.2
	Zn	2.1 ± 0.9	1.4 ± 0.2	2.2 ± 0.3
Liver	*Fe*	1.8 ± 0.3	2.9 ± 0.1	2.4 ± 0.2
	Cu	2.1 ± 0.1	2.8 ± 0.1	2.5 ± 0.3
	Zn	1.2 ± 0.5	2.4 ± 0.5	1.5 ± 0.1

Changes of EGG: The frequency, amplitude range, activity index, reactive region and phase position in electrogastrogram from goats with spleen-asthenia, changed remarkably before experiment and after treatment. The parameters in the cases are markedly lower than those in normal goats, which come back to normal level after treatment. The parameters of goats that recovered naturally changed slowly. See Tab. 6 – 12.

Tab. 6 – 12 Changes of EGG (M ±SD)

	HCG	SCG	*TMG*	NRG
F (c/min)	2. 84 ±0. 05	2. 76 ±0. 04	2. 85 ±0. 05	2. 8 ±0. 09
A (μv)	384 ±26. 8	201 ±26. 2	336 ±32. 8	244 ±25. 8
AI (C×μv /3min)	1 923 ±279	1 075 ±126	1 690 ±204	1 339 ±261
RA (mm^3/3min)	1 376 ±157	779 ±72	1 323 ±128	974 ±115
Dp (sec)	3. 3 ±1. 9	–0. 9 ±3. 7	3. 6 ±2. 5	2. 3 ±0. 7

C. Observation of EGG of Clinical Cases:

A. type of degeneration has been observed in changes of EGG in clinical unhealthy cattle, which includes slow frequency, lower amplitude index, disorder in rhythm, disappeared active region, repressed active index and reversed phase position. The changes are closely correlated with development of illness together with self-regulations in different symptom.

The symptom with weakness/asthenia in stomach and spleen (*Pi Wei Xu Ruo* Spleen/Stomach deficiency)[10] showed an EGG at lower frequency and amplitude index. A. linear fiat appears after three continuously low amplitude peaks. The case with asthenia and cold in stomach and spleen (*Pi Wei Xu Han*)[10] represent an EGG at alternate appearance of Sole wave, Double waves and triplicate waves, among them a. plane and slow equipotential line connected, which are called typical reverse transmit.

The cases with asthenia in spleen and recessive in stomach (*Pi Xu Wei Shi*)[10] showed EGG with a. character in bad rhythmicity, varied Amplitude, shorten up-phase, longer down-phase in slow wave, plat form electric position has some low position in wave. The cases with wet and distress in spleen and stomach (*Pi Wei Shi Kun*)[10] showed an EGG at had. The rhythmicity irregular Amplitude with small wave below 30, and the big wave about 1200, the similar waves hardly was found. See Tab. 6 – 13.

Tab. 6 – 13 EGG in clinical spleen-asthenia cattle before and after treatment (M ±SD)

		N	F (c/min)	A (μv)	AI (C× μv /3min)	RA (mm^2/ 3min)	Dp (sec)
Normal		150	2. 75 ±0. 05	725 ±111. 7	49 878 ±1 007	1 286 ±182	0. 43 ±0. 41
Pi Wei Xu Ruo	*before*	61	2. 6 ±0. 19	147 ±24. 9	10 774 ±1 128	517 ±198	–9. 3 ±3. 8
	after	23	2. 7 ±0. 37	625 ±86. 7	43 214 ±919	1 076 ±471	–2. 6 ±4. 3
Pi Wei Xu Han	*before*	55	2. 65 ±0. 21	288 ±59. 2	22 915 ±1 035	840 ±488	–7. 4 ±5. 1
	after	19	2. 74 ±0. 13	710 ±95. 4	47 316 ±1 713	1 130 ±507	–2. 1 ±6. 2
Pi Xu Wei Shi	*before*	39	2. 7 ±0. 14	487 ±63. 2	31 124 ±113	1 048 ±369	–3. 2 ±3. 7
	after	13	2. 77 ±0. 18	738 ±79. 2	46 112 ±1 268	1 311 ±315	–1. 3 ±2. 1
Pi Wei Shi Kun	*before*	46	2. 81 ±0. 12	838 ±86. 1	61 340 ±1 799	1 577 ±245	5. 1 ±4. 2
	after	17	2. 72 ±0. 26	690 ±68. 5	50 762 ±145	1 369 ±409	2. 5 ±1. 6

Discussion

The spleen asthenia. pattern in cattle and goat is a. kind of integrated pathological conditions of vegetative nervous system. The symptoms include indigestion, poor absorption, and poor metabolic function due to the disorder in the vegetative nervous system.

The research indicates that AchE is an indicator to reflect the level of Ach inside body, and is a. chemical transmitter in peripheral parasympathetic nerve system. cGMP participates in synapse transmit among cholinergic neurons. Those two substances reflect indirectly functional state of parasympathetic nerve[11,12]. In our study, the goats with spleen asthenia. have a. higher concentration of cGMP and AchE, indicating hyperfunction in parasympathetic nerve system. Until now, there is not any hormone that has been found to control directly secretion of saliva. The processing of saliva. secretion is controlled by vegetative nervous system. When stimulated by citric acid, amylase in saliva. in goats with spleen asthenia. decreased. This result indicated reduction of reaction of reactive function in parasympathetic nervous system.

Studies indicate that the skin electric potential reflects the functional state of sympathetic nervous system[13]. We found that the skin electric potential of cases with spleen asthenia. is lower than that in the normal animals. This result indicated the animal patients with spleen asthenia. might have a. hyper-excitement of sympathetic central nerve system.

The level of VMA. is one of indicators of functional state of peripheral sympathetic nervous-adrenal medulla[12]. Our experiment demonstrated that the concentration of VMA. in spleen asthenia goats was lower than normal state, indicating hyperfunction of peripheral sympathetic nervous-adrenal medulla. cAMP is the second messenger of communication of the azotohormone and the noradrenalin nerve cells. When sympathetic excites and catecholamine manifolds, cAMP may be manifold through uniting catecholamine with target cell β accepter. cAMP in plasma. of spleen asthenia. goats was lower than normal one's. It shows partly that the sympathetic target cell excitement is lower in spleen asthenia. animal. This result may indicate that the electric potential of skin , VMA. and cGMP in plasma. in spleen asthenia. animals dropped, which is closely related with reductions including sympathetic nervous center peripheral sympathetic nervous and its target cells. In a. word, the key of spleen asthenia. in animal is hyperfunction in sympathetic nervous system. There are deeply material and adjusting changes in vegetation in sympathetic nervous system.

There are deeply found from our research that mainly pathological changes occurred in mitochondrion of cells of stomach, spleen and liver by electronic microscope. The result from analyzing trace elements also demonstrated that the concentrations of Cu, Fe, Zn in mitochondrion of spleen asthenia. animal are lower than normal, which showed the obstructs in energy metabolism and reduction in catalysis and oxidizing reaction. The pathological change in cellular metabolism and from spleen asthenia. animal, is morphological features of disorders in mitochondrion is probably material base of symptoms of spleen asthenia. animals.

EGG is regulated chiefly by vagus, sympathetic nerve, gastrointestinal hormones and vasoactive intestinal peptide. When an animal is in a. state of spleen asthenia, the pathological changes on function of vegetative nervous and general level of gastrointestinal hormones of spleen-asthenia. animal can be known by analyzing all parameters on EGG.. Therefore, EGG can be used an objective indicator to differentiate spleen-asthenia. in cattle and goats. EGG possesses a. unique advantage in veterinary clinical medicine.

The EGG can provide objective information to distinguish and diagnose a. disease including the spleen asthema. pattern of cattle and goat. Fixing electrodes on skin of animals do not only protect internal environment from being disturbed m order to reveal characters of illness, but also are proven to be a. safe and simple diagnostic tool in veterinary clinical practice.

REFERENCES

[1] Zheng Jifang, Qu Ziming, Tang Liping. Characters of Electrogastrogram (EGG) Slow Waves of Dairy Cattle. *Chinese Journal of Veterinary Science and Technology*. 1991, 21 (7): 34 ~ 35. (in Chinese)

[2] Zheng Jifang, Qu Ziming. Observation on electrogastrogram of sick goat model. *Chinese Journal of Veterinary Science and Technology*. 1993, 23 (2): 26 ~ 28. (in Chinese)

[3] Medical Test Institute of Shanghai. *Biochemistry Test for clinic*. Science and Technology Publishing Company of Shanghai. 1984, 118: 345. (in Chinese)

[4] Zhang Wenlong, Xu Zuoliang, Niu Qingshu, Pan Wenxian. Mensuration on 4 indexes of rumen liquid in yellow cattle. *Shangdong Animal Husbandry & Veterinary*. 1980, (3): 22. (in Chinese)

[5] Guangzhou College of Traditional Chinese Medicine. Study on activity of ptyalin at eating glucagons in the spleen-asthenia. patients. *Chinese Medicine Journal*. 1980, (5): 10. (in Chinese)

[6] Chen Jiewen, Ji Qihui. On the condition of sympathetic nerve in the spleen-asthenia. patients. *Journal of Guangzhou College of Traditional Chinese Medicine*. 1984, (1): 23. (in Chinese)

[7] Jin Jingshan. The simple determine method of D-*xylose in urine*. *Journal of Traditional Chinese Medicine*. 1981, (3): 6. (in Chinese)

[8] Wei Yanming, Qu Ziming, Zheng Jifang. Neuroendocrine Pattern in RabbitS With "Insufficiency Of the Spleen" and Effects of Sijunzi Decoction on It. *Chinese Journal of Veterinary Science*. 1999, (5): 486 ~ 488. (in Chinese)

[9] Zheng Jifang, Qu Ziming. Observation of trace elements in super-micro stracture of spleen debility goat. *Studies of trace elements and Health*. 1992, (4): 36 ~ 37. (in Chinese)

[10] The editorial department. *The Volume of Traditional Chinese Veterinary Medicine, Chinese Encyclopedia. of Agriculture*. Agricultural Publishing Company. 1991: 183 ~ 185. (in Chinese)

[11] Wei Muxin, Pei Shuying. Study on functional state of autonomic nerve in spleen yin deficiency. *Chinese Journal of Integrated Traditional and Western Medicine*. 1988, (4): 202. (in Chinese)

[12] Davenport H W. *Enteron Physiology*. Science Publishing Company. 1976: 123. (in Chinese)

[13] WarWick R. *Grays Anatowy*. 35ed: 1224. Jarrold and Sons Ltd, longman, 1973. (in Chinese)

第一讲

中兽医基础理论

中兽医学是一个博大精深的文化宝库，是灿烂的中国古代文化的一部分。中兽医学既汲取了中国古代深邃的哲学、文化和科学思想，又对中华民族数千年来与畜禽疾病作斗争的经验进行了总结，所以，它不但具有极其丰富的理论思辨性和创造性，而且临床实用性极强。即使在现代医学十分发达的今天，中兽医学依然具有很强的生命力，其重要原因之一，就在于其卓越的临床疗效。近几十年来，中兽医学的独特优势正在逐渐被国际兽医学界所认识，这一古老的传统医学正一步步走向世界。

1 兽医学的基本特点

中兽医学有两个基本特点。一是整体观念，二是辨证论治。它体现在对畜体生理功能和病理变化的认识，以及对疾病的诊断和治疗等各个方面认知。

1.1 整体观念

整体观念，即指畜体是一个统一的整体，以及畜与自然相互关联的整体的思想。

1.1.1 畜体是一个统一的整体

中兽医学认为，畜体由许多组织器官所构成，包括脏腑经络、四肢百骸等。但彼此之间在结构上、生理上、病理上却有着密切的联系，是一个统一的整体。因此，在临床上诊治疾病时，必须从整体出发，通过五官、形体、色脉等外在的变化，来分析和判断内脏的病变，从而在整体观的指导下制定正确的治疗原则和方法。

1.1.2 畜与自然相互关联

畜体生活在自然界之中，自然界存在着各种畜体赖以生存的必要条件，如阳光、空气、水源等。同时自然界的各种变化又可以直接或间接地影响着畜体，使畜体产生生理或病理上的相应反应。不仅季节气候变化对畜体有影响，地理环境的不同也影响着畜体的生理活动和病理状态。

1.2 辨证论治

1.2.1 辨证论治的含义

辨证论治，就是将通过四诊（望、闻、问、切）所收集的症状、体征等临床资料，进行分析、归纳和综合，判断、概括为某种性质的证，然后再根据辨证的结果，确定相应的治疗方法。辨证论治是中兽医学认识疾病和治疗疾病的基本原则。所谓“证”，即“证候”，是对疾病发生、发展过程中某一阶段的病理变化的概括，包括了疾病的部位、原因、性质以及邪正关系等内容，它比单一的症状能更全面、更深刻、更正确地反映出疾病的本质。例如，病畜表现出恶寒、发热、疼痛、无汗、脉浮紧等临床症状，通过分析归纳，辨清病因为风寒之邪，病位在表，疾病性质为寒，邪正关系是实，于是就概括判断为风寒表实证，治以疏风散寒，辛温解表。可见辨证是决定治疗的前提和依据，论治是辨证的目的，也是检验辨证是否正确的方法和手段，辨证与论治是诊治疾病过程中相互联系而不可分割的两个方面。

1.2.2 辨证与辨病的关系

在临床上，疾病与证候之间既有内在的联系，又有一定的区别。通常认为，疾病是包括其整个病理过程在内的，而证候则是疾病过程中某一阶段的病理概括，因此同一种疾病可表现出不同的证，而不同的疾病在发展过程中又可出现相同的证，所以中兽医认识疾病和治疗疾病，主要是着眼于辨证，着眼于在辨病的过程中找出证的共性或差异性。如同一种疾病，由于病畜的体质不同，或发病的时间、地区不同，或处于不同的发展阶段等，可以表现出不同的证候，因而治法也不一样。以感冒为例，由于病畜感受的邪气不同，临床上有风寒证与风热证的区别，前者治以辛温解表，后者治以辛凉解表，中兽医学把这种情形叫做“同病异治”。又如不同的疾病，在其发展过程中，只要出现相同的证候，便可采用相同的治疗方法，如痢疾与黄疸，是两种不同的疾病，但如果都表现为湿热证，就都可用清利湿热的方法来进行治疗，中兽医学把这种情形叫做“异病同治”。由此可见，辨证论治能够从本质上理解病和证的关系，强调“证”在治疗中的首要作用，认为“证同治亦同，证异治亦异”。

2　阴阳五行学说

阴阳五行本属中国古代的哲学思想。中国古代医学家，将阴阳五行学说运用于医学领域，借以阐明畜体生理功能和病理变化，指导诊断与治疗，成为中兽医学理论体系的重要组成部分。

2.1　阴阳学说的概念

阴阳代表着事物相互对立又相互联系的两个方面，一般说，凡是活动的、上升的、明亮的、进行的、机能亢进的或属于功能方面的皆为阳；凡是沉静的、下降的、隐晦的、退行性的、机能衰退的、或属于器质方面的皆属于阴；天在上故属阳、地在下故属阴；水性寒而下走故属阴，火性热而上炎故属阳。

事物阴阳两方面的对立性并不是绝对的，而是相对的。阴阳的相对性，一方面表现为在一定条件下，阴阳可以相互转化，阴可以变为阳，阳也可以变为阴；另一方面则体现在阴阳之中可以再分阴阳，例如昼为阳、夜为阴，而上午为阳中之阳，下午为阳中之阴。

2.2　五行学说

中国古代劳动人民在长期的生活和生产实践中，认识到木、火、土、金、水五种物质是人类生活中不可缺少的东西。人们把这五种物质的属性加以抽象推演，用来说明整个物质世界，并认为这五种物质既具有相互资生、相互制约的关系，而且是在不断运动、变化之中，故称之为“五行学说”。

2.2.1　五行的特性

古代医家运用五行学说，对畜体的脏腑组织、生理、病理现象，以及与有关的事物，作了广泛的联系和研究，并用“取象比类”的方法，按照事物的不同性质、作用与形态，分别归属于木火土金水“五行”之中，借以阐述畜体脏腑组织之间生理、病理的复杂联系，以及畜体与外界环境之间的相互关系。

2.2.2　五行生克乘侮

相生相克

相生，是指五行之间的相互资生和助长。相生的次序是：木生火、火生土、土生金、金生水、水生木。在五行的相生关系中，任何一行都有“生我”“我生”两方面的关系，生我者为“母”，我生者为“子”，所以五行的相生关系也称为“母子关系”。以木为例，由于木生火，故木为火之母，而木由水所生，故木又为水之子。

相克，是指五行之间的相互克制和约束。克的次序是：木克土、土克水、水克火、火克金、金克木。在五行的相克关系中，任何一行都有“克我”“我克”的两方面的关系，我克者为“所胜”，克我者为“所不胜”。以木为例，克木者为金，金为木之所不胜；木克者为土，土为木之所胜。

相乘相侮

相乘，含有乘虚侵袭的意思，是事物间的关系失去正常协调的一种表现。如木气偏亢，而金又不能对木正常克制时，太过的木便去乘土，使土更虚。

相侮，含有恃强欺侮的意思，即五行中某一行本身太过，使原来克它的一行，不仅不能去制约它，反而被它所克制，故又称“反克”。例如：正常的相克关系是金克木，若木气偏亢，或金气不足，木就会反过来侮金。五行相侮恰与五行相克的次序相反。

3 脏腑学说

中兽医学将畜体的内脏分成五脏、六腑。五脏，即心、肝、脾、肺、肾，它们的共同特点是具有化生贮藏精气的功能。六腑，即胆、胃、大肠、小肠、膀胱、三焦，它们的共同特点是具有受盛和传化水谷的功能。

3.1 五脏

3.1.1 心

心位于胸中，有心包围护于外。它的生理功能是主血脉，为畜体血液运行的动力所在。主神，开窍于舌。

3.1.2 肺

肺位于胸中，上通喉咙，开窍于鼻。它的生理功能是司呼吸，主一身之气，有宣发与肃降的作用，而且能通调水道，外合皮毛。

3.1.3 脾

脾位于中焦，它的生理功能有主运化，统血，主升清，主肌肉及四肢的作用，开窍于口，其华在唇。

3.1.4 肝

肝位于胁部，它的生理功能是主疏泄、藏血、主筋、开窍于目、其华在目。

3.1.5 肾

肾位于腰部，左右各一。它的生理功能是藏精、主骨，为生殖发育之源；又主纳气、主水、开窍于耳及二阴。

3.2 六腑

3.2.1 胆

胆附于肝，内藏“精汁”，精汁即胆汁，来源于肝。胆汁注入肠中，有促进饮食物的消化作用，是脾胃运化功能得以正常进行的重要条件之一。由于胆的主要生理功能是贮存和排泄胆汁，胆汁直接有助于草料的消化，故为六腑之一；因胆本身并无转化草料的生理功能，且藏精汁，与胃、肠等腑有别，故属奇恒之腑。

3.2.2 胃

胃位于膈下，其上口名曰贲门，与食道相接；下口名幽门，与小肠相通。胃受纳，腐熟水谷。草料入口，经过食道，容纳于胃，所以又称胃为“水谷之海”。容纳于胃中的水谷，经过胃气腐熟、消磨、形成食糜，并下移于小肠。如果胃的这一功能障碍，可出现食欲不振、食少、消化不良、胃胀等。当然，胃受纳、腐熟水谷的功能，必须与脾的运化功能相配合，才能顺利完成。

3.2.3 小肠

小肠的主要功能是主化物，分清别浊。小肠上接于胃，接受胃中转化来的水谷，作进一步的消化、吸收，即所谓化物作用。然后分别清浊，其精微物质（清）由脾转送全身；其糟粕（浊）通过“阑门”下注大肠；其无用之水液渗入膀胱，以保证水液、精微、糟粕各行其道，小便正常，大便通利。

3.2.4 大肠

大肠上接阑门，与小肠相连，下端为肛门。大肠接受小肠下注的内容物，吸收其中剩余的水液后，使之变化成为粪便，最后经肛门排出体外。所以大肠是传送糟粕的通道。大肠有病，可见传送失常的种种病变，如大便溏泄或大便秘结等症。

3.2.5 膀胱

膀胱的主要生理功能是贮藏、排泄尿液。在畜体水液代谢过程中，水液通过肺、脾、肾等脏的作用，布散全身，发挥润泽肌体的作用。其被畜体利用之后，下归于肾，经过肾的气化作用，升清降浊，其清者回流体内，浊者变成尿液，下输膀胱，当其达到一定容量时，通过膀胱的气化功能及时而自主地将尿液排出体外。如膀胱气化不行，可见小便不利或癃闭；膀胱失其约束，则见尿频、小便失禁；膀胱湿热，常见尿频、尿痛等。

3.2.6 三焦

三焦是上、中、下焦的总称。上焦包括心、肺等脏器，中焦包括脾、胃等脏器，下焦包括肝、肾、大肠、小肠、膀胱等脏器。三焦的生理功能，为总司畜体的气化，是水谷精微生化和水液代谢的通路。食物自受纳、腐熟，到精气的敷布、代谢产物的排泄都与三焦有关。

3.3　脏与腑之间的关系

脏与腑主要是表里关系。脏为阴，腑为阳；阴主里，阳主表。心与小肠、肺与大肠、脾与胃、肝与胆、肾与膀胱、一脏一腑、一阴一阳、一里一表，它们通过经脉互相络属，相互配合，组成脏腑表里关系。这种关系，不仅说明它们在生理上相互联系，而且也决定了它们在病理上的相互影响。

3.3.1　心与小肠

心的经脉属心而络小肠，小肠的经脉属小肠而络心。由于心与小肠的经脉互为络属，便构成一脏一腑的表里关系。心与小肠的内在联系，表现在病理上较为明显。如心火下移小肠，熏蒸水液，常可引起尿少、尿赤、尿热等小肠实热的病证。反之，如果小肠有热，顺经脉上熏于心，则可出现口舌糜烂等症状。

3.3.2　肺与大肠

肺的经脉与大肠的经脉互为络属，所以肺与大肠构成了表里关系。由于它们经脉互相沟通，所以在生理上和病理上都是互相影响的。肺气肃降，则大肠功能正常，大便通畅；若大肠积滞不通，也能反过来影响肺气的肃降。临床上，如肺气肃降失职，津液不能下达，则可见大便困难；若大肠实热，又可影响肺气不降。

3.3.3　脾与胃

脾胃同居中焦，有经脉互相络属，构成表里关系。胃主受纳，脾主运化，脾为胃行其津液；胃气宜降，脾气宜升；胃喜润恶燥，脾喜燥恶湿，二者一升一降，一润一燥，分工合作，共同完成草料的消化、吸收与水谷精微的转输。在病理上，二者亦互相影响。如脾脏虚寒，常同时兼有胃阳不足，中兽医常称之为“中焦虚寒”，治疗亦脾胃同治，如温中散寒法。

3.3.4　肝与胆

胆附于肝，经脉相互络属，互为表里。胆有贮藏和排泄胆汁的功能，以帮助胃肠消化。胆汁来源于肝。若肝的疏泄失常，则会影响胆汁的正常排泄；反之，胆汁的排泄失常，又会影响到肝，故肝胆证候往往同时并见。

3.3.5　肾与膀胱

肾与膀胱的经脉互相络属，互为表里。膀胱的气化功能，取决于肾气的盛衰，肾气有助膀胱的气化津液和司膀胱的开合以约束尿液的作用。肾气充足，固摄有权，膀胱开合有度，以维持水液的正常代谢。如果肾气不足，失去固摄与司膀胱开合作用，就可出现小便失禁、多尿等病证。故有关尿的贮存与排泄，除膀胱本身的病变外，多与肾脏有关。

4　气血津液

气、血、津、液是构成畜体的基本物质，是脏腑经络等器官组织进行生理活动的物质基础。机体的脏腑经络等器官组织进行生理活动所需要的能量，来源于气、血、津液；它们的生成和代谢，又依赖于脏腑、经络等器官组织的正常生理活动。因此，无论是在生理还是在病理上，气、血、津、液和脏腑、经络等器官组织之间，都存在着密切的关系。气、血、津液禀受于先天，或化生于后天的水谷精微，都是维持畜体生命活动的物质基础；脏腑、经络依赖这些物质的滋养和补充，而发挥其正常生理活动。

在中兽医学中，“精”也是构成畜体的一种基本物质。有广义与狭义之分。狭义之“精”即为肾所藏之生殖之精，作为肾脏功能的一部分。广义之“精”包括一切精微物质，即气、血、津液及水谷精微等，总称为“精气”。

4.1　气

4.1.1　气的概念

中兽医学中气的概念来源于中国古代哲学中的“气”。在古代哲学家认为，“气”是构成世界的最基本物质，宇宙间的一切事物都是由气的变化运动而产生。“气”为万物之本，是构成畜体的基本物质和维持生命活动的基本元素。

4.1.2　气的功能

4.1.2.1　推动作用

气是活力很强的精微物质，它对于畜体的

生长发育，各脏腑、经络等器官组织的生理活动，血的生成和运行，津液的生成、输布和排泄等，均起着推动和激活的作用。如果气的推动、激活作用不足，则机体的生长、发育，脏腑、经络等的生理活动，以及血与津液的生成和运行等均会受到影响。

4.1.2.2 温煦作用

气是畜体热量的来源。畜体的体温是依靠气的温煦作用而维持恒定。气的温煦作用也与脏腑、经络等器官组织的生理活动，血和津液的正常循行密切相关。

4.1.2.3 防御作用

机体在防御病邪方面主要依赖畜体的“正气”。畜体的正气包括气、血、津液和脏腑、经络的功能活动，但是其中最重要的是“气”。气的防御作用，主要体现于护卫全身的肌表和抵御外邪的入侵。

4.1.2.4 固摄作用

气的固摄作用主要表现在其防止血、津液等物质的流失。例如固摄血液（主要是脾气的功能），防止其逸出脉外而产生出血；固摄汗、尿、唾液及精液等（分别是卫气、肾气、脾气等的功能），控制分泌量，以防其无故流失。气的固摄作用和气的推动作用是相互协调、相辅相成。气的推动作用为血的运行和津液的代谢提供动力，气的固摄作用则保证了其运行和代谢循常道进行。

4.1.2.5 气化作用

气化，是指通过气的运行而产生的各种变化，气、血、津液和精各自的新陈代谢及由于气的作用而发生的相互转化。例如气、血、津液的生成，都需要将草料转化为水谷精微，然后再化生成气、血、津液等，津液经过代谢，转化为汗液和尿液，草料经过消化吸收后，其残渣转化为糟粕等，都是气化的具体表现。所以气化的过程，实际上是体内物质代谢的过程，是物质和能量转化的过程。

4.2 血

4.2.1 血的概念

血，是红色的液态物质，它循行于血脉之中，是构成畜体生命活动的基本物质之一。

4.2.2 血的功能

血，具有营养和滋润全身的功能。其营养和滋润作用主要体现于口色的红润、肌肉的丰满和壮实、皮肤和毛发的润泽有华和运动的灵活自如等方面。

4.3 津液

4.3.1 津液的概念

津液，是机体一切正常水液的总称，包括各脏腑组织器官的内在体液及其正常的分泌物，如胃液、肠液和泪等。津液，同气和血一样，是构成和维持畜体生命活动的基本物质。

津和液，同属于水液，都来源于草料，有赖于脾胃的运化功能而生成。由于在其性状、功能及其分布部位等方面有所不同，因而两者也有一定的区别。一般说，性质较清稀、流动性较大，分布于体表、皮肤、肌肉和孔窍，并能渗注于血脉，起滋润作用的，称为津；性质较稠厚、流动性较小，灌注于骨节、脏腑、脑髓等组织，起濡养作用的，称为液。津和液之间也可互相转化，故津和液常同时并称。

4.3.2 津液的功能

津液有滋润和濡养的生理功能。如：布散于肌表的津液，具有滋润皮毛的作用；流注于孔窍的津液具有滋润和保持眼、口、鼻等孔窍润泽的作用。

5 经络学说

经络学说，是研究畜体经络系统的组成、生理功能、病理变化，以及与脏腑、气血等相互关系的理性知识。它同脏腑学说紧密结合，是中兽医学基础理论中的重要组成部分，对于中兽医临床各科，特别是对针灸治疗，都具有指导意义。

5.1 经络的概念

经络是运行全身气血，联络脏腑肢节，沟通上下内外的通路，是经脉和络脉的总称。经脉是主干，络脉是分支。经，有路径的意思；络，有网络的意思。经脉大多循行于深部，络脉循行于较浅的部位，有的络脉还显现于体表。

经脉有一定的循行径路，而络脉则纵横交错，网络全身，把畜体所有的脏腑、器官、孔窍以及皮肉筋骨等组织联结成一个统一的有机整体。

5.2　经络的生理功能

5.2.1　沟通上下表里，联系全身各部

畜体是由五脏六腑、四肢百骸、五官九窍、皮肉筋骨等组成。它们虽各有不同的生理功能，但又共同进行着有机的整体活动，使机体内外、上下保持协调统一，构成一个有机的整体。这种有机配合、相互联系，主要是依靠经络的沟通、联络作用实现的。由于十二经脉及其分支的纵横交错，入里出表，通上达下，相互络属于脏腑；奇经八脉联系沟通于十二正经；十二经筋、十二皮部联络筋脉皮肉，从而使畜体的各个脏腑组织器官有机的联系起来，构成了一个表里、上下彼此间紧密联系，协调共济的统一体。

5.2.2　运行气血，濡养全身

畜体的各个脏腑组织器官均需要气血的温养濡润，才能够发挥其正常作用。气血是畜体生命活动的物质基础，必须依赖经络的传注，才能输布周身，以温养濡润全身各脏腑组织器官，维持机体的正常功能，如营气之和调于五脏，洒陈于六腑，为五脏藏精、六腑传化的功能活动提供了物质条件。所以说经脉具有运行气血，调节阴阳和濡养全身的作用。由于经络能行血气而营阴阳，营气运行于脉中，卫气行于脉外，使营卫之气密布于周身，加强了机体的防御能力，所以，起到了抗御外邪，保卫机体的作用。

5.2.3　感应传导

经络有感应刺激、传导信息的作用。当畜体的某一部位受到刺激时，这个刺激就可沿着经脉传入体内有关脏腑，使其发生相应的生理或病理变化。而这些变化，又可通过经络反应于体表。针刺中的“得气”、“行气”，就是经络感应、传导功能的具体体现。

5.2.4　调节机体平衡

经络在正常情况下能运行气血和协调阴阳，在疾病情况下，出现气血不和及阴阳偏胜偏衰的证候，即可运用针灸等治法以激发经络的调节作用，以泻其有余，补其不足，调节机体，维持平衡。实验证明，针刺有关经络的穴位，可对各脏腑机能产生调整作用，即原来亢进的可使之抑制，原来抑制的可使之兴奋。

6　病因学

导致疾病发生的原因是多种多样的，如六淫、草料、劳逸等，在一定条件下都能使畜发生疾病。中兽医学认为，临床上没有无原因的证候，任何证候都是在原因的影响和作用下，患病机体所产生的一种病态反映。中兽医认识病因，主要是以病证的临床表现为依据，通过分析疾病的症状、体征来推求病因，为治疗用药提供依据，这种方法称为“辨证求因”。所以，中兽医学的病因学，不但研究病因的性质和致病特点，同时也探讨各种致病因素所致病证的临床表现，以便更好地指导临床诊断和治疗。

六淫　即“风、寒、暑、湿、燥、火”六种外感病邪的统称。“风、寒、暑、湿、燥”，是自然界六种不同的气候变化，在正常的情况下称为“六气”，人们在生活实践中逐步认识了它们的变化特点，产生了一定的适应能力，所以六气不至于使畜体生病。只有气候异常急骤的变化或畜体的抵抗力下降时，六气才能成为致病因素，侵犯畜体发生疾病，这种情况下的六气就称为“六淫”。淫，有太过和浸淫之意。由于六淫是不正之气，所以又称“六邪”，属于外感病的一类病因。

6.1　风

风为春季的主气，但四季皆有风。故风邪致病虽春季多见，但不限于春季，其他季节亦均可发生。风邪外袭多从皮毛肌腠而入，从而产生外风病证。

6.2　寒

寒为冬季的主气。寒邪为病有外寒、内寒之分。外寒是指寒邪外袭，其致病又有伤寒、中寒之别。寒邪伤于肌表，郁遏卫阳，称为“伤寒”；寒邪直中于里，伤及脏腑阳气，则为

"中寒"。内寒则是机体阳气不足，失却温煦的病理反映。外寒与内寒虽有区别，但它们又是互相联系，互相影响的。阳虚内寒之体，容易感受外寒；而外来寒邪侵入畜体，积久不散，又常能损及畜体阳气，导致内寒。

6.3 暑

暑为夏季的主气，乃火热所化。暑邪有明显的季节性，独见于夏季。暑邪纯属外邪，无内暑之说。

6.4 湿

湿为长夏的主气，夏秋之交，为一年中湿气最盛的季节。湿邪为病有外湿、内湿之分。外湿多由于气候潮湿、涉水淋雨、居处潮湿等外在湿邪侵袭畜体所致。内湿多由于脾失健运、水湿停聚而生。外湿和内湿虽有不同，但在发病中又常相互影响。伤于外湿，湿邪困脾，脾失健运，则湿从内生；而脾阳虚损，水湿不化，亦易招致外湿的侵袭。

6.5 燥

燥为秋季的主气，此时气候干燥，水分亏乏，故多燥病。燥邪感染途径，多从口鼻而入，侵犯肺卫。燥邪为病又有温燥、凉燥之分，初秋有夏热之余气，燥与温热结合而侵犯畜体，则多见温燥病证；深秋又有近冬之寒气，燥与寒邪结合侵犯畜体，故有时亦见凉燥病证。

6.6 火

火热为阳盛所生，故火热常可并称。但火与温热，同中有异，热为温之渐，火为热之极，热多属外淫，如风热、暑热、湿热之类病邪；而火常由内生，如心火上炎、肝火亢盛、胆火横逆等类病变。

7 防治原则

7.1 预防

预防，就是采取一定的措施，防止疾病的发生与发展。中兽医学历来重视疾病的预防，早在两千多年前，《黄帝内经》中就已提出了"治未病"的预防思想。所谓"治未病"，包括未病先防和既病防变两个方面的内容。

7.1.1 未病先防

未病先防，就是在未病之前，做好各种预防工作，以防止疾病的发生。

疾病的发生，一是因畜体的正气不足或功能紊乱；二是因邪气侵入畜体。邪气是导致疾病发生的重要条件，而正气不足是疾病发生的内在原因和根据，外邪通过内因而起作用。因此，预防疾病的发生亦须从这两方面着手。

7.1.2 既病防变

未病先防，是最理想的预防措施。但如果疾病已经发生，就应该争取早期诊断、早期治疗，以防止疾病的发展与传变。

7.2 治则

治则，即治疗疾病的法则。它是在整体观念和辨证论治基本精神指导下制订的，对临床治疗立法、处方、用药，具有普遍的指导意义。扶正祛邪为治疗总则，在总则的指导下采用益气、养血、滋阴、补阳等方法，就是扶正的具体方法。发汗、涌吐、攻下等法，就属于祛邪的具体方法。

由于疾病的证候表现是多种多样的，病理变化是极为复杂的，因此，必须善于从复杂多变的疾病现象中，抓住疾病的本质，治病求本；采取相应的扶正祛邪，调整阴阳的措施；按发病的不同时间、地点和不同的病变个体，因时、因地、因畜制宜。

7.2.1 治病求本

治病求本，就是寻找出疾病的根本原因，并针对其根本原因进行治疗。这是辨证论治的一个基本原则。"本"是对"标"而言的。标本是一个相对的概念，有多种含义，可用以说明病变过程中各种矛盾的主次关系。从邪正双方来说，正气为本，邪气为标；从病因与症状来说，病因为本，症状为标；从疾病先后来说，旧病、原发病为本，新病、继发病为标。

7.2.2 扶正祛邪

疾病的过程，在一定意义上，就是正气与邪气双方相互斗争的过程。正邪斗争的胜负，决定着疾病的进退。邪胜于正则病进，正胜于邪则病退。因此，治疗疾病，就要扶助正气，祛除邪气，改变邪正双方的力量对比，使之有利于疾病向痊愈方向转化。所以扶正祛邪是临床治疗的一个重要法则。

7.2.3　调整阴阳

疾病的发生从根本上说就是畜体阴阳的相对平衡遭到破坏，出现偏盛偏衰的结果。因此，调整阴阳，补偏救弊，恢复阴阳的相对平衡，促成阴平阳秘，是临床治疗的根本法则之一。

7.2.4　因时、因地、因畜制宜

因时、因地、因畜制宜，是指治疗疾病要根据季节、地区以及患者的体质、性别、年龄等不同而制定适宜的治疗方法。

第二讲

中兽医药物学

中药的性能是以中兽医学的阴阳、脏腑、经络、治疗法则等理论为基础，从医疗实践中予以归纳总结出来的，是分析药物及临床用药的基本依据。中药药性理论主要有四气五味、升降浮沉、归经及毒性等。

1 四气

指药物的“寒、热、温、凉”四种药性，也称为四性。寒凉与温热是两种不同的属性，寒凉属阴，温热属阳，而寒与凉、热与温仅是程度上的不同。寒凉之性的药物有清热、泻火、解毒等作用，主要适用于热性病证；温热之性的药物有散寒、助阳的作用，主要适用于寒性病证。另外，有些药物寒热之性不明显，而称为平性。所谓“平性”，实际上仍有偏凉、偏温的区别，所以仍属四气的范围。

2 五味

五味是指药物“辛、甘、酸、苦、咸”五种不同的滋味。药物的滋味主要是通过口尝得出来的。随着药性理论的发展，不少药物的味是从药物的临床功效归纳出来的，药物的味与实际口尝之味也就不尽相同。味相同的药物，它们的作用大多相近；而味不同的药物，所表现出的治疗作用也不同。

3 升降浮沉

“升”是指药物作用趋势向上，具有升提作用，适用于病势向下的疾病。如黄芪、升麻能升提中气，治疗久泻久痢、脱肛、子宫脱垂、胃下垂等中气下陷之证。“降”是指药物作用趋于向下，具有降逆作用，适用于病势向上的疾病。如代赭石、沉香、石决明能降上逆之气火，潜降肝阳，降肺、胃气逆，适用于气火炎上之出血、牙龈肿痛、口舌生疮，肺气上逆之咳嗽气喘，胃气上逆之恶心呕吐、嗳气呃逆等证。“浮”是指药物作用趋势向外、向上，具有上浮、发散功能，适用于病位在表、在上的疾病。如麻黄、紫苏、防风、独活能发散风寒湿邪，用治外感表证、风湿痹痛等证，“沉”是指药物作用趋势向下、向内，具有下行、泄利功能，适用于病位在下、在里的疾病。如大黄、木通分别具有泻下、利尿作用，治疗大便不通、腹胀腹痛、小便不利等证。

4 归经

归经是指某药对畜体某脏腑经络病变有明显和特殊的治疗作用，而对其他脏腑经络起的作用较小或不起作用，体现了药物对畜体各部位治疗作用的选择性。归经是以脏腑经络理论为基础，以主治病证为依据的。如麻黄、杏仁用于咳嗽气喘等肺经病证有效，便将其归入肺经；青皮、香附用于乳房、胁肋胀痛及疝气痛等肝经病证有效，便将其归入肝经。故一般而论，某一药物能治何经之病，即可归入何经。如果某味药能归数经，就说明该药治疗范围较广，对数经病证都能发挥治疗作用。由此可见，中药的归经，是从临床疗效观察中总结出来，经过反复实践，逐步发展而上升为理论。

5　毒性

毒性是指药物对畜体的危害作用，有毒性的药物称有毒药。有毒药物及性质猛烈的药物使用不当，轻者可引起机体的损害，重者可导致死亡。为了保证用药安全，必须认识中药的毒性。

有毒药物的治疗剂量与中毒剂量比较接近或相当，因而安全度小，易引起中毒。无毒药物治疗量与中毒量相距较大，安全度亦较大，但也并非绝对不会引起中毒反应。

为了保证用药安全并发挥其治疗作用，避免毒性反应的产生，应用有毒药物应注意以下几点：严格炮制、控制剂量、注意用法。

6　中药的炮制

炮制是药物在使用前，根据中兽医药理论，按照医疗、调剂、制剂和贮藏的需要，对药材进行加工处理的工艺过程。它包括对药材进行的一般修治和部分药材的特殊处理。由于中药大多是生药，有的在采集过程中混有杂质，有的易变质而不能久贮，有的毒烈之性较强而不能直接服用，还有的需要特定的方法处理才能符合治疗需要。因此，中药在应用或制成制剂以前，都必须经过加工处理，才能充分发挥药效，保证用药安全。

6.1　水制

水制是用水处理药物的一种方法。水制的目的主要是清洁、软化药材，以便于切制或调整药性，以及减低毒性，使矿物类药物纯净、细腻等。常用的方法有：洗、淘、浸、润、渍、腌、提、水飞等。

6.2　火制

火制是用火加热处理药物的方法。常用的有炒、炙、煅、煨等。

7　中药的配伍

中药的配伍就是根据病情需要和药物性能，有目的地将两种以上的药物配合在一起使用。配伍是临床用药的主要方式，也是组成方剂的基础。

药物通过配伍之后，可发生复杂的变化。有的增强药效，有的减低药效，有的能抑制和消除原有的毒副作用，而有的却能产生毒性和不良反应。前人把中药配伍后产生的不同效应总结为“七情”，即单行、相须、相使、相畏、相杀、相恶、相反七个方面。

8　配伍禁忌

配伍禁忌是指某些药物不能配伍应用，否则会产生毒副作用，危害病畜的健康，甚则危及生命，也就是指相反的药物，具体是指“十八反”和“十九畏”。

“十八反”的内容是：乌头反半夏、瓜蒌、贝母、白蔹、白及；甘草反海藻、大戟、芫花、甘遂；藜芦反党参、沙参、丹参、玄参、苦参、细辛、芍药。“十九畏”的内容是：硫黄畏牵牛子，丁香畏郁金，川乌、草乌畏犀角，牙硝畏三棱，官桂畏五灵脂。必须说明，这里所说的“畏”是相恶之意，与中药配伍关系中的“相畏”含义不同。

“十八反”和“十九畏”中的药物不宜同用，但有部分内容与实际应用并不完全相同，历代医家也有所论及，“畏”、“反”药物同用的方剂也不乏其例。现代实验研究，结论亦不完全一致。因此，尚难给予肯定和否定，还有待于进一步研究。但在未得出结论之前，应慎重应用，一般情况下，仍遵循传统习惯，避免配伍同用。

9　妊娠用药禁忌

母畜妊娠期间，有些药物不能使用，而又有些药物应谨慎使用，否则会损伤胎元或引起堕胎。妊娠用药禁忌一般分为两类：一类为禁用药，多是毒性强烈或药性峻猛药；一类为慎用药，主要是活血祛瘀药、行气药、攻下药、温里药中的部分药，这些药物应用不当亦可损伤胎气，引起堕胎。

禁用药：水银、砒霜、雄黄、轻粉、斑蝥、蜈蚣、马钱子、蟾酥、川乌、草乌、藜芦、胆矾、瓜蒂、巴豆、甘遂、大戟、芫花、牵牛子、商陆、麝香、干漆、水蛭、虻虫、三棱、莪术等。

慎用药：牛膝、川芎、红花、桃仁、姜黄、牡丹皮、枳实、枳壳、大黄、番泻叶、芦荟、天南星、芒硝、附子、肉桂等。

10 剂量

剂量，即用药量。主要是指每一味药的成畜一日量；其次是指方剂中药与药之间的比较用量，即相对剂量。一般而言，各药下所标明的用量是指干燥后的生药在汤剂中的成畜一日内服量及干燥后生药研作末后成畜一次服用量。中药的用量，大多以重量计算，个别以数量、容量表示。现在中国内地规定的重量计算单位是以公制克为单位。

11 煎药法

煎药的器皿宜用砂锅或砂罐。煎药用水必须无异味，洁净澄清。将药物倒入器皿中加入洁净的水，以淹没药物为度，浸泡半小时左右即可煎煮。火候的控制可根据药物的性质和质地而定，气味芳香者，宜武火急煎，煮沸数分钟后，改用文火略煮即可，否则药效减弱；滋补药物质地滋腻，宜文火久煎，否则有效成分难以煎出。一般每日 1 剂，煎煮 2 次，滋补药可煎 3 次。

各种药物的质地、性质往往有显著差异，因此，煎煮方法或煎煮时间常不相同。

12 投药法

中药汤剂一般宜温服，每日 1 剂，煎 2 次，每服 1 煎。病情急重者可 1 日 2 剂，甚或 3 剂，即每 4 小时煎服一次；慢性病可隔日 1 次，或 1 剂分 2 日服，即每日服 1 煎。止吐药宜少量多次服；发汗药宜热服，以助药力，以出汗为宜；泻下药以下为度，否则泻下太过损伤正气。丸、散等药可用温开水送服。若用从治法，则温热药冷服，或寒凉药热服。

13 解表药

凡以发散表邪，解除表证为主要作用的药物，称解表药，又叫发表药。使用解表药时，应控制用量，中病即止，否则发汗太多，易耗伤阳气，损及津液。表虚自汗，阴虚盗汗，以及疮疡日久、淋证、失血者，虽有表证，均当忌用或慎用。解表药多含挥发油，入汤剂不宜久煎，以免有效成分挥发而降低疗效。

13.1 麻黄

麻黄是麻黄科植物草麻黄、木贼麻黄和中麻黄的草质茎。主产于河北、山西、内蒙古自治区（全书称内蒙古）、甘肃等地。秋季采收，阴干，切段。生用、蜜炙或捣绒用。

【药性】味辛、微苦，性温。归肺、膀胱经。

【功效】发汗解表，宣肺平喘，利水消肿。

【临床应用】

1. 用于外感风寒证。本品发汗作用最强，多用于外感风寒表实证，症见恶寒发热，无汗，脉浮紧，常与桂枝相须配伍，如麻黄汤。

2. 用于咳喘实证。邪壅于肺，肺气不宣之咳嗽气喘，无论寒、热、痰、饮，有无表证均可应用。风寒郁肺之咳喘，与杏仁、甘草配伍，如三拗汤。肺有寒饮，痰多清稀者，可配伍干姜、细辛、半夏等，如小青龙汤。肺有郁热，喘息痰黄稠者，也可以与石膏、杏仁、甘草配伍以清肺平喘止咳，如麻黄杏仁甘草石膏汤。

【用法用量】水煎服，马、牛用 15 ~ 30 克，猪、羊 3 ~ 10 克。发汗解表宜生用，止咳平喘多蜜炙用。

【使用注意】不可过量使用，表虚自汗、阴虚盗汗、虚喘病畜禁用。

13.2 桂枝

桂枝指樟科植物肉桂的嫩枝。主产于广东、广西壮族自治区（全书称广西）及云南。春、夏季割取嫩枝，晒干或阴干，切片或切段。生用。

【药性】味辛、甘，性温。归心、肺、膀胱经。

【功效】发汗解表，温通经脉，通阳化气。

【临床应用】

（1）用于风寒表证，无论表实或表虚证均可用。感受风寒，表虚自汗，常与白芍药等同用，以调和营卫，如桂枝汤。表实无汗者，常与麻黄同用。

（2）用于寒凝血滞诸痛证。胸阳不振，心脉瘀阻，胸痹心痛，常与枳实、薤白等同用，如瓜蒌薤白桂枝汤；还可用于风寒湿邪痹阻，关节疼痛，常配附子同用，如桂枝附子汤。

（3）用于小便不利、水肿者，与猪苓、泽泻等同用，如五苓散。

【用法用量】水煎服，马、牛用 15 ~ 45 克，猪、羊 3 ~ 10 克。

【使用注意】本品辛温，易伤阴动血，凡温热病热盛伤阴及杂病阴虚阳亢，血热妄行诸证忌用。孕畜慎用。

【说明】桂枝与麻黄均为发散风寒药，桂枝发汗力不及麻黄，作用缓和，可用于表虚自汗者，多与白芍药配伍，并有温通经脉，通阳化饮利水等功效。麻黄为肺经主药，发汗之力较强，多用于风寒表实无汗者，还能宣肺平喘，利水消肿。

13.3　柴胡

柴胡为伞形科植物柴胡（北柴胡）和狭叶柴胡（南柴胡）的根。前者主产于辽宁、甘肃、河北、河南等地；后者主产于湖北、江苏、四川等地。春秋两季采挖，晒干，切段。生用或醋炙用。

【药性】味苦、辛，性微寒。归肝、胆经。

【功效】和解退热，疏肝解郁，升阳举陷。

【临床应用】

（1）用于外感发热，寒热往来。善于疏散少阳半表半里之邪，为治疗邪在少阳、寒热往来、少阳证之要药。常与黄芩、半夏等同用，如小柴胡汤。治外感发热，可与葛根、黄芩、石膏等同用，如柴葛解肌汤。

（2）用于气虚下陷诸证。本品善于升举脾胃清阳之气，可治气虚下陷、脱肛、子宫脱出、短气乏力等，可与升麻、黄芪等升提中气之品同用，如补中益气汤。

【用法用量】水煎服：马、牛用 15 ~ 45 克，猪、羊 6 ~ 15 克。

【使用注意】肝风内动，阴虚火旺及气机上逆者忌用。

14　清热药

凡以清解里热为主要作用的药物，称为清热药。

清热药药性寒凉，有清热泻火、燥湿、凉血、解毒、清虚热等功效。主要用于外感热病，高热烦渴，湿热泻痢，温毒发斑，痈肿疮毒及阴虚发热等病证表邪已解，里热炽盛，而无积滞的各种里热证候。

清热药性均寒凉，易伤脾胃，脾胃气虚，食少便溏者慎用；热病易伤津液，清热燥湿药性味苦寒，易化燥伤阴伤津，故阴虚津伤者也应当慎用。

14.1　石膏

石膏为硫酸盐类矿物硬膏族石膏，含结晶水硫酸钙。主产于湖北、甘肃及四川。全年可采，研细。生用或煅用。

【药性】味辛、甘，性大寒。归肺、胃经。

【功效】清热泻火，除烦止渴，收敛生肌。

【临床应用】

（1）用于壮热烦渴。清热泻火之力强，并能除烦止渴，为清泻肺胃二经气分实热的要药。用于温病邪在气分，壮热、烦渴、汗出等实热证，常与知母相须为用。

（2）用于热郁于肺，咳嗽痰稠，发热口渴。常与麻黄、杏仁等配伍以增强止咳平喘之力，如麻杏石甘汤。

（3）用于胃火上炎，牙龈红肿，常配伍黄连、升麻等清泻胃热，如清胃散。

（4）外用收敛生肌，用于疮疡，湿疹，烧烫伤及外伤出血。可单用，也可配伍青黛、黄柏等同用，以清热解毒，收湿敛疮。

【用法用量】水煎服，马、牛用 30 ~ 120 克，猪、羊 15 ~ 30 克。

【使用注意】脾胃虚寒及阴虚内热病畜忌用。

14.2 黄芩

黄芩为唇形科植物黄芩的根。主产于河北、山西、内蒙古等地。春秋季采挖。蒸透或开水润透切片。生用，酒炙用或炒炭用。

【药性】味苦、性寒。归肺、胃、胆、大肠经。

【功效】清热燥湿，泻火解毒，凉血止血，安胎。

【临床应用】

(1) 用于湿温病，湿热黄疸，泻痢，淋证。本品尤善清中上二焦湿热。治湿温病发热汗出，苔腻，与滑石、通草、白豆蔻等同用，如黄芩滑石汤。治湿热黄疸，常配伍茵陈、栀子、大黄等，增强利胆退黄之功。治湿热泻痢，配黄连、葛根同用，如葛根芩连汤。治膀胱湿热，小便淋漓涩痛，与木通、滑石、车前子等清热利湿药同用。

(2) 用于气分实热，肺热咳嗽。本品善清肺火及上焦实热。若肺热壅盛，肺失清降，咳嗽痰稠，可单用，即清金丸，也可配伍桑白皮、知母、麦门冬等同用。如气分实热，壮热烦渴，尿赤便秘，多配伍大黄、栀子等药同用以泻火通便，如凉膈散。

(3) 用于外疡内痈及其他内科、外科、五官科热毒证，可与黄连、连翘、蒲公英等配伍。

(4) 用于邪热迫血妄行之吐血、咯血、尿血、便血等。可单用，或与三七、槐花、白茅根等同用。

(5) 用于胎热所致胎动不安，可与白术、当归等药物同用，方如当归散。

【用法用量】水煎服，马、牛用 15～90 克，猪、羊 9～15 克。清热多生用，安胎多炒用，止血多炒炭用。清上焦热多酒炒使药力上行。

【使用注意】本品苦寒伤胃，脾胃虚寒病畜不宜用。

14.3 苦参

苦参为豆科植物苦参的根。产于中国各地。春秋两季采挖，切片，晒干。生用。

【药性】味苦，性寒。归心、肝、胃、大肠、膀胱经。

【功效】清热燥湿，利尿杀虫。

【临床应用】

(1) 用于湿热泻痢，黄疸尿赤。治湿热泻痢，可单用，或配伍木香、甘草同用，如香参丸；治黄疸，常配伍栀子、茵陈、龙胆等以利湿退黄；治湿热淋证，小便不利，常与车前子、泽泻等利水药同用。

(2) 用于阴痒，湿疹，疥癣。既能清热燥湿，又能杀虫止痒，为治上述皮肤病的常用药。内服外用均可，常配伍黄柏、蛇床子等同用。

【用法用量】水煎服，马、牛用 15～60 克，猪、羊 6～15 克。

【使用注意】苦寒之性较甚，脾胃虚寒者忌用。反藜芦。

14.4 生地黄

生地黄为玄参科植物地黄的根。主产于河南、河北、内蒙古及东北。全国大部分地区有栽培。秋季采挖，鲜用或干燥切片生用。

【药性】味甘、苦，性寒。归心、肝、肾经。

【功效】清热凉血，养阴生津。

【临床应用】

(1) 用于温热病热入营血。善清血分之热，又能滋阴生津。用于温热病热入营血，发斑，舌质红绛，配伍玄参、麦门冬等清热生津凉血之品同用，如清营汤。

(2) 用于血热妄行之吐血、衄血、便血、尿血等，常与侧柏叶、生荷叶、生艾叶等同用，如四生丸。用于血热发斑，可与牡丹皮、赤芍药等同用。

(3) 用于阴虚津伤证。热病伤阴，舌红口干，可配伍沙参、麦门冬、玉竹等。治热伤津液而肠燥便秘，常与玄参、麦门冬同用，如增液汤。

【用法用量】水煎服，马、牛用 30～60 克，猪、羊 10～15 克。

【使用注意】本品寒凉滋腻，脾虚有湿及便溏病畜忌用。

14.5 金银花

金银花为忍冬科植物红腺忍冬、山银花或毛花柱忍冬的花蕾。产于中国各地。夏季花含苞未放时采收，阴干。生用，炒炭或蒸馏制露用。

【药性】味甘，性寒。归肺、心、胃经。

【功效】清热解毒，疏散风热，清热解暑。

【临床应用】

（1）用于温热病及外感风热表证。温热病无论邪在卫分、气分，或入营血均可应用。温病初起，邪在卫分，或外感风热表证，发热微恶寒者，常与荆芥、薄荷、牛蒡子等同用，如银翘散；若热在气分壮热烦渴，可与石膏、知母等同用。热入营血，斑疹隐隐，舌绛而干，则须配伍生地黄、牡丹皮等。

（2）用于痈肿疮疡。热毒阳证最适宜。痈肿未成者可消，已成者可溃，初溃热毒未消者也可用。内服或鲜品捣烂外敷。或配伍蒲公英、野菊花、紫花地丁等。治肠痈，可配薏苡仁、黄芩、当归等同用；治肺痈，可与鱼腥草、芦根、桃仁等同用，以解毒排脓。

（3）用于热毒血痢，便脓血，可单味生品浓煎；亦可与黄芩、黄连、白头翁等药同用，以增强止痢效果。

（4）用于暑热烦渴、咽喉肿痛、热疮等。可加水蒸馏制成金银花露，供内服、外用。

【用法用量】水煎服，马、牛用 15 ~ 60 克，猪、羊 9 ~ 15 克。

14.6 板蓝根

板蓝根为十字花科植物菘蓝的根。秋季采挖，除去泥沙，晒干。生用。

【药性】味苦，性寒。归心、胃经。

【功效】清热解毒，凉血利咽。

【临床应用】

用于温热病发热、斑疹、大头瘟、痈肿疮毒等。清热解毒之力较强，多用于热毒壅盛之证，又以解毒利咽散结见长。如外感风热发热或温病初起有上述证候者，常与金银花、连翘、荆芥等同用；治大头瘟毒，常配伍玄参、连翘、牛蒡子等，如普济消毒饮。现多用于病毒感染性疾病。

【用法用量】水煎服，马、牛用 30 ~ 90 克，猪、羊 15 ~ 30 克。

【使用注意】脾胃虚寒者忌用。

14.7 青蒿

青蒿为菊科植物黄花蒿的地上部分。产于中国各地。秋季花盛开时采割。鲜用或阴干，切段生用。

【药性】味苦、辛，性寒。归肝、胆、肾经。

【功效】清虚热，解暑热，截疟。

【临床应用】

（1）用于阴虚发热，骨蒸劳热。本品入阴分清虚热，兼有透散之功，尤宜于骨蒸无汗者。常与银柴胡、胡黄连、知母、鳖甲等同用，如清骨散；温热病后期，余热未清，夜热早凉，热退无汗，或热病后低热不退等。常与鳖甲、牡丹皮、生地黄等同用，如青蒿鳖甲汤。

（2）用于暑热证，暑热挟湿及湿温病湿热交蒸者。治暑热发热汗出，可与西瓜翠衣、金银花、荷叶等清解暑热之品同用。如治暑热挟湿或湿温证可配藿香、佩兰、滑石等化湿、利湿之品。

（3）用于疟疾寒热。本品为治疟疾常用之品。可单用，或随证配伍黄芩、滑石、青黛等同用。现代从其中提取青蒿素，制成片剂、注射剂等应用，进一步提高疗效。

【用法用量】水煎服，马、牛用 15 ~ 60 克，猪、羊 6 ~ 15 克。不宜久煎；或鲜用绞汁饮。

15 泻下药

凡能引起腹泻，或润滑大肠，促使排便的药物称泻下药。

泻下药有通利大便、清泄热邪、攻逐水饮的作用。主要适用于大便秘结、实积停滞、水饮滞留，以及实热壅滞等证。

根据泻下作用强弱及适应范围的不同，本章药物可分为攻下、润下、峻下逐水 3 类。其中，攻下药和峻下逐水药作用峻猛，尤以后者为甚，润下药润滑肠道，作用缓和。

攻下药与逐水药，对久病正虚，体弱，以及母畜妊娠、产后，均应慎用或忌用。中病即止，不可过量。里实而正虚者，当与补益药同用。

15.1 大黄

大黄为蓼科植物掌叶大黄、唐古特大黄或药用大黄的根及根茎。掌叶大黄和唐古特大黄称为北大黄，主产于青海、甘肃等地；药用大黄称南大黄，主产于四川。秋末茎叶枯萎或次年春发芽前采挖。除去须根，刮去外皮，切块，干燥。生用，或酒炒，酒蒸，炒炭用。

【药性】味苦，性寒。归脾、胃、大肠、肝、心经。

【功效】泻下攻积，清热泻火，凉血止血，解毒，活血祛瘀。

【临床应用】

（1）用于胃肠积滞，大便秘结。为治疗积滞便秘要药，尤宜于热结便秘。治实热内结便秘，腹痛拒按者，常与芒硝、枳实、厚朴同用，如大承气汤；如兼见气血虚者，再配党参、当归等，以扶正攻邪，如黄龙汤；热结阴伤者，配生地黄、麦门冬等药，如增液承气汤；至于脾阳不足，冷积便秘，也可以配附子、干姜等同用，如温脾汤。

（2）用于积滞泻痢。治湿热痢疾初起，腹痛里急后重，常与黄连、木香等同用，如芍药汤；治食积腹痛，泻痢不畅，可与青皮、木香等同用。

（3）用于火热上炎之目赤、口舌生疮等，无论有无便秘均可应用。常与黄连、黄芩、牛黄等配伍，如泻心汤。

（4）用于血热出血。既能泻火止血，又兼活血作用，止血而不留瘀。凡血热妄行之吐血、衄血，或瘀血血不归经所致的出血均可应用。可单用，或与其他清热凉血止血药同用。

（5）用于热毒疮疡，烧烫伤。既可内服，又能外用。治热毒痈肿，常与金银花、蒲公英、连翘等同用；治疗肠痈，常与牡丹皮、桃仁等同用，如大黄牡丹皮汤；治烧烫伤，可单用研粉，或配地榆粉、麻油调敷患处。

（6）用于产后瘀血腹痛，癥瘕积聚，跌打损伤及蓄血证。治母畜产后血瘀腹痛，可与桃仁等同用，如下瘀血汤。治跌打损伤，瘀血肿痛，可与桃仁、红花等同用，如复元活血汤。

【用法用量】水煎服，马、牛用 18 ~ 45 克，猪、羊 6 ~ 12 克。外用，适量。生用泻下作用较强，欲攻下者生用，入煎剂后下或开水泡服，不宜久煎。酒制大黄泻下力减弱，活血作用较好，宜用于瘀血证。大黄炭多用于出血证。

15.2 火麻仁

火麻仁为桑科植物大麻的成熟种子。产于中国各地。秋季果实成熟时采收，晒干，打碎。生用。

【药性】味甘，性平。归脾、大肠经。

【功效】润肠通便。

【临床应用】

用于肠燥便秘及体弱津血不足而致的便秘，可与当归、熟地黄等同用。热邪伤阴或素体阴虚，大便秘结，可与大黄、厚朴配伍，如麻子仁丸。

【用法用量】水煎服，马、牛用 120 ~ 180 克，猪、羊 12 ~ 30 克。

15.3 甘遂

甘遂为大戟科植物甘遂的块根。主产于陕西、山西、河南等地。秋末或春初采挖，除去外皮，晒干。醋制用。

【药性】味苦、甘，性寒。有毒。归肺、肾、大肠经。

【功效】泻下逐饮，利水消肿。

【临床应用】

（1）用于水肿，鼓胀，胸胁停饮。本品泻水逐饮之力峻，前证正气未衰者，均可用之。可单用研末服，或与大戟、芫花等同用，如十枣汤；若水饮与热邪结聚而致的结胸证，可与大黄、芒硝同用，如大陷胸汤。现多用于腹水、渗出性胸膜炎、肠梗阻等。

（2）用于疮疡肿毒。可用甘遂末水调外敷。

【用法用量】入丸散服，马、牛用 6 ~ 15

克，猪、羊1～3克。内服须醋制以减低毒性。

【使用注意】孕畜及体虚者忌用。反甘草。

16　祛湿药

凡能祛除体内之湿邪，治疗湿邪所致诸证的药物，称为祛湿药。其中以祛除风湿，解除痹痛为主要作用的药物称祛风湿药。气味芳香，具化湿健脾作用的药物称芳香化湿药。有通利水道，渗泄水湿作用的药物称利水渗湿药。

本类药物适用于风寒湿痹、水肿、淋证、黄疸及湿阻中焦证等。

某些药物性温燥，或渗泄之性强，易耗伤阴血，故阴亏血虚、津伤者慎用。

16.1　独活

独活为伞形科植物重齿毛当归的根。主产于四川、湖北、安徽等地。秋末或春初采挖，晒干，切片。生用。

【药性】味辛、苦，性温。归肝、膀胱经。

【功效】祛风湿，止痹痛，解表。

【临床应用】

（1）用于风寒湿痹证。对湿邪偏重腰背尤为适宜。若治全身性关节痹痛，可配羌活、秦艽一起使用；痹证日久，肝肾不足，气血亏虚者，配桑寄生、杜仲、牛膝等补肝肾之晶，如独活寄生汤。

（2）用于风寒表证挟湿，症见恶寒发热。可配羌活、防风、荆芥使用。

【用法用量】煎服，马、牛用15～45克，猪、羊3～9克。

【使用说明】独活与羌活均有祛风湿、散寒止痛作用，都可以治疗风湿痹痛，风寒湿表证。但羌活偏于解表，发汗散寒力量强，用于祛风湿，多治上半身痹痛；而独活解表力较弱，偏于祛风湿，除痹痛，并偏治下半身痹痛。对于全身关节疼痛，两药配伍同用，效果更好。

16.2　藿香

藿香为唇形科植物广藿香的地上部分。主产于广东。夏、秋季采割，鲜时切段用，或阴干。生用。

【药性】味辛，性微温。归脾、胃经。

【功效】化湿，解暑，止呕。

【临床应用】

（1）用于中焦湿阻气滞，多与苍术、厚朴等同用。

（2）用于暑湿证及湿温证初起。暑季外感风寒，内伤湿冷而致恶寒发热，呕吐泄泻，常配伍紫苏叶、厚朴等同用，如藿香正气散。湿温病初起，湿热并重，可与黄芩、滑石、茵陈等同用，如甘露消毒丹。

（3）用于呕吐。为治湿浊呕吐要药，常与半夏配伍。偏热者，再配黄连、竹茹等。

【用法用量】水煎服，马、牛用15～45克，猪、羊6～12克。鲜品用量加倍。

16.3　茯苓

茯苓为多孔菌科真菌茯苓的菌核。多寄生于松科植物赤松或马尾松等的树根上。主产于云南、湖北、四川等地。7～9月采挖，反复堆置、晒干。生用。

【药性】味甘、淡，性平。归心、脾、肾经。

【功效】利水渗湿，健脾安神。

【临床应用】

（1）用于水肿，小便不利。可用于各种水肿证。膀胱气化不利所致，可与桂枝、猪苓、白术、泽泻等同用，如五苓散；气虚者配防己、黄芪；脾肾阳虚者，配附子、干姜等同用，如真武汤。

（2）用于脾虚诸证。对脾虚有湿者尤宜，可与党参、白术配伍，如四君子汤。若脾虚饮停，可与桂枝、白术等同用，如苓桂术甘汤；若脾虚泄泻，可与山药、白术、薏苡仁等同用。

【用法用量】水煎服，马、牛用18～60克，猪、羊9～18克。

17　温里药

凡能温散里寒，治疗里寒证的药物，称为温里药。

本类药物性温热，能祛散里寒，振奋阳气，适用于寒邪内侵，阳气受抑；或阳气衰弱，阴寒内生之里寒证，症见脘腹冷痛、呕吐、呃逆、泄泻痢疾、畏寒肢冷、小溲清长等。部分药物能回阳救逆，适用于亡阳证。此外，有的能温肺化饮，温肝治疝，降逆止呃。可治肺寒咳喘，疝气痛及胃寒呕吐，脘腹冷痛。

17.1 附子

附子为毛茛科植物乌头的子根的加工品。主产于四川、湖北、湖南等地。6月下旬至8月上旬采收，加工炮制为盐附子、黑顺片、白附片、淡附片、炮附片。

【药性】味辛、甘，性热。有毒。归心、肾、脾经。

【功效】回阳救逆，补火助阳，散寒止痛。

【临床应用】

（1）用于亡阳证。本品为治亡阳证主药，症见汗出清冷、呼吸气微、四肢厥冷、脉微欲绝者，常配干姜、甘草同用，如四逆汤；若阳虚气脱，可配党参益气固脱，如参附汤。

（2）用于阳虚诸证。若治肾阳不足，尿频尿多，可与肉桂、山茱萸、熟地黄等同用，如右归丸；若治脾肾阳虚，寒湿内盛的脘腹冷痛，纳少，大便泄泻，可与党参、白术、干姜等同用；治阳虚水肿，小便不利，多与白术、茯苓、生姜同用；治阳虚外感配麻黄同用；治阳虚自汗配黄芪同用。

（3）用于诸寒痛证。若治风寒湿痹，周身骨节疼痛，属寒湿盛者，可与桂枝、白术、甘草同用；寒凝气滞腹痛，配伍丁香、高良姜等同用。

【用法用量】水煎服，马、牛用15～30克，猪、羊3～9克。至口尝无麻辣感为度。

【使用注意】本品辛热燥烈，阴虚阳亢及孕畜忌用。反半夏、瓜蒌、贝母、白蔹、白及。内服须炮制后用，且须久煎，不可过量久服。

17.2 肉桂

肉桂为樟科植物肉桂的树皮。主产于广东、广西、云南等地。秋季剥取，刮去栓皮，阴干，切片或研末。生用。

【药性】味辛、甘，性热。归脾、肾、心经。

【功效】补火助阳，散寒止痛，温经通脉。

【临床应用】

（1）用于肾阳不足证。治肾阳不足之畏寒、尿频等，多与附子、熟地黄、山茱萸等同用，如肾气丸；若治下元虚衰，虚阳上浮之面赤、虚喘，可用本品与附子同用以引火归原。

（2）用于寒凝气滞或寒凝血瘀所致的诸痛。治脾胃虚寒，脘腹冷痛，可单用，或与干姜、高良姜等同用；治寒疝腹痛，常配吴茱萸、小茴香同用

（3）用于阴疽及虚寒性疮疡，久不愈合者。可与鹿角胶、炮姜、麻黄等同用，如阳和汤；也可与黄芪、当归等同用，如托里黄芪散。

【用法用量】水煎服，马、牛用15～30克，猪、羊3～9克，宜后下或开水泡服。

【使用注意】血热出血忌用。

【说明】肉桂与桂枝来源于同一植物。肉桂用树皮，桂枝用嫩枝。两者均能散寒助阳，桂枝主上行而偏散寒解表，肉桂主温里且入下焦，偏于温肾阳。

18 理气药

凡能疏理气机，以治疗气滞或气逆证为主的药物，称为理气药。

理气药大多气香性温，其味辛、苦，具有调气健脾、行气止痛、顺气降逆、疏肝解郁或破气散结等功效。适用于肺失宣降，肝失疏泄，脾胃升降失司。

本类药物辛温香燥者居多，易耗气伤阴，故气虚及阴虚者慎用。

18.1 青皮

青皮为芸香科橘及其同属植物的幼果或未成熟果实的果皮。5～6月间采集，晒干。生用或醋炒用。

【药性】味苦、辛，性温。归肝、胆、胃经。

【功效】疏肝破气，散结消滞。

【临床应用】

（1）用于肝气郁滞治乳痈肿痛，常配瓜蒌、金银花、蒲公英、甘草等同用；寒疝腹痛，

则配乌药、小茴香、木香等以散寒理气止痛，如天台乌药散。

（2）用于食积胃脘胀痛，常与山楂、麦芽、神曲等配伍，如青皮丸。

【用法用量】水煎服，马、牛用 15 ~ 30 克，猪、羊 6 ~ 12 克。

【使用注意】性较峻烈，易耗气、破气，气虚病畜慎用。

18.2　香附

香附为莎草科植物莎草的根茎。主产于广东、河南、四川等地。9 ~ 10 月采收，挖取根茎，晒干，烧去须根。生用或醋炒用。

【药性】味辛、微苦、微甘，性平。归肝、三焦经。

【功效】疏肝理气，调经止痛。

【临床应用】

（1）用于肝气郁滞所致的胁肋作痛、脘腹胀痛及疝痛等证。治胁痛，可与柴胡、白芍药、枳壳等配伍，如柴胡疏肝散；肝气犯胃、寒凝气滞之胃脘冷痛，可配高良姜，即良附丸；治寒疝腹痛，则与小茴香、乌药等同用。

（2）用于乳房胀痛。尤宜用于因肝气郁结所致者，常配当归、川芎、白芍药、柴胡等以疏肝行滞，调和气血。

【用法用量】水煎服，马、牛用 15 ~ 45 克，猪、羊 9 ~ 15 克。

19　消食药

凡以消食化积，恢复脾胃运化为主要功效，治疗食积证的药物，称为消食药。

消食药除能消化草料积滞外，多数具有开胃和中的作用。可以治疗饮食积滞、脘腹胀满、嗳腐吞酸、恶心呕吐、不思水草、大便失常等消化不良证。

使用时，常配理气药，可助消食化滞，并根据病情分别配伍温中、清热、化湿药。食积因脾胃虚弱者，应以补养脾胃为主，辅以消导。

19.1　山楂

山楂为蔷薇科植物山里红或山楂的果实。主产于河南、江苏、浙江、安徽等省。秋末冬初采收，晒干。生用或炒用。

【药性】味酸、甘，性微温。归脾、胃、肝经。

【功效】消食化积，活血散瘀。

【临床应用】

1. 用于食滞不化，腹痛泄泻。常与神曲、麦芽等配伍；脘腹胀痛明显者，可配木香、枳壳等；伤食腹痛泄泻，可用焦山楂研末，开水调服。

2. 用于产后瘀阻腹痛、恶露不尽，疝气偏坠胀痛。治产后血瘀证，常与当归、川芎、益母草等配伍；治疝气，可与小茴香、橘核等同用。

【用法用量】水煎服，马、牛用 18 ~ 60 克，猪、羊 9 ~ 15 克。生山楂偏于消食散结，焦山楂偏于止泻止痢。

19.2　神曲

神曲为面粉和杏仁泥、赤小豆粉、鲜青蒿、鲜苍耳、鲜辣蓼自然汁等混合后经发酵而成的加工品。中国各地均产。长出菌丝（生黄衣）后，取出晒干。生用或炒用。

【药性】味甘、辛，性温。归脾、胃经。

【功效】消食和胃。

【临床应用】

用于食积不化，脘腹胀满，不思水草及肠鸣泄泻。常与山楂、麦芽等同用。又本品略有解表之功，故外感食滞者有之尤宜。

【用法用量】水煎服，马、牛用 20 ~ 60 克，猪、羊 10 ~ 25 克。

19.3　麦芽

麦芽为禾本科植物大麦的成熟果实经发芽干燥而成，中国各地均产。随时制备。生用或炒黄用。

【药性】味甘，平。归脾、胃、肝经。

【功效】消食和中，回乳。

【临床应用】

用于食积不化、脘闷腹胀等。擅于助淀粉类食物消化。常与山楂、神曲、鸡内金等同用。脾胃虚弱而运化不良者，可以在补气药中配用本品。

【用法用量】水煎服，马、牛、用20～60克，猪、羊9～15克。

【使用注意】喂乳期不宜用。

20 止血药

凡以制止体内外出血为主要作用的药物，称为止血药。

止血药分别具有凉血止血、收敛止血、化瘀止血、温经止血等不同作用。主要适用于全身各部分出血，如咯血、吐血、衄血、尿血、便血及外伤出血。

临床使用时，须根据出血的原因和具体的证候，辨证选用，并进行适当的配伍，以增强疗效。如血热妄行者，应选用凉血止血药，并配伍清热凉血药；阴虚阳亢者，应与滋阴潜阳药同用；瘀血阻滞而出血不止者，应选用化瘀止血药为主，并酌配行气活血药；虚寒性出血，应与温阳、益气、健脾等药同用；出血过多而导致气虚欲脱者，应急予大补元气之药，以益气固脱。使用凉血止血和收敛止血药时，必须注意有无瘀血。瘀血未尽之出血，应酌加活血去瘀药，不能单纯止血，更不宜选用收敛止血及凉血止血药，以免有留瘀之弊。

20.1 大蓟

大蓟为菊科植物蓟的根及全草。中国各地均产。夏、秋花期时采集全草，秋末采挖根部，晒干，切段。生用或炒用。

【药性】味甘、苦，性凉。归心、肝经。

【功效】凉血止血，散瘀消痈。

【临床应用】

（1）用于血热咯血、衄血、尿血等。可单味应用，也可与小蓟、侧柏叶等同用。

（2）用于疮痈肿毒。无论内服、外敷，皆有效，以鲜品为佳。也可配金银花、大青叶、赤芍药等同用。

【用法用量】水煎服，马、牛用18～60克，猪、羊9～18克。

20.2 槐花

槐花为豆科植物槐的花蕾，中国大部分地区均有栽培。6～7月采摘花蕾，晒干。生用或炒用。

【药性】味苦，性微寒。归肝、大肠经。

【功效】凉血止血，清肝降火。

【临床应用】

（1）用于血热出血证。尤善于治下部出血。多炒炭用，用于便血，常与地榆相须配伍；治咯血、衄血，则多与仙鹤草、白茅根、侧柏叶等同用。

（2）用于肝火上炎之目赤。可单用煎汤代茶，或配夏枯草、菊花等同用。

【用法用量】水煎服，马、牛用15～30克，猪、羊6～12克。

21 活血化瘀药

凡以通畅血行，消除瘀血为主要作用的药物，称活血化瘀药，或称活血祛瘀药。其中活血逐瘀作用较强者，又称破血药。

活血化瘀药，味多辛、苦，主归肝、心经，入血分。善于走散，通过活跃血行，消散瘀血，而能达到通经、利痹、消肿、定痛、疗伤等功效，适用于血行失畅，瘀血阻滞之证。症见疼痛（痛处固定不移）；身体内外部肿块，出血色暗，夹有紫黯色血块；皮肤、黏膜或舌质瘀斑等。形成瘀血证的原因颇多，在运用活血祛瘀药时，应辨证审因，选择适当的药物，并作适宜的配伍。如属寒凝气滞者，须配温里祛寒药同用；属热伤营血，瘀血内阻者，配清热凉血药同用；风湿痹痛，须与祛风湿药合用；如跌打损伤，宜与行气和络之品配伍；对癥瘕痞块，应与化痰软坚散结药配用；兼有正气不足之证者，又当配伍相应的补虚药同用。

畜体气血之间有着密切的关系，气行则血行，气滞则血凝，故在使用活血祛瘀药时，常配合行气药，以增强行血散瘀的作用。

21.1　丹参

丹参为唇形科植物丹参的根及根茎。中国大部分地区均产。主产于河北、安徽、江苏、四川等地。春、秋季采挖，晒干。生用或酒炒用。

【药性】味苦，性微寒。归心、心包、肝经。

【功效】活血祛瘀，调经止痛，凉血消痈，清心安神。

【临床应用】

（1）用于血滞，产后瘀滞腹痛，心腹疼痛，癥瘕积聚以及肢体疼痛等证。因其性属寒，故对瘀血兼热的病证尤为适宜。对母畜经产诸证因瘀血者，常与红花、桃仁、益母草等配伍；用于血瘀气滞所致的心腹、胃脘疼痛，可与檀香、砂仁配伍，如丹参饮；治癥瘕积聚，可与三棱、莪术、泽兰、鳖甲等配伍；跌打损伤、瘀滞作痛，常与当归、红花、川芎同用；风湿热痹，关节红肿疼痛，与忍冬藤、赤芍药、秦艽同用。

（2）用于疮痈肿痛。与清热解毒药相配，有助于消除痈肿。如乳痈肿痛与乳香、金银花、连翘等同用，即消乳汤。

（3）用于治温热病，邪入营血，常与生地黄、玄参、竹叶心等同用，如清营汤；

【用法用量】水煎服，马、牛 15～45 克，猪、羊 6～15 克。

【使用注意】反藜芦。

21.2　红花

红花为菊科植物红花的花冠。主产于河南、湖北、四川、浙江等地。夏季花色由黄转为鲜红时采摘，阴干。生用。

【药性】味辛，性温。归心、肝经。

【功效】活血祛瘀，通经止痛。

【临床应用】

（1）用于产后瘀阻腹痛、癥瘕积聚、跌打损伤瘀痛以及关节疼痛等证。常与桃仁、当归、川芎、赤芍药等配用。

（2）用于斑疹色暗，因热郁血滞所致者。可与当归、紫草、大青叶等配伍，如当归红花饮。

【用法用量】水煎服，马、牛 10～30 克，猪、羊 3～6 克。

22　化痰止咳平喘药

凡具祛痰或消痰作用，以治疗“痰证”为主要作用的药称化痰药；以减轻或制止咳嗽和喘息为主要作用的药物，叫止咳平喘药。化痰药多兼止咳、平喘之功，止咳平喘药亦多兼化痰之效。所以，两类药合于一章，总称为化痰止咳平喘药。化痰药，主要用于痰多咳嗽或痰饮气喘，咯痰不爽之证。止咳平喘药，主要用于内伤、外感所引起的咳嗽和喘息。瘿瘤瘰疬、阴疽流注等证，在病机上均与痰有密切的关系，故亦可用化痰药治之。

外感、内伤均可引起咳喘或多痰，因而在应用时除根据各药的特点加以选择外，还须根据致病的原因和证型作适当的配伍。例如，兼有表证者配解表药，兼有里热者配清热药，兼有里寒者配温里药，虚劳咳喘者配补益药。

22.1　半夏

半夏为天南星科植物半夏的块茎。中国各地均产，长江流域产量最多。夏、秋间采挖，去皮及须根，晒干。生用或用生姜、明矾等炮制后用。

【药性】味辛，性温。有毒。归脾、胃、肺经。

【功效】燥湿化痰，降逆止呕，消痞散结；外用消肿止痛。

【临床应用】

（1）于湿痰、寒痰证。为治痰湿证要药。治痰湿壅肺之咳嗽气喘，痰多，常与橘皮、茯苓同用，如二陈汤；兼见寒象，痰多清稀，可加配细辛、干姜；若见热象，痰稠色黄者，则需与黄芩、知母、瓜蒌同用。

（2）用于胃气上逆呕吐。常与生姜同用。胃热呕吐，则可配黄连、竹茹等。

（3）用于瘿瘤痰核，痈疽发背及乳疮。治瘿瘤痰核，可与昆布、海藻、浙贝母等软坚散结药同用；治痈疽，可将生半夏研末，用鸡蛋清调敷患处。此法也可用于治疗毒蛇咬伤。

【用法用量】水煎服，马、牛 15～45 克，猪、羊 3 ~ 10 克。

【使用注意】其性温燥，对阴亏燥咳、血证、热痰等证，当忌用或慎用。反乌头。

【说明】内服一般制用，主要有姜半夏、法半夏、竹沥半夏等，姜半夏长于降逆止呕，多用于治呕吐；法半夏长于燥湿且温性较弱，多用于湿痰证；竹沥制半夏，使药性由温变凉，能清化痰热，主治热痰、风痰之证。

22.2 杏仁

杏仁为蔷薇科植物山杏、东北杏、西伯利亚杏或杏的成熟种子。产于东北、华北、西北、新疆及长江流域。夏季果实成熟时采收种子，除去果肉及核壳，晒干。生用。

【药性】味苦，性微温。有小毒。归肺、大肠经。

【功效】止咳平喘，润肠通便。

【临床应用】

（1）用于咳嗽气喘。为治咳喘要药，随证配伍可用治各种咳喘证。治风寒咳喘，可配麻黄、甘草，如三拗汤；治风热咳嗽，每与桑叶、菊花等配伍；治燥热咳嗽，与桑叶、贝母、沙参等同用；治肺热咳喘，与麻黄、生石膏等合用，如麻杏石甘汤。

（2）用于肠燥便秘。常与火麻仁、当归、枳壳等同用，如润肠丸。

【用法用量】水煎服，马、牛 15 ~ 30 克，猪、羊 3 ~ 12 克，宜打碎入煎。

【使用注意】本品有毒，用量不宜过大。

23 平肝熄风药

具有平肝潜阳或平熄肝风作用，主治肝阳上亢或肝风内动的药物，称为平肝熄风药。

本类药物，主入肝经，多为介类、昆虫等动物药及矿物药，主要用于肝风内动，抽搐惊痫。

应用平肝熄风药时，应根据不同病因和兼证，予以不同配伍。肝风内动，多由火热炽盛所致；肝阳上亢亦每兼肝热，故须与清热泻火、清泄肝热药同用。阴虚血少，肝失滋养，以致肝风内动与肝阳上亢，则又当与滋肾养阴或补血药同用。

本类药物多偏于寒凉，但也有偏于温燥者，应区别使用。凡脾虚慢惊，非寒凉药所宜；而阴虚血亏者，又当慎用温燥之品。

牡蛎

牡蛎为牡蛎科动物长牡蛎、大连湾牡蛎或近江牡蛎的贝壳。中国沿海地区均产，冬、春采集，去肉留壳，洗净晒干。捣碎生用，或煅用。

【药性】味咸，性微寒。归肝、肾经。

【功效】平肝潜阳，软坚散结，收敛固涩。

【临床应用】

（1）用于热病伤阴，肝风内动，四肢抽搐，常配龟板、鳖甲、地黄等药，如大定风珠。熟地黄等同用，如石决明丸。

（2）用于痰火郁结之瘰疬，痰核及癥瘕积聚。常与浙贝母、玄参配伍，如消瘰丸。近来临床用以治胁下癥块，常与丹参、泽兰、鳖甲等配伍使用。

（3）用于滑脱诸证。治自汗、盗汗，可与黄芪、麻黄根配伍。

【用法用量】水煎服，马、牛 18 ~ 40 克，猪、羊 6 ~ 12 克，先煎。收敛固涩、制酸煅用，余均生用。

24 补益药

凡能补益畜体气血阴阳不足，改善脏腑功能，提高抗病能力，消除各种虚弱证候的药物，称为补益药，亦称为补虚药或补养药。

虚证临床表现非常复杂，但概括起来有气虚、阳虚、血虚、阴虚之分。根据其作用和适应范围，补益药也相应分为补气、补阳、补血、补阴四类。由于畜体的气血阴阳有着相互依存的关系，阳虚多兼有气虚，气虚也易导致阳虚，气虚和阳虚表示畜体生理功能不足；阴虚多兼有血虚，血虚也易导致阴虚，血虚和阴虚表示畜体精血津液的损耗。因此，补气和补阳、补血和补阴药常相须为用。如果气血不足、阴阳俱虚的证候同时出现，又当气血双补或阴阳并补。

补益药不宜用于实邪未尽而正气不虚者，以免影响邪气的驱除而加重病情，但若病邪未清而正气已虚的，可于祛邪药中酌加补益药，以增强抗病能力，扶正祛邪。

补益药一般服药时间较久，且某些药物有碍消化，故常配伍健脾胃药物，以加强疗效。补益药虽可补虚，但不可滥用，以防变生他证。

24.1　党参

党参为桔梗科植物党参、素花党参或川党参的根。野生者习称野台党，栽培者习称潞党参。主产于山西、陕西、甘肃。秋季采挖。晒干，切段。生用。

【药性】味甘，性平。归脾、肺经。

【功效】补中益气，生津养血。

【临床应用】

（1）用于脾虚倦怠乏力、食少便溏及各种原因所致的气虚体弱之证。常与白术、茯苓、甘草同用。

（2）用于肺气亏虚之气短咳喘。常与黄芪、五味子同用，如补肺汤。

（3）用于热病气津两伤，气短口渴。多配伍麦门冬、五味子等同用。

【用法用量】水煎服，马、牛 18～60 克，猪、羊 6～12 克。

【使用注意】热证及阴虚阳亢证不宜用。反藜芦。

24.2　黄芪

黄芪为豆科植物蒙古黄芪或膜荚黄芪的根。主产于内蒙古、山西、甘肃、黑龙江等地。春、秋两季采挖，除掉须根及根头。切片，晒干。生用或蜜炙用。

【药性】味甘，性微温。归脾、肺经。

【功效】补气升阳，益肺固表，利水消肿，托毒生肌。

【临床应用】

（1）用于脾肺气虚及中气下陷证。病后气虚体弱，倦怠乏力，常与党参配伍，如参芪膏；兼阳虚而见畏寒，体倦多汗者，可与附子同用，如芪附膏；气虚血亏，可与当归同用；脾气虚弱，食少便溏或泄泻者，可与白术同用，如芪术膏；脾阳不升，中气下陷，内脏下垂，子宫脱垂，久泻脱肛，可与党参、白术、升麻、柴胡等配伍，如补中益气汤；若气虚不能摄血之便血等，又可与党参、酸枣仁、桂圆肉等配伍，如归脾汤。

（2）用于肺虚咳喘，肌表不固之自汗、盗汗。治肺气虚弱，咳喘气短，常与紫菀、五味子等同用；治体虚多汗，容易感冒，常与白术、防风同用，如玉屏风散；盗汗可与生地黄、黄柏等同用，如当归六黄汤。

（3）用于气虚水湿失运之面目浮肿，小便短少。每与防己、白术等同用，如防己黄芪汤；对慢性肾炎浮肿，尿中蛋白长期不消者，用之有良好疗效。

（4）用于气血不足之痈疽不溃或溃久不敛。治痈疽久不溃，常与当归、皂角刺等同用，如透脓散；溃久脓水清稀，久不收口，可与当归、党参、肉桂等配伍，如十全大补汤。

【用法用量】水煎服，马、牛 15～60 克，猪、羊 6～10 克。

【使用注意】表实邪盛，气滞湿阻，食积内停，阴虚阳亢，痈疽初起或溃久热毒尚盛者，均不宜用。

24.3　巴戟天

巴戟天为茜草科植物巴戟天的根。主产于广东、广西、福建等地。全年均可采挖。晒干，再经蒸透，除去木心者，称“巴戟肉”。切段，干燥。生用或盐水炙用。

【药性】味甘、辛，性微温。归肾、肝经。

【功效】补肾阳，强筋骨，祛风湿。

【临床应用】

（1）用于肾阳不足的不孕，少腹冷痛。治不孕，常配党参、山药、肉苁蓉等；治少腹冷痛，常与高良姜、肉桂、吴茱萸等同用，如巴戟丸。

（2）用于肾阳虚兼有风湿之腰膝疼痛或软弱无力。可与萆薢、杜仲等同用，如金刚丸。

【用法用量】水煎服，马、牛 12～30 克，猪、羊 3～9 克。

【使用注意】阴虚火旺及有湿热者不宜用。

24.4 当归

当归为伞形科植物当归的根。主产于甘肃、陕西等地。以产于甘肃岷县的当归质量最好。秋末采挖后除去须根，用微火熏干，切片。生用或酒炒用。

【药性】味甘、辛，性温。归肝、心、脾经。

【功效】补血，活血，止痛，润肠。

【用法用量】

（1）用于血虚唇舌色淡，趾甲苍白。常与黄芪同用，如当归补血汤。

（2）用于各种瘀血疼痛及风湿痹痛。肢体瘀血作痛，可配丹参、乳香、没药等，如活络效灵丹；跌打损伤，瘀肿疼痛，常配大黄、桃仁、红花等，如复元活血汤；治风湿肩臂疼痛，多配羌活、防风、姜黄等，如蠲痹汤。与桂枝、白芍药、生姜同用，还可用治中焦虚寒腹痛。

（3）用于痈疽疮疡。初起红肿疼痛而尚未化脓者，与金银花、赤芍药、天花粉等配伍，如仙方活命饮；中期脓成未溃者，应与皂角刺、黄芪等配用，如透脓散；若溃后因气血不足，脓水不尽，久不收口者，常配党参、黄芪、熟地黄等，如十全大补汤。

（4）用于血虚肠燥便秘。尤其适宜于久病体虚及产后血虚津亏便秘，常配生何首乌、火麻仁、肉苁蓉等同用。

此外，还可以用5%当归注射液注入肺俞、膻中等穴位，治疗慢性支气管炎。

【用法用量】水煎服，马、牛15~60克，猪、羊10~15克。

24.5 熟地黄

熟地黄为玄参科植物地黄的根，经加工炮制而成。主产于河南、河北、内蒙古等地。生地黄干燥后，以黄酒、砂仁、橘皮为辅料，经反复蒸晒，至内外色黑、质地柔软黏腻为度。切片用。

【药性】味甘，性微温。归肝、肾经。

【功效】养血滋阴，补精益髓。

【临床应用】

（1）用于血虚萎黄，常与当归、白芍药、川芎等同用，如四物汤。

（2）用于肝肾阴虚，常与山药、山茱萸等同用，如六味地黄丸。如肾阴不足，虚火偏旺，低热、盗汗明显者，还可配滋阴清火之知母、黄柏、龟板等，如大补阴丸。

（3）用于精血亏虚，常与何首乌、女贞子、旱莲草、山茱萸等同用。

【用法用量】水煎服，马、牛200~400克，猪、羊50~100克。

【使用注意】本品性质黏腻，有碍消化，故凡气滞痰多、脘腹胀满、食少便溏者均不宜服用。

24.6 麦门冬

麦门冬为百合科植物麦门冬的块根。中国各地均产，夏季采挖，除去须根，洗净，晒干。生用。

【药性】味甘、微苦，性微寒。归心、肺、胃经。

【功效】养阴润肺，益胃生津。

【临床应用】

（1）用于肺燥干咳，痰黏及劳嗽咯血。治温燥之邪犯肺干咳，常与桑叶、杏仁、生石膏同用，如清燥救肺汤；治肺阴亏损，劳嗽咯血，常与天门冬配伍，如二冬膏。

（2）用于胃肠阴虚证。胃阴不足，舌干口渴，常与沙参、生地黄、玉竹等同用，如益胃汤；治肠燥便秘，可与生地黄、玄参同用，即增液汤。

【用法用量】水煎服，马、牛15~45克，猪、羊2~18克。

【使用注意】外感风寒及痰湿阻肺的咳嗽，或脾胃虚寒泄泻者，均不宜服用。

25 收敛固涩药

凡以收敛固涩为主要作用的药物，称为收敛固涩药。

本类药物大多性味酸涩，分别具有敛汗、止泻、固精、缩尿、止血、止嗽等作用，适用于体虚精气耗散所致的自汗、盗汗、久泻、久痢、尿频，以及久咳虚喘不止等证。

收涩药有恋邪之弊，所以表邪未解、内有湿滞，以及实邪盛者，均不宜用。

浮小麦

浮小麦为禾本科植物小麦未成熟的颖果。中国各地均产。以水淘小麦，取浮起者，晒干。生用或炒用。

【药性】味甘，性凉。归心经。

【功效】益气，除热，止汗。

【临床应用】

（1）用于自汗、盗汗。治盗汗及虚汗不止，可单用本品；治体虚自汗不止，常与牡蛎、麻黄根、黄芪同用，即牡蛎散。

（2）用于骨蒸劳热。多与生地黄、麦门冬、地骨皮等养阴清虚热药同用。

【用法用量】水煎服，马、牛 30 ~ 120 克，猪、羊 12 ~ 18 克。或炒焦研末服。

第三讲

中兽医方剂学

方剂学是研究治法与方剂配伍理论及其临床运用的一门学科，与临床各科紧密相连，起着沟通基础与临床的桥梁作用。

方剂是由药物组成的，是在辨证审因、确定治法之后，按照组方原则配伍而成，是中兽医临床防治疾病的主要工具。

方剂的应用也经历了一个由经验上升到理论的过程。最初人们只是针对病证来选药用方，随着方剂数量的增加以及大量的临床实践，逐渐总结出了方剂功用的一些规律性的认识，在此基础上产生了治法理论。治法就是针对不同的病证，通过辨证求因、审因论治而确定的治疗指导原则。例如针对里热证候要采用清热法，针对里寒证候要采用温里法，针对血瘀证候要采用活血化瘀法等等。治法理论一经形成之后，便成为指导人们运用成方或创制新方的重要理论依据。例如，一病畜，临床表现为恶寒发热，无汗而喘，舌苔薄白，脉浮而紧，兽医师经过四诊合参，确定其为外感风寒引起的表寒证。根据表证宜用汗法，治寒当用温药的原则，确定使用辛温解表法治疗，这时候就可以拟定治疗处方了；或者是按照治法的要求选用相应的辛温解表成方加减（如麻黄汤），或者是自行选择合适的药物，根据方剂的组成原则组成辛温解表剂，病畜如法服用，便能汗出病解，邪去畜安。由此可见，治法是组方的理论依据，方剂是治法的具体体现，即“方从法出，以法统方”，二者密切相连，构成了中兽医学辨证论治过程中的两个重要环节。

1　方剂的组成与变化

目前临床使用的方剂中，除了极少数单味药方（俗称“单方”）之外，大多是由两味或两味以上的中药所组成的复方。这是因为单味中药的作用是有限的，有些对畜体还会产生一些副作用甚至毒性反应。如果将若干味中药配合起来使用，相互之间扬长抑短，显然较之仅用一味药物治疗疾病有着更多的优越性。这种优越性具体表现在以下 3 个方面：其一，将功效相近的药物配伍同用，可以增强疗效，以适应较为严重的病证，例如大黄与芒硝合用，可以加强泻下逐邪的作用，治疗热结重证，方如大承气汤；其二，将功效不同的药物配伍同用，可以扩大治疗范围，适应较为复杂的病变，例如人参补气，麦门冬滋阴，两者合用。则有气阴双补作用，治疗气阴两虚证候，方如生脉散；其三，在使用药性峻烈或有毒性的药物时，配伍一些能够减轻或消除其毒副作用的药物，则可以避免或减轻畜体正气的损伤以及毒性反应，例如甘遂泻下逐水，但药性峻猛，且有毒性，使用时配伍大枣则能够缓和其对畜体的不利影响，方如十枣汤。由此可见，方剂通过合理妥善的配伍，可以最大限度地发挥药物的治疗作用，最大限度地降低乃至消除药物的毒性和副作用，这就是复方被广泛使用的主要原因。要达到上述要求，使得所拟方剂尽可能地切合临床病证，就必须在方剂组成原则的指导下遣药制方，并且针对具体证候加以灵活的变化。

1.1　方剂的组成

如何将一些各不相同的中药合理配伍组成方剂呢？除了准确的辨证、立法以及合理选择药物，权衡用药剂量之外，还必须遵循方剂特有的组成原则，即“君臣佐使”。其具体含义如下。

君药：即针对病因或主证，起主要治疗作用的药物，是方剂组成中不可缺少的药物。

臣药：是协助君药以加强治疗作用的药物。

佐药：有三种意义。①佐助药，即配合君药、臣药以加强治疗作用，或治疗兼病与兼证的药物；②佐制药，即起到制约君药、臣药的峻烈之性，或减轻与消除君药、臣药毒性反应的药物；③反佐药，即病重邪甚，可能拒药时，配用与君药性味相反而又能在治疗中起相成作用的药物。

使药：有两种意义。①引经药，即能引导方中诸药达到病所的药物；②调和药，即具有调和方中诸药性味的药物。

兹举麻黄汤为例对上述组成原则加以说明：麻黄汤由麻黄30克、桂枝30克、杏仁30克、甘草21克组成，主治外感风寒表实证，症见恶寒发热，无汗而喘，舌苔薄白，脉象浮紧。此证病因是外感风寒，主证为风寒束表，兼证为肺气失宣。治疗宜用散寒解表，宣肺平喘之法。方中麻黄与桂枝均味辛性温，可散寒解表，但麻黄发汗散邪力强，且药量较重，因而是在本方中针对病因和主证起主要治疗作用的药物，即君药；桂枝协助麻黄加强发汗散寒解表作用，故为臣药；杏仁降气止咳平喘，专门针对肺气失宣、咳嗽气喘的兼证而设，故为佐药（佐助药），甘草可以调和药性，属于使药中的调和药，因其味甘性缓，又能缓和麻黄、桂枝辛温发散可能导致发汗太过之弊，兼作佐药。兹将上述麻黄汤的组方意义概括如下。

麻黄汤

君药——麻黄：发汗解表，宣肺平喘。

臣药——桂枝：助麻黄发汗、解表、散寒。

佐药——杏仁：合麻黄宣降肺气，止咳平喘。

使药——甘草：调和药性，并防麻黄、桂枝过汗伤正（兼佐药）。

上述“君臣佐使”的组方原则告诉我们：①方剂中药物的作用有主次之分。其中君药至为重要，臣药次之，佐、使药物又再次之。②方刑中药物之间存在着多方面的联系，如君药与臣药之间的相互配合与协助，佐药与君药、臣药之间的协同或制约，通过相辅相成或相反相成的配伍关系，使方剂发挥最佳的治疗效应。③并非每首方剂均包含君、臣、佐、使各类药物，也不一定每味药只专任一职。这是因为“君臣佐使”是根据治疗的需要而设，除君药必不可缺外，其余类型药物并不一定必须具备，若君药药力足够，则不必以臣药辅之；若君、臣药无毒亦不峻烈时，亦无须以佐药制之；若主病药物能直达病所，则不必再加引经的使药；有的臣药兼有佐药之职，有的佐药兼有使药之能，如麻黄汤中的甘草既为使药又兼佐药之功，所以切不可机械地理解“君臣佐使”的组方原则。

遵循“君臣佐使”的组方原则配伍组方，能够使方中各药主从有序，既有明确的分工，又有密切的配合，相互之间协调制约，使方剂成为一个配伍法度严谨的有机整体，就能取得临床预期的疗效。

1.2　方剂的变化

临床在运用成方时，必须在君、臣、佐、使的组方原则指导下，结合病畜的病情、体质、年龄、性别与季节、气候等，予以加减运用，灵活化裁，所谓“师其法而不泥其方”，就是指原则性与灵活性的统一，才能使方药与病证相吻合，达到预期的治疗目的。方剂的组成变化，归纳起来主要有以下3种方式。

1.2.1　药味增减的变化

药味的增减变化有两种情况，一种是佐使药的加减，即在原方的主证与现证基本相同而兼证不同时，减去原方中某些不适宜的药物，或加上某些原方中没有但现证又需要的药物，以适应兼证的治疗要求；由于佐使药在方中的作用较为次要，其变化不至于引起原方功效的根本改变，故又称为“随证加减”。如四君子汤主治脾胃气虚证，症见气短乏力，食少便溏，舌淡苔白，脉细弱，该方由人参、白术、茯苓、炙甘草组成，功在益气补脾，若在上述症状基础上又兼见腹胀，则为脾虚不运，又兼气滞之征，可在四君子汤中加入陈皮以行气消胀（此方亦名异功散），这就是根据兼证的变化，临床上予以随证加减的运用。

另一种是君臣药的加减，或者君臣药虽然

仍保留在方剂中，但由于其他药物的增减使方中的君药及其配伍关系发生了改变，从而使方剂的功效发生根本变化。例如将麻黄汤中的桂枝换成石膏，就成为麻黄杏仁甘草石膏汤。前者以麻黄为君药，与桂枝配伍以发汗散寒，治疗风寒表实证；后者以麻黄与石膏共为君药，两药配伍共同发挥清肺平喘作用，治疗肺热咳喘证。由此可见，虽然两方仅一药之差，但因为改变了君药及其配伍关系，结果使方剂的主要功用亦随之发生改变，由辛温解表之方一变而成为辛凉解表之剂。

在临床运用成方时，可以根据不同的需要选用相应的药味增减。一般来说，"随证加减"方法易于掌握，较为符合临床兽医师用方的思路，因而在临床实践中被广泛应用。

1.2.2 药量增减的变化

这种变化是指方剂的药物组成不变，仅通过增加或减少方中药物的剂量，以改变其药效的强弱乃至配伍关系，进而影响方剂的功用。如四逆汤和通脉四逆汤均由附子、干姜、炙甘草3味药组成，且均以附子为君，干姜为臣，炙甘草为佐使。但前方附子、干姜用量相对较小，功能回阳救逆，主治阴盛阳微而致的四肢厥逆，恶寒踡卧，下利清谷，脉沉微细的证候；后方附子、干姜用量较前方俱有增加，温里回阳之功增大，能够回阳通脉，主治阴盛格阳于外而致四肢厥逆，身反不恶寒，其畜口色赤，下利清谷，脉微欲绝的证候。又如，小承气汤和厚朴三物汤，都由大黄、枳实、厚朴3味药物组成，但小承气汤中大黄用量较大，作为君药，枳实为臣药，厚朴用量较小，是大黄的1/2，为佐使，功能泻热通便，主治阳明腑实轻证；厚朴三物汤中厚朴用量独重，为君药，枳实为臣药，用量亦较小承气汤中枳实为大，大黄为佐使，用量是厚朴的1/2，全方功能行气通便，主治气滞便秘证。

由上可见，四逆汤和通脉四逆汤的药量虽有轻重之异，但其剂量的改变并未影响原方君臣佐使的配伍关系，结果其作用仅有强弱的差别，主治证候亦是轻重之异；而小承气汤和厚朴三物汤则由于药量的增减导致了配伍关系改变，因而两方的功用和主治证发生了质的改变。

2 方剂的常用剂型

将药物配伍成方之后，再根据病情的需要、药物的性质以及给药的途径，将原料药进行加工制成的型态，称为剂型。适宜的剂型是方剂治疗作用和药效发挥不可缺少的条件。

现将临床常用的方剂剂型简介如下。

2.1 汤剂

汤剂是将药物饮片混合加水浸泡，再煎煮一定时间，去渣取汁而成的液体剂型。汤剂主要供内服，如麻黄汤、桂枝汤等。外用的多作洗浴、熏蒸。汤剂的特点是吸收较快，能迅速发挥药效，特别是便于根据病情的变化而随证加减使用，适用于病证较重或病情不稳定的病畜。汤剂有利于满足辨证论治的需要，是中兽医临床运用最广泛的一种剂型。汤剂的不足之处是服用量大，某些药物的有效成分不易煎出或易挥发散失，煎煮费时而不利于危重病畜的抢救，难以服用，亦不便于携带等。

2.2 散剂

散剂是将药物粉碎，混合均匀而制成的粉末状制剂。根据其用途，分内服和外用两类。内服散剂一般是研成细粉，以温开水冲服，如七厘散、行军散等。亦有制成粗末，临用时加水煎煮去渣取汁服的，称为煮散，如银翘散、败毒散等。外用散剂一般作为外敷、掺撒疮面或患病部位，如金黄散、生肌散等；亦有作点眼、吹喉等外用的，如八宝眼药、冰硼散等。散剂的特点是制备方法简便，吸收较快，节省药材，性质较稳定，不易变质，便于服用与携带。

2.3 丸剂

丸剂是将药物研成细粉或用药材提取物，加适宜的赋形剂制成的圆形固体剂型。丸剂与汤剂相比，吸收较慢，药效持久，节省药材，体积较小，便于携带与服用。适用于慢性、虚弱性疾病，如六味地黄丸、香砂六君子丸等；也有取峻药缓治而用丸剂的，如十枣丸、抵当丸等；还有因方剂中含较多芳香走窜药物，不宜入汤剂煎煮而制成丸剂的，如安宫牛黄丸、苏合香丸等。

2.4　片剂

片剂是将药物细粉或药材提取物与辅料混合压制而成的片状制剂。片剂用量准确，体积小，易于服用。

2.5　注射剂

注射剂是将药物经过提取、精制、配制等步骤而制成的灭菌溶液、无菌混悬液或供配制成液体的无菌粉末，供皮下、肌肉、静脉注射的一种制剂。具有剂量准确，药效迅速，不受消化系统影响的特点以上剂型各有特点，临证应根据病情与方剂中药物特性酌情选用。此外，胶囊剂、灸剂、熨剂、灌肠剂、气雾剂等亦在临床广泛应用，目前中成药剂型已达 60 种左右，还有为数不少的中兽药传统产品，通过剂型改进研制成新剂型，进一步提高了临床药效。

3　常用处方

3.1　麻黄汤

【组成】麻黄 30 克，桂枝 30 克，杏仁 30 克，炙甘草 21 克。

【用法】水煎服。

【功用】发汗解表，宣肺平喘。

【临床应用】适用于外感风寒表实证。症见恶寒发热，无汗而喘，舌苔薄白，脉浮紧。感冒、流行性感冒、急性支气管炎、支气管哮喘等以恶寒无汗、咳嗽或气喘为主要表现，属风寒表实证者，可用本方治疗。若兼挟湿邪而见关节疼痛、肢体困重者，可加白术以祛湿，即麻黄加术汤；若恶寒不甚而以咳喘为主者，可去桂枝以专于宣肺平喘，即三拗汤；若恶寒无汗身痛较甚并兼里热烦躁者，可倍用麻黄以加强发汗散邪之力，再加石膏以清泄里热，即大青龙汤。

【方解】本方证由风寒束表，卫阳被遏，腠理闭塞，肺气失宣所致，治宜发汗解表，宣肺平喘之法。方中麻黄发汗解表，宣肺平喘，为君药。桂枝温经散寒，助麻黄发汗解表，为臣药。杏仁降肺气，与麻黄配伍，一宣一降，可加强止咳平喘作用，为佐药。炙甘草调和药性，又可益气补中，防止麻黄、桂枝可能发汗太过而耗伤正气，为佐使药。

【注意事项】本方为辛温发汗峻剂，风寒表证而有汗者禁用；素体阴虚、血虚、内热较重者慎用。

3.2　银翘散

【组成】金银花、连翘、芦根各 45 克，桔梗、薄荷、竹叶、荆芥穗、牛蒡子各 30 克、淡豆豉各、生甘草各 21 克。

【用法】汤煎服。

【功用】辛凉透表，清热解毒。

【临床应用】适用于温病初起，表热重证。症见发热无汗，或有汗不畅，微恶风寒，口渴，咳嗽，舌尖红，苔薄白或微黄，脉浮数。流行性感冒、急性扁桃体炎，以及流行性乙型脑炎、流行性脑脊髓膜炎、腮腺炎等初起属风热表证，见有发热，表郁无汗者，可用本方治疗。若热毒较重，高热口渴者，可加石膏、黄芩、大青叶以清热泻火解毒。

【方解】本方证为外感风热，卫气被郁，肺失清肃所致，治宜疏风透表，清热解毒之法。方中重用金银花、连翘，清热解毒，并兼透表之功，共为君药。薄荷、牛蒡子辛凉疏散风热，解毒利咽；荆芥穗、淡豆豉辛温发散，且温而不燥，既可加强本方散邪解表之力，又无温燥伤津之弊，四药共助君药以加强解表散邪之功，同为臣药。芦根、竹叶清热生津，桔梗宣肺止咳，皆为佐药。生甘草清热解毒，调和药性，合桔梗擅清利咽喉，为佐使药。

【注意事项】风热表证虽见有恶寒无汗，但发热不甚、口不渴者，不宜使用本方。

3.3　白虎汤

【组成】石膏 180 克（先煎），知母 60 克，炙甘草 45 克，粳米 30 克。

【用法】水煎至米熟汤成，去渣分服。

【功用】清热生津。

【临床应用】适用于阳明经证。症见壮热口赤，烦渴引饮，汗出恶热，脉洪大有力。感染性疾病，如大叶性肺炎、流行性乙型脑炎、流行性出血热、牙龈炎等出现阳明气分热盛的“四大”（大热、大渴、大汗、脉洪大）特征者，可用本方治疗。若热盛耗伤气津，兼见脉大无力，饮不解渴者，可加党参以益气生津；若温热病气血两燔，身发斑疹者，可加生地黄、水牛角以清热凉血；若热盛动风，兼四肢抽搐者，可加羚羊角、钩藤以熄风止痉；若里热挟湿，如湿温病热重于湿，或风湿热痹，关节红肿疼痛者，可加苍术以燥湿，即白虎加苍术汤。

【方解】本方证由气分邪热炽盛，灼伤津液所致，治宜清热生津之法。方中石膏辛甘大寒，清热泻火而不伤津液，为君药。知母苦寒质润，助君药清热泻火，又能滋阴生津，为臣药。粳米、炙甘草益胃护津，又可防石膏大寒伤胃，为佐药。炙甘草兼能调和药性，为使药。

【注意事项】若脉象洪大而重按无力，月.舌质淡者，乃血虚发热，禁用本方。

3.4 黄连解毒汤

【组成】黄连 45 克，黄芩 60 克、黄柏各 30 克，栀子 45 克。

【用法】水煎服。

【功用】泻火解毒。

【临床应用】适用于三焦火毒热盛证。症见高热，口渴引饮，或身热下利，或痈肿疔毒，舌红苔黄，脉数有力。败血症、脓毒血症、痢疾、肺炎、泌尿系感染、流行性脑脊髓膜炎、流行性乙型脑炎以及其他感染性疾病属火毒炽盛者，均可用本方治疗。若兼便秘者，可加大黄以通便泻火，导热毒下行；若兼吐血、衄血、发斑者，可加生地黄、玄参、牡丹皮以清热凉血；若出现黄疸者，可加茵陈蒿、大黄以清热祛湿退黄。

【方解】本方证由热毒壅盛于三焦所致，治宜大寒泻火解毒之法。方中黄连长于泻心火，兼泻中焦之火，为君药。黄芩泻上焦之火，黄柏泻下焦之火，栀子通泻三焦，导热下行，均为臣、佐药。四药皆大苦大寒之品，配伍同用，相得益彰，直折火毒之功颇著。

【注意事项】本方苦寒之药群聚，苦燥劫阴，苦寒败胃，非火毒炽盛者不可服用，且应中病即止。若阴伤较著，舌质光绛者禁用。

3.5 理中汤

【组成】党参 60 克，干姜 60 克，炙甘草 30 克，白术 45 克。

【用法】水煎服。

【功用】温中祛寒，补气健脾。

【临床应用】适用于脾胃虚寒证。症见腹疼痛，喜温喜按，畏寒肢冷，食欲不振，呕吐腹泻，或阳虚失血，出血不多，血色黯淡，舌淡苔白，脉象沉细。急性胃肠炎、慢性胃肠炎、胃及十二指肠溃疡等以吐、利、冷、痛为特征，属脾胃虚寒者，可用本方治疗。若寒甚者，重用干姜，或加附子以助温中祛寒之力，即附子理中丸；若用于阳虚失血证，宜将干姜改为炮姜；若虚甚者，重用党参以益气补脾；若下利重者，重用白术以健脾助运止利；若呕吐甚者，可加吴茱萸、生姜以温胃止呕；若脾虚不运，聚湿生痰者，可加半夏、茯苓以温化痰饮，即理中化痰丸。

【方解】本方证由中阳不足，脾胃虚寒，运化失司所致，治宜温中祛寒，补气健脾之法。方中干姜温中助阳祛寒，为君药。党参益气补脾助运，为臣药。白术补气健脾燥湿，合干姜以温运脾阳，配党参可益气补脾，为佐药。炙甘草甘温补中，既可助三药温补脾胃之力，又能调和药性，为佐使药。

3.6 大承气汤

【组成】大黄 60 克，厚朴 90 克，枳实 60 克，芒硝 120 克。

【用法】枳实、厚朴先煎，大黄后下，汤成去滓，溶入芒硝。

【功用】峻下热结。

【临床应用】适用于阳明腑实证。症见大便不通，腹痛拒按，午后热甚，口渴引饮，舌红苔黄燥起刺或焦黑燥裂，脉沉实。急性单纯性肠梗阻、粘连性肠梗阻、急性胆囊炎、急性胰腺炎，以及热病过程中出现高热惊厥，甚则发狂而见大便不通，苔黄脉实者，均可用本方治疗。若热盛耗气，兼少气倦怠者，可加党参以补气，防止峻下而致气脱；若热结伤阴，大渴引饮，舌干红少苔者，可加玄参、生地黄以滋阴生津，润燥通便。

【方解】本方证由邪热积滞结于大肠，腑气不通所致，治宜泻下热结，攻积通腑之法。方中大黄泻热通便，荡涤肠胃积滞，为君药。芒硝软坚润燥，助大黄以加强泻热通便之功，为臣药。枳实、厚朴行气除满，并助大黄、芒硝推荡积滞，加速热结排泄，均为佐药。四药配伍，泻下热结之力颇为峻猛。若减去芒硝则成为轻下热结的小承气汤，减去枳实、厚朴，加甘草则成为缓下热结的调胃承气汤，两者与上方合称三承气汤，临床常根据阳明腑实证候的轻重缓急酌情选用。

【注意事项】阳明腑实而热结不甚，或素体气虚阴亏等慎用本方。

3.7　小柴胡汤

【组成】柴胡 45 克，黄芩 45 克，党参 30 克，炙甘草 20 克，半夏 45 克，生姜 30 克，大枣 20 枚。

【用法】水煎服。

【功用】和解少阳。

【临床应用】适用于伤寒少阳证。症见往来寒热，舌苔薄白，脉弦。若有感冒、流行性感冒、慢性肝炎、急性胆囊炎、慢性胆囊炎、胸膜炎、肾盂肾炎、产后感染、胃溃疡等上述病症，属伤寒少阳证者，可用本方治疗。若心烦而不呕吐者，为热聚于胸，宜去半夏、党参，加瓜蒌以清热理气宽胸；若热伤津液，兼口渴者，宜去半夏，加天花粉以生津止渴；若肝气乘脾，兼腹痛者，宜去黄芩，加白芍药以柔肝缓急止痛。

【方解】本方证由邪犯少阳，邪正相争于表里之间所致，治宜和解少阳之法。方中柴胡透达半表之邪，疏畅少阳气机之壅滞，为君药。黄芩清泄半里之热，为臣药。柴胡合黄芩一散一清，共解少阳之邪，为和解少阳的要药。半夏、生姜和胃止呕，党参、炙甘草、大枣益气扶正，实里以防邪气内传，均为佐药。甘草调和药性，兼作使药。

【注意事项】方中柴胡升散，黄芩、半夏性燥，故少阳证而见阴虚血少者慎用。

3.8　四君子汤

【组成】党参、白术、茯苓各 45 克，炙甘草 30 克。

【用法】水煎服。

【功用】益气健脾。

【临床应用】适用于脾胃气虚证。症见口色萎白，气短乏力，食少便溏，舌淡苔白，脉细弱。慢性胃炎、胃及十二指肠溃疡等见有上述症状，属脾胃气虚者，可用本方治疗。若兼气滞者，可加陈皮以行气化滞，即异功散；若兼痰湿内阻，咳嗽痰多者，可加陈皮、半夏以燥湿化痰，即六君子汤；若兼寒湿中阻，胃气失和，气机不畅者，可加木香、砂仁、陈皮、半夏以行气畅中，和胃止呕，即香砂六君子汤。

【方解】本方证由脾胃气虚，运化乏力所致，治宜益气健脾之法。方中党参益气健脾养胃，为君药。白术益气健脾燥湿，为臣药。茯苓渗湿健脾助运，为佐药。炙甘草益气补中，调和药性，为使药。方中党参、白术、甘草合用以益气补脾养胃；白术、茯苓相伍以健脾祛湿助运。四药配伍，补而不滞，温而不燥，共成甘温平补脾胃之剂。

3.9　四物汤

【组成】熟地黄 45 克，当归 45 克，白芍药 45 克，川芎 30 克。

【用法】水煎服。

【功用】补血和血。

【临床应用】适用于营血虚滞证。症见口舌色淡，脉细。慢性湿疹、荨麻疹等慢性皮肤病，骨伤科疾病等属营血虚滞者，均可用本方治疗。若兼气虚，神倦气短者，可加党参、黄芪以补气生血，即圣愈汤；若血瘀明显，经行腹痛较甚者，可将白芍药改为赤芍药，再加桃仁、红花以加强活血祛瘀之力，即桃红四物汤；若血虚有寒，腹痛喜温者，可加肉桂、炮姜、吴茱萸以温通血脉；若血虚有热，口干咽燥者，可将熟地黄改为生地黄，再加黄芩、牡丹皮以清热凉血。

【方解】本方证由营血亏虚，血行不畅所致，治宜补血和血之法。方中熟地黄滋阴养血，为君药。当归补血养肝，和血调经，为臣药。白芍药养血柔肝和营，川芎活血行气，为佐药。方中熟地黄、白芍药专于滋补阴血，当归、川芎补中有行，四药配伍，补中有行，补血而不滞血，行血而不伤血，既有养血补虚之功，又具和血调经之效。

3.10 平胃散

【组成】苍术45克，厚朴、陈皮各45克、生姜30克、甘草30克、大枣30枚。

【用法】水煎服。

【功用】燥湿健脾，行气和胃。

【临床应用】适用于湿滞脾胃证。症见腹胀，草料减少，呕吐，嗳气，肢体沉重，怠惰嗜卧，舌苔白腻而厚，脉缓。慢性胃炎、消化道功能紊乱、胃及十二指肠溃疡等以腹胀、苔白腻为主症，属湿滞脾胃者，可用本方治疗。若湿蕴化热，舌苔黄腻者，可加黄连、黄芩以清热燥湿；若兼里寒，腹冷便溏，畏寒喜温者，可加干姜、草豆蔻以温化寒湿；若兼脾胃不和，饮食难消者，可加神曲、山楂、麦芽以消食化滞；若湿盛泄泻者，可加茯苓、泽泻以利湿止泻。

【方解】本方证由湿阻中焦，气机不畅，脾失健运，胃失和降所致，治宜燥湿健脾，行气和胃之法。方中重用苍术燥湿运脾，为君药。厚朴燥湿行气，既可协苍术加强燥湿作用，又能理气以除腹胀，为臣药。陈皮行气和胃，合厚朴以助行气除胀之力，合苍术更增燥湿和中之功，为佐药。炙甘草调和药性，为使药。用时少加生姜以和胃，加大枣以补脾，从而进一步加强了本方调和脾胃的作用。

【注意事项】脾胃虚弱或素体阴虚者，慎用本方。

3.11 五苓散

【组成】猪苓45克，泽泻45克，白术30克，茯苓45克，桂枝30克。

【用法】水煎服。

【功用】利水渗湿，温阳化气。

【临床应用】适用于水湿内停证。症见水肿，泄泻，小便不利，舌淡，苔白滑，脉濡。肾炎所引起的水肿，急性肠炎、尿潴留等属水湿内停者，可用本方治疗。若水湿壅盛者，可与五皮散（陈皮、茯苓皮、生姜皮、桑白皮、大腹皮）合用以增强利水消肿之效。

【方解】本方证由脾失健运，水湿内停，膀胱气化不利所致，治宜利水渗湿，温阳化气之法。方中重用泽泻利水渗湿，为君药。茯苓、猪苓甘淡利湿，茯苓又能健脾，两药协助君药加强利水消肿之功，同为臣药。白术合茯苓则健脾运湿之力更著，桂枝助膀胱之气化，均为佐药。原书以本方治疗太阳表邪未解，内传膀胱以致气化不行，小便不利的蓄水证，故用桂枝还有解表散邪之意。诸药配伍，重在健脾利水，故凡脾失健运，水湿内停之证皆可治之。

【注意事项】本方药性偏温，水湿化热者不宜使用。

3.12 二陈汤

【组成】半夏30克，橘皮30克，茯苓30克，炙甘草24克。

【用法】水煎服。

【功用】燥湿化痰，理气和中。

【临床应用】适用于湿痰咳嗽或呕吐。症见咳嗽，痰多色白易咯，恶心呕吐，肢体倦怠，舌苔白腻，脉滑。慢性支气管炎、肺气肿、慢性胃炎、妊娠呕吐、神经性呕吐等见有痰白量多，属湿痰为患者，可用本方治疗。本方又是治痰的基本方，可加减应用于多种痰证。若为热痰，可加黄芩、胆星以清热化痰；若为寒痰，可加干姜、细辛以温化寒痰；若为风痰，可加天南星、竹沥以熄风化痰；若为食痰，可加莱菔子、神曲以消食化痰。

【方解】本方证由脾失健运，湿邪内停，聚而成痰，气机失畅所致，治宜燥湿化痰，理气和中之法。方中半夏辛温性燥，善能燥湿化痰以止咳，并可降逆和胃以止呕，为君药。橘皮助半夏化痰与和胃之力，并可行气以使气顺痰消，为臣药。茯苓健脾渗湿，治生痰之源，与君、臣药相伍可收标本兼治之功，为佐药。炙甘草和中益脾，调和药性，为使药。煎药时加生姜，取其降逆化饮，助半夏、橘皮行气消痰，和胃止呕，并制半夏之毒；再用少许乌梅收敛肺气，与君、臣药相伍，散中有收，可使祛痰而不伤正。诸药合用，燥湿化痰为主而兼健脾理气，体现了治疗痰病的基本大法，故本方又为祛痰的通用方剂。方中半夏、橘皮宜选用较陈久者以减其燥散之性，故方以“二陈”为名。

【注意事项】本方药性温燥，故阴虚肺燥及咯血者禁用。

第四讲

中兽医针灸学

兽医针灸疗法是古代中国人民创造的一种治病方法，其广泛使用的历史已有数千年之久，是中兽医治病的主要手段之一，对中国畜禽的医疗保健起着重大的作用。

1 腧穴总论

腧穴的“腧”与“输”义通，即有输注的含义；“穴”含有“孔”、“隙”的意思。腧穴是畜体脏腑经络气血输注出入于体表的部位。腧穴的名称说明了腧穴的两个基本特征：其实质是脏腑经络之气的输注部位，其部位所在多为分肉筋骨之间的空隙之处。

1.1 腧穴分类

根据腧穴的不同特点，通常可将其分为经穴、奇穴、阿是穴三大类。

1.1.1 经穴

经穴又称十四经穴，是指分布于十二经脉及任、督两脉上的腧穴。

1.1.2 奇穴

奇穴又称经外奇穴，是指十四经以外，有一定的位置和名称，对某些病证有专门的治疗作用的腧穴。

1.1.3 阿是穴

阿是穴是指病痛局部或与病痛有关部位的压痛点。其特点是：无固定的位置，无穴位名称、无归经。

1.2 腧穴作用

腧穴具有近治、远治和特殊作用三大特点。

1.3 取穴方法

准确地定取腧穴是兽医针灸治疗疾病的前提，而要快速准确地定取穴位，必须采用恰当的取穴方法。常用的定位方法体表标志法、同身寸法、简便取穴法和揣穴法。

1.3.1 体表标志法

包别体表固定标志法和体表活动标志法两种。

1.3.2 简便取穴法

这是古今医家在多年的临床实践中总结出的简便、快捷的取穴方法，如两耳尖直上与头部前后正中线的交点取百会；两手虎口交叉取列缺；两手自然下垂于大腿外侧，当中指尖抵达处取风市等。

1.3.3 揣穴

揣穴是医者用手指在穴位部位上下左右按压，以揣摸腧穴的方法。腧穴大多位于骨缝、肌肉间隙及一些凹陷中，定取腧穴时，就要在相应的部位揣摸，以找到骨缝、间隙等；腧穴具有反应病候的功能，在畜体发生病变时，寻找这些反应点进行兽医针灸治疗，往往能取得满意的疗效。

2 针具

临床常用的针具有圆利针、毫针、三棱针、宽针和火针等，多由不锈钢制成。见图 4－1。

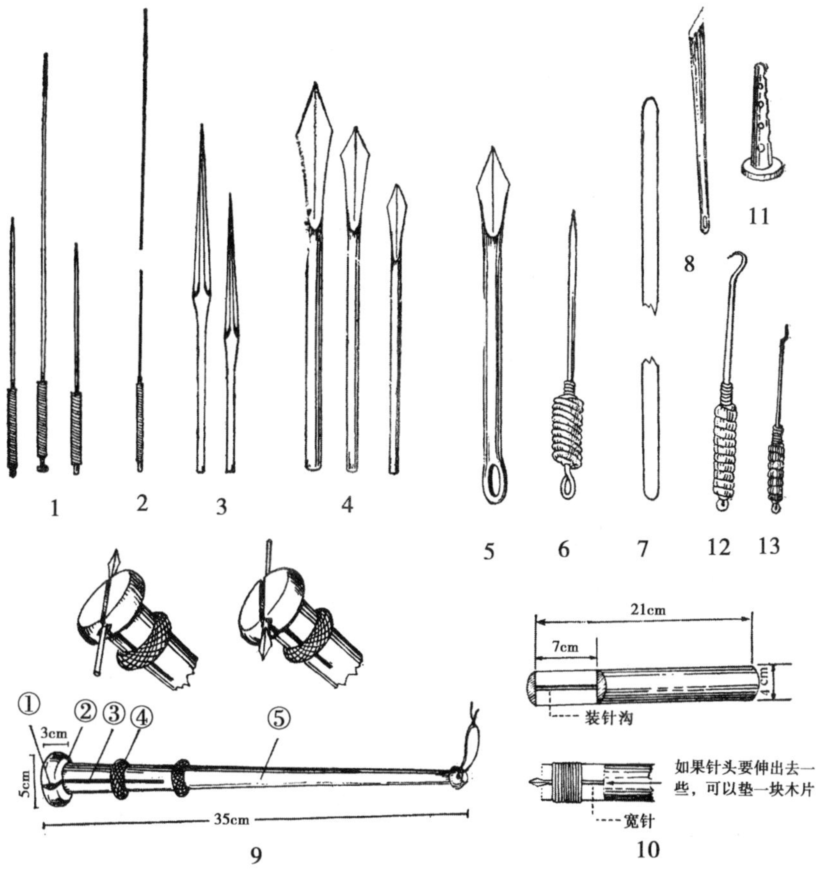

图 4－1　针具

1. 圆利针；2. 毫针；3. 三棱针；4. 宽针；5. 穿黄针；6. 火针；7. 夹气针；8. 眉针刀；9. 持针器（1. 针锤；2. 针棒；3. 锯缝；4. 活动箍；5. 锤柄）；10. 其他针治用具；11. 宿水管；12. 玉堂钩；13. 三弯针

3　兽医针灸操作方法

3.1　保定

在兽医针灸疗法前，要根据扎针部位情况，对病畜进行适当的保定。

3.2　消毒

包括针具、针刺部位以及兽医双手的消毒。

3.3　针刺方法

3.3.1　毫针、圆利针刺法

3.3.1.1　进针方法

针刺施术时，一般是右手持针（称“刺手”），进行针刺的主要操作；左手配合（称“押手”），切按所刺部位或夹持针身，以帮助右手进针，双手协同操作将针刺入皮肤。临床上常用的进针方法有以下几种。

指切进针法：用左手拇指或食指的指甲切按在穴位处，右手将针紧靠左手指甲缘刺入腧穴。此法适用于短针的进针（图 4－2.1）。

夹持进针法：用左手的拇指和食指拿消毒干棉球夹住针身的下端，将针尖轻置于穴位的皮肤表面，然后右手与左手同时用力，将针刺入皮肤，此法适用于长针的进针（图 4－2.2）。

舒张进针法：以左手的拇指和食指将穴位处的皮肤向两侧撑开而使皮肤绷紧，右手将针从两指之间刺入。此法用于皮肤松弛部位的进针（图 4－2.3）。

提捏进针法：用左手拇指、食指将穴位处的皮肤捏起，右手将针从捏起处刺入。此法用于皮薄肉少部位的进针（图 4－2.4）。

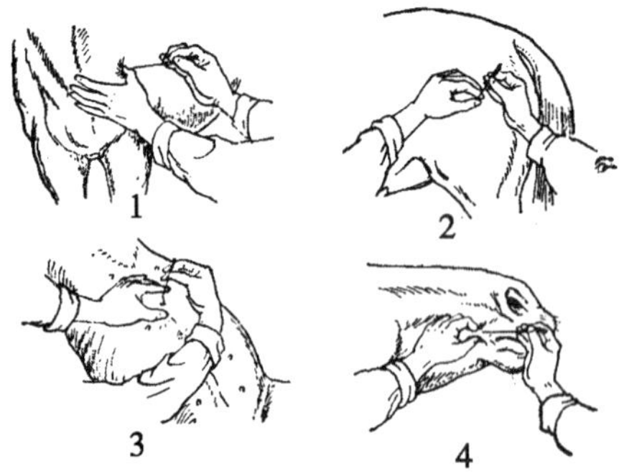

图4-2　进针方法

3.3.1.2　针刺的角度、深度

对同一个腧穴，如果针刺的角度、深度不同，所刺及的组织、产生的针刺感应和治疗效果会有明显的差异；对不同的腧穴，由于所在部位的解剖特点各异，针刺的角度和深度就应有所区别。所以，掌握正确的针刺角度与深度，对于获得预期的针刺治疗效果、防止意外事故的发生等，都具有重要的意义。

针刺角度：指进针时针身与皮肤表面间的夹角。一般分为三种角度。

直刺：指针身与皮肤表面呈90°角左右垂直刺入，适用于大部分腧穴。

斜刺：指针身与皮肤表面呈45°角左右斜向刺入，适用于内有重要组织或脏器，或不宜直刺、深刺的腧穴。

平刺：指针身与皮肤表面呈15°角左右沿皮刺入，又称“横刺”、“沿皮刺”。适用于皮薄肉少部位的腧穴。

针刺深度：指针身刺入肌肤的深浅程度。临床上主要根据腧穴部位、病畜体质、疾病情况等因素决定针刺的深浅。

基本手法

捻转法：进针至一定深度后，（以右手的拇指和食、中二指持针柄）来回捻转针柄。

提插法：进针至一定深度后，将针身在肌肉中进行上提、下插的反复操作。

捻转法和提插法，是构成针刺手法的基本操作方法，两者既可单独使用，也可结合运用。捻转的角度、提插的幅度，以及操作的频率、持续时间等量度，要根据病畜的体质、病情和所刺部位而定。

3.3.1.3　留针和出针

留针：即在施用针刺手法后，将针留置于腧穴内。留针能加强针刺的作用，对得气较慢较弱者还可起到候气的作用。留针与否、留针时间的长短，主要依病情而定。一般病证，施用一定针刺手法后即可出针，或留针10～20分钟；某些特殊病证，可适当延长留针时间，并在留针过程中间歇行针，以增强疗效。

出针：出针时，左手用消毒干棉球轻轻压住针孔周围的皮肤，右手微捻针柄，慢慢将针上提至皮下，而后迅速拔出，随即用消毒干棉球按住针孔，以防出血。最后，医者应检查针数，以防遗漏。

3.3.1.4　异常情况的处理

滞针：是指医者感觉针下非常紧涩，难以进行捻转和提插等操作，处理方法须视不同原因而定，如果是由于病畜紧张致局部肌肉过度收缩者，适当延长留针时间，或在局部轻柔按压，或在邻近处再刺一针；如果是因医者单向捻针所致者，应反方向捻针，再轻轻提插，即可消除滞针。

弯针：如果在针刺后出现针身弯曲，不可再行捻转等操作，应将针顺着弯曲方向缓缓退出。医者针刺操作要轻柔，不可用力过猛；留针时针柄不得受外物碰撞和压迫，以避免发生弯针。

断针：若出现针身断于体内的情况，医者须镇静。如果针身尚有部分露出于体外，可用镊子取出；如果断端与皮肤相平或稍陷，可用左手拇、食两指在针旁垂直向下按压皮肤，以使针体露出，右手用镊子将针取出；如果针体已进入深部，应手术取出。针刺前要做好预防工作，应认真检查针具；医者针刺操作要轻柔，不可用力过猛等，以防断针。

血肿：是指针刺部位出现皮下出血而致肿痛。如果出针后局部青紫或疼痛，应立即用消毒干棉球按压针孔片刻，即能止血；若是微量的皮下出血所致局部小块青紫，可自行消退；若局部肿胀疼痛较剧，青紫面积大，须即时冷敷止血，待血止后改用热敷或轻轻揉按以助消散。

气胸：在针刺缺盆、胸、背、腋、胁等部位的腧穴时，如果直刺过深，刺伤胸膜和肺脏，则使空气进入胸腔而致气胸。表现为病畜突感胸闷、胸痛、气短；甚则呼吸困难，唇发绀、出汗、血压下降等。体检时胸部叩诊过度反响，肺泡呼吸音减弱或消失，甚者气管向健侧移位。X线胸部透视可确诊气胸的程度。对轻者可给予镇咳、抗菌药物等对症处理，并严密观察；重者应立即采取抢救措施。

3.3.2　三棱针和宽针刺法

三棱针是一种针身呈三棱形、针尖锋利的针具，古代称“锋针”，用于刺络放血。宽针的针头呈矛尖状，针刃锋利，分为大、中、小三种，多用于马牛放血疗法。

3.3.3　火针刺法

火针疗法是用特制的针具，烧热后刺入一定的穴位，以治疗疾病的一种方法，是兽医临床广泛使用的一种传统的治疗技术。因为在针刺的同时使穴位的局部组织发生较深的灼伤灶，所以在一定的时间内保持对穴位的刺激作用。

实践证明，火针疗法对虚寒证有特殊的治疗效果，如风湿症及慢性腰肢病等。

3.3.4　艾灸方法

灸法，是利用灸火刺激畜体以防治疾病的一种疗法，为针灸疗法的重要组成部分。施灸的材料多用艾叶，其味香而易燃，临床所用是由干燥艾叶加工制成的艾绒。

3.3.5　电针

电针是在刺入畜体的（毫）针通以微量的脉冲电流，利用针、电两种刺激的综合作用治疗疾病的——种疗法。应用器材主要是电针器，种类较多，多以半导体元件装制，采用振荡发生器，输出接近畜体生物电的低频脉冲电流。

3.3.6　穴位注射

穴位注射也称“水针”，是将药液注入穴位或阳性反应点，通过针刺和药物的双重作用治疗疾病的一种疗法。

4　选穴与配穴

针灸的治疗效应是通过针刺、艾灸刺激腧穴而产生的，所以腧穴的选择及其配合运用与针灸治疗效果密切相关。正确地选穴和配穴，是达到针灸治病目的的重要环节。

4.1　选穴方法

选穴的主要原则和方法是循经取穴，其次是根据病情选用有针对主治作用的腧穴。

4.1.1　近部选穴

近部选穴是在靠近病变的部位选取腧穴的方法。因为腧穴都能够治疗其所在部位和邻近周围处的病证。

4.1.2　远部选穴

远部选穴是在距病变处的较远部位选取腧穴的方法，多是取四法是基于经络的联系。例如：腹泻取后三里等。

4.1.3　左右选穴

左右选穴或称交叉选穴，是选取与患侧相反的一侧的腧穴，所谓“左病取右，右病取左”方法。例如：左侧踝关节扭伤，可选用右侧踝部腧穴。

4.1.4　随证选穴

随证选穴是根据病证选取具有相应主治作用的腧穴。主要用于全身性病证，其临床表现不限于某一局部。根据腧穴的主治特点，以及中兽医辨证论治的理论方法来运用。例如：高热取大椎，感冒取天门穴，咳嗽取肺俞或肺攀穴，便秘取支沟等。

4.2　配穴组方

针对病证将两个以上的腧穴配合使用，组成针灸治疗的用穴处方。配穴的目的主要有两方面：一是加强主治作用，二是具有综合、全面的治疗作用。因此，其配穴方法也就有两大类：一是将主治作用相同或相近的腧穴配合使用，二是根据病情或辨证配用有相应主治作用的腧穴。

5　影响针灸疗效的主要因素

影响针灸疗效的因素是多方面的。广义上讲，构成针灸诊治过程的各个环节、各个部分，都会对针灸治疗的效果产生影响。在诸多的影响因素中，治法选择恰当、正确用穴组方和刺灸操作等，对提高针灸疗效具有更为重要的、直接的影响。

6 马的针灸穴位

6.1 头部穴位

头部穴位共36个，其中大风门、三江、分水、锁口、通关、耳尖等穴，临床较为常用（图4-3、图4-4、图4-5、图4-6）。

#1. 大风门

【位置】头顶部，主穴在门鬃下缘，二副穴各斜下旁开约3厘米处，共三穴。

【解剖】主穴在顶骨外矢状嵴分叉处，二副穴在顶外嵴上，与主穴成正三角形，皮下为耳肌，有颞浅动、静脉和耳睑神经分布。

【取穴】主穴在门鬃下方的骨棱上，两副穴各沿骨棱向下3厘米处。

【操作】毫针由下而上沿皮下刺入3厘米，火针刺入2厘米；艾灸或火烙至皮肤呈焦黄色。

【主治】破伤风，脑黄，脾虚湿邪，心热风邪。

#13. 三江

【位置】内眼角下方约3厘米处的血管上，左右侧各一穴。

【解剖】内眼角下方约3厘米处的眼角静脉上。有眼角动脉伴行，并有面神经分布。

【取穴】使马低头，在内眼角下方的血管即可显露。

【操作】三棱针或小宽针由下而上沿血管方向刺入1厘米，出血。

【主治】冷痛，肚胀，月盲，肝热传眼。

#21. 分水

【位置】上唇外面，正中旋毛处，一穴。

【解剖】穴部皮下是口轮匝肌，有上唇动、静脉，上唇神经和颊上神经分布。

【取穴】上唇外面，正中旋毛处是穴。

【操作】紧握上唇，小宽针或三棱针直刺2厘米，出血；毫针直刺3厘米。

【主治】冷痛，黑汗风，歪嘴风，中暑。

#29. 锁口

【位置】口角后上方2厘米处，左右侧各一穴。

【解剖】口角后上方，口轮匝肌外缘，颊肌与口轮匝肌相接处，有颊肌动、静脉和颊上神经分布。

【取穴】口角后上方一指取穴。

【操作】毫针顺口角向后上方透刺开关穴；火针刺入3厘米，或烧烙3厘米长。

【主治】破伤风，歪嘴风。

#34. 通关

【位置】舌下两旁血管上，左右侧各一穴。

【解剖】舌系带两侧黏膜下的舌下静脉上，有舌下动脉和舌下神经分布。

【取穴】将舌拉出口外，向上翻转，紧握舌体，舌系带两侧舌下静脉就显露出来。

【操作】三棱针或小宽针刺入1厘米，出血，并用冷水冲洗。

【主治】舌疮，木舌，胃热，黑汗风。

#35. 耳尖

【位置】耳尖部血管上，左右耳各一穴。

【解剖】耳背侧，耳大静脉的内、中、外支汇合处，有耳后动脉伴行以及耳后神经分布。

【取穴】紧握耳根，耳背部静脉即可显露，在耳尖部取之。

【操作】小宽针或三棱针顺血管刺入1厘米，出血。

【主治】冷痛，感冒，中暑。

6.2 躯干及尾部穴位

躯干及尾部穴位共50个，其中颈脉、九委、大椎、百会、肾俞，关元俞、脾俞、带脉、后海等穴，临床较为常用。

#39. 九委（上上委、上中委、上下委、中上委、中中委、中下委、下上委、下中委、下下委）

【位置】颈部两侧各九穴，九委第一穴称上上委，位于伏兔穴后方（耳后）约3厘米处，距鬃下缘约3.5厘米处；最后一穴称下下委，在肩胛骨前角（膊尖穴）前方约4.5厘米，距鬃下缘约5厘米处。在此二穴之间作八等分，为其余七穴，排列于颈的背侧部，各穴都在肌沟中，相距约6厘米，与鬃线距离两端几穴稍近，中间穴稍远，呈弧线状分布在颈部两侧。

【解剖】上上委和上中委在头后斜肌内，上下委及中上委在菱形肌下缘，夹肌深部及头半棘肌中，中中委至下中委均在斜方肌下、菱形肌腹侧缘的夹肌中，各穴沿菱形肌下缘分布成弧形，九穴之间距离相等。各穴都有颈深动、静脉和颈神经背侧支分布。

【取穴】按穴位的自然位置取穴。

【操作】毫针直刺4.5～6厘米；火针刺入2.5～3厘米。

【主治】项脊怯。

#40. 颈脉（鹘脉）

【位置】颊下6厘米的大血管上，左右侧各一穴。

【解剖】颈静脉沟上、中1/3交界处的颈静脉上，背侧是臂头肌，腹侧是胸头肌，深部是肩胛舌骨肌，有颈总动脉并行和颈神经腹支分布。

【取穴】在下颌骨角后方四指处取之。

【操作】将马头高抬，在颈基部拴一细绳，缠个活扣，使颈静脉怒胀，用装有大宽针的针锤，对准穴位，急刺1厘米，出血，术后松开绳扣，血流停止。

【主治】热性病，肺黄，脑黄，中暑，中毒，五攒痛，舌疮，遍身黄。

#43. 大椎

【位置】肩胛前缘的颈背侧，向前数第一个骨间隙中，一穴。

【解剖】第七颈椎与第一胸椎棘突间的凹陷中，项韧带索状部，棘上韧带深部的棘间肌和棘间韧带内，有颈横动、静脉及胸神经背侧支分布。

【取穴】当家畜作颈部上、下运动时，颈椎可随颈部运动而活动，胸椎则不活动，从而确定第一胸椎，在第一胸椎与第七颈椎之间的凹陷处，即是穴。

【操作】毫针或圆利针稍向前下方刺入6～9厘米。

【主治】感冒，咳嗽，腰背风湿，发热，癫痫。

#59. 脾俞

【位置】倒数第三肋间，距脊梁12厘米的凹陷中，左右侧各一穴。

【解剖】第十五肋间，针刺在背阔肌深部、背最长肌与髂肋肌之肌沟中，有肋间动、静脉和胸神经的背侧支的外支分布。

【取穴】在三焦俞后方同高位的第一个凹陷中取之。

【操作】同厥阴俞穴。

【主治】胃冷吐涎，肚胀，结症，泄泻，冷痛。

#62. 关元俞

【位置】最后肋骨后缘，距脊梁12厘米的凹陷中，左右侧各一穴。

【解剖】第十八肋骨后缘与第一腰椎横突顶端之间。针刺在胸腹皮肌、背最长肌与髂肋肌之肌沟中，有最后肋间动、静脉和最后肋间神经分布。

【取穴】在大肠俞后方同高位的第一个凹陷中取之。

【操作】同厥阴俞穴。

【主治】结症，肚胀，泄泻，冷痛。

#68. 百会

【位置】腰椎后、荐椎（节脊骨）前的凹陷中，一穴。

【解剖】最后腰椎棘突和第一荐椎棘突顶端之间的凹陷处，有腰动、静脉及腰神经背侧支分布。

【取穴】在腰荐十字部结合处取之。

【操作】火针或圆利针直刺3～4.5厘米，毫针直刺6～7.5厘米。

【主治】腰胯风湿，闪伤腰胯，肚胀，风症，泄泻，虚劳。

#69. 肾俞

【位置】百会穴正旁6厘米处，左右侧各一穴。

【解剖】臀部，刺入臀中肌肉，有臀前动、静脉和臀前神经分布。

【取穴】在百会侧方6厘米处取之。

【操作】火针或圆利针直刺3～4.5厘米；毫针刺入6厘米，亦可向肾棚、肾角透刺，且三穴可互相透刺。

【主治】腰胯风湿，腰痿。

#73. 带脉

【位置】肘后约 6 厘米处的血管上，左右侧各一穴。

【解剖】胸侧壁，肘后与第 7 肋骨相对的胸外静脉上，有胸外动脉、胸外神经伴行。

【取穴】在肘后 6 厘米处的血管上取之。

【操作】以大宽针或中宽针顺血管刺入 1 厘米，出血。

【主治】肠黄，冷痛，黑汗。

#78. 脿俞

【位置】右侧肷窝部，距脊梁约 21 厘米，距最后肋骨 4.5 厘米的交点上，一穴。

【解剖】腹壁右侧，髋关节至最后肋骨之间水平线上的中点处，穿通皮肤、胸腹皮肌售腹内外斜肌、腹横肌及腹膜，刺入盲肠底部。腹壁有旋髂深动、静脉和髂腹股沟神经分布。

【取穴】从最后肋骨到髋结节中部水平线的正中点取之。

【操作】巧治，术部剪毛消毒后，用大宽针将皮肤直切 1 厘米创口，用套管针向对侧肘头方向迅速刺入，拔出内针，使气体徐徐排出，起针前插入内针，压定针孔处的皮肤拔针，消毒。

【主治】肚胀。

#82. 后海

【位置】肛门上方，尾根下的凹陷中，一穴。

【解剖】肛门与尾根之间的凹陷处，刺入肛门外括约肌与尾肌之肌间隙内，有尾动、静脉及直肠后神经分布。

【取穴】提起尾根，在肛门与尾根中间陷中处取之。

【操作】圆利针或火针向前上方刺入 9 厘米；毫针刺入 12 ~ 18 厘米。

【主治】结症，肚胀，腹泻，直肠麻痹。

6.3 前肢穴位

前肢穴位共 32 个，其中膊尖、抢风、胸堂、肘俞、前缠腕，前蹄头等穴，临床较为常用。

#87. 膊尖

【位置】肩胛软骨前角的凹陷中，左右侧各一穴。

【解剖】肩胛软骨与肩胛骨前角结合部的凹陷中，刺入颈斜方肌深部、菱形肌与颈下锯肌之间的肌间隙内。穴位分布着颈横动、静脉，颈神经背侧支和肩胛上神经。

【取穴】沿肩胛软骨前上缘向下按取（一般用食指取穴），在肩胛骨前缘，正当肩胛软骨与肩胛骨前角结合部的凹陷中。

【操作】圆利针或火针沿肩胛骨内缘稍向后下方刺入 3 ~ 5 厘米，毫针刺入 10 ~ 12 厘米。

【主治】肺气把膊，寒伤肩膊痛。

#94. 抢风

【位置】臂骨节后下方 15 厘米处凹陷中，左右侧各一穴。

【解剖】臂骨三角肌隆起后上方的凹陷处，刺入三角肌深部、臂三头肌长头与外头之间的肌间隙内，有旋臂后动、静脉及桡神经、腋神经分布。

【取穴】以中指按触肩端部（膀尖），并以同手拇指向后按取，在有较大的深坑，即方处取之。

【操作】圆利针或火针直刺 3 ~ 4 厘米；毫针刺入 8 ~ 10 厘米。

【主治】闪伤夹气，前肢风湿，前肢麻木。

#103. 胸堂

【位置】胸骨两旁，腋窝前方血管上，左右侧各一穴。

【解剖】胸前两侧，胸外静脉沟下部，桡骨上端水平位处，臂皮下静脉上，有肌皮神经的皮支分布。

【取穴】头高位，取正常站立姿势，在胸前血管怒张处取之。

【操作】使马头高抬，用大宽针或小宽针安装在针锤上，或徒手持针顺血管急刺 1 厘米，出血。

【主治】前肢闪伤，胸膊痛，五攒痛，心经积热。

#104. 肘俞

【位置】肘头前方的凹陷中，左右侧各一穴。

【解剖】臂骨外上髁上缘与尺骨肘突前缘之间凹陷处，针刺臂三头肌的外头，有臂深动、静脉及桡神经分布。

【取穴】在肘端直前方的凹陷中。

【操作】火针或圆利针直刺3～4厘米，毫针直刺6厘米。

【主治】肘头肿胀，肩膊麻木，肘部风湿。

#113. 前缠腕

【位置】球节后上缘内、外侧筋前骨后血管上，每肢内外侧各一穴。

【解剖】球节上方两侧，指深屈肌腱与骨间中肌之间凹陷中，指内、外侧静脉上，其深部为滑液囊，有指内、外动脉及指神经伴行。

【取穴】在球节上缘二指处的掌心浅外（内）侧静脉上，按压有波动处是穴。

【操作】以中宽针顺血管刺1厘米，出血。

【主治】缠腕痛，板筋肿痛。

#116. 前蹄头

【位置】前蹄蹄缘（毛边）上约1厘米处，从正中向外旁开2～3厘米处血管上。左右侧各一穴。

【解剖】蹄背侧稍外方，蹄缘与皮肤交界处，皮下是蹄冠动脉、静脉丛，有指背侧神经分布。

【取穴】按穴位自然位置取之。

【操作】以中宽针向蹄内刺入1厘米，出血。

【主治】缠腕痛，五攒痛，蹄头痛，冷痛，结症。

6.4　后肢穴位

后肢穴位共36个，其中巴山、环跳、邪气、汗沟、后三里、后缠腕、后蹄头等穴，临床较为常用。

#119. 巴山

【位置】髂骨翼（雁翅骨）正中部上方约9厘米处，左右侧各一穴。

【解剖】臀部，百会穴与股骨大转子连线的中点处。皮下为臀筋膜和臀浅肌，深层为臀中肌，有臀前动、静脉及臀前神经分支。

【取穴】从股骨大转子向百会穴方向按压，在其中途感有凹陷处是穴。

【操作】圆利针或火针直刺3～4.5厘米，毫针直刺10～12厘米。

【主治】腰胯风湿，闪伤腰胯，后肢麻木。

#124. 环跳

【位置】股骨大转子（大胯尖）前方约6厘米凹陷处，左右侧各一穴。

【解剖】大转子前方约6厘米凹陷处。皮下是臀肌与股阔筋膜张肌、股四头肌之间的间隙。有股神经和臀前神经及臀前动、静脉分支。

【取穴】在髋关节前缘的凹陷中取穴。

【操作】圆利针或火针直刺3～4.5厘米；毫针直刺6～7.5厘米。

【主治】雁翅肿痛，后肢风湿，后肢麻木。

#127. 大胯

【取穴】与尾根同位的股二头肌沟起始部凹陷中按取，

【操作】火针或圆利针直刺3～4.5厘米，毫针直刺6～8厘米。

【主治】股胯风湿，闪伤腰胯，后肢麻木。

#132. 邪气

【位置】尾根旁开9厘米肌沟中，左右侧各一穴。

【解剖】坐骨弓上方约6厘米处、股二头肌与半腱肌之间的肌沟中，有臀后动、静脉和臀后神经分支。

【取穴】与肛门平位约四指处的股二头肌与半腱肌之间的肌沟中。

【操作】圆利针或火针直刺4.5厘米，毫利针直刺6～8厘米。

【主治】后肢风湿，股胯闪伤。

#133. 汗沟

【位置】邪气穴下方约7厘米处肌沟中，左右侧各一穴。

【解剖】股部，大转子后下方、股二头肌与半腱肌之间的肌沟中，有股深动、静脉及胫神经近侧支的分布。

【取穴】在邪气下方约四指处的股二头肌与半腱肌之间的肌沟中。

【操作】同邪气穴。

【主治】同邪气穴。

#139. 后三里

【位置】膝盖骨下缘外下方约7.5厘米处肌沟中，左右侧各一穴。

【解剖】小腿上部外侧，腓骨小头下部，

趾长伸肌与趾外侧伸肌之间的肌沟中。有胫前动、静脉与腓神经分布。

【取穴】在掠草穴下方，穴位与胫骨和腓骨小头成三角形。

【操作】圆利针或火针斜向下方刺入 2 ~ 4 厘米；毫针刺入 4. 5 ~ 6 厘米。

【主治】脾胃虚弱，后肢风湿。

#148. 后缠腕

【位置】球节（后寸腕骨）后上缘内，外侧筋前骨后血管上，每肢内外侧各一穴，共四穴。

【解剖】球节上部两侧，近侧籽骨上缘处的趾内外侧静脉上，有同名动脉及足底外侧神经伴行。

【取穴】在球节上缘二指处的血管上，以手指按压有波动处是穴。

【操作】同前缠腕穴。

【主治】同前缠腕穴。

#151. 后蹄头

【位置】后蹄毛边正中上方约 1 厘米处血管上，左右侧各一穴。

【解剖】蹄缘背侧正中稍外，皮下是蹄冠动、静脉丛，有趾背侧神经分支。

【取穴】按穴位自然位置取穴。

【操作】同前蹄头穴。

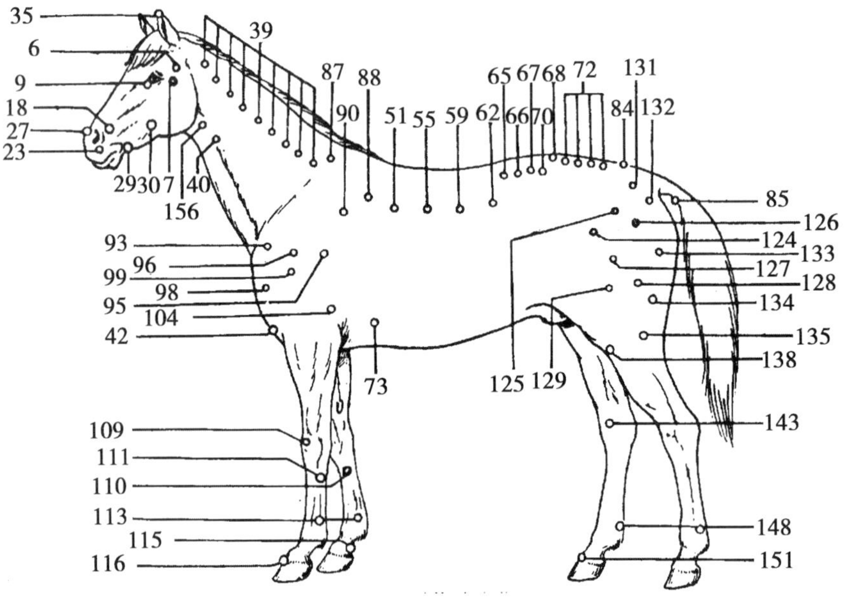

图 4 – 3　马体表穴位

6. 上关　7. 下关　9. 睛明　18. 血堂　23. 姜牙　27. 抽筋　29. 锁口　30. 开关　35. 耳尖　39. 九委　40. 颈脉　42. 穿黄　51. 肺俞　55. 胃俞　59. 脾俞　62. 关元俞　65. 腰前　66. 腰中　67. 腰后　68. 百会　70. 肾棚　72 八窌　73. 带脉　84. 尾根　85. 尾本　87. 膊尖　88. 膊栏　90. 肺攀　93 肩井　95. 冲天　96. 肩贞　98. 肩颙　99. 肩外颙　104. 肘俞　109. 膝眼　110. 膝脉　111. 攒筋　113. 前缠腕　115. 前蹄门　116. 前蹄头　124. 环跳　125. 环中　126. 环后　127. 大胯　128. 小胯　129. 后伏兔　131. 会阳　132. 邪气　133. 汗沟　134. 仰瓦　135. 牵肾　138. 掠草　143. 曲池　148. 后缠腕　151. 后蹄头　156. 喉门

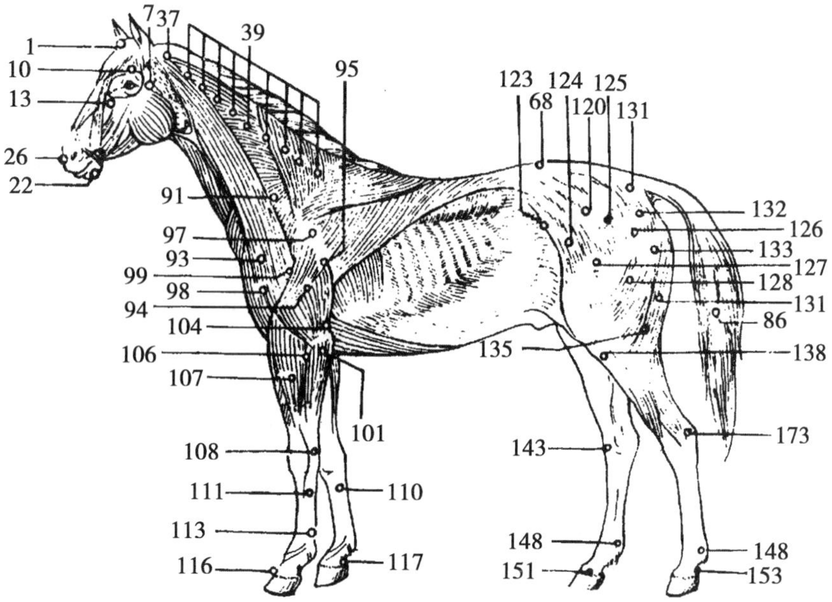

图 4－4　马的肌肉及穴位

1. 大风门 7. 下关　10. 睛俞　13. 三江　22. 承浆　26. 外唇阴　37. 风门　39. 九委　68　百会 86. 尾尖　91. 膊中　93. 肩井　94. 抢风　95. 冲天　97. 天宗　98. 肩髃　99. 肩外髃　101. 乘蹬　104. 肘俞　106. 乘重　107. 前三里　108. 过梁　110. 膝脉　111. 攒筋　113. 前缠腕　116. 前蹄头　117. 前臼　120. 路股　123. 居髎　124. 环跳　125. 环中　126. 环后　127. 大胯　128. 小胯　131. 会阳　132 邪气　133. 汗沟　134. 仰瓦　135. 牵肾　138. 掠草　143. 曲池　148. 后缠腕　151. 后蹄头　153. 后臼　173. 昆仑

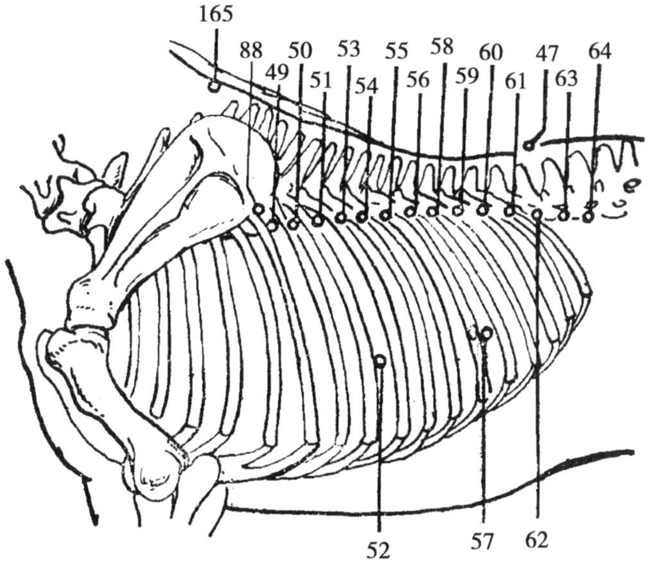

图 4－5　马的前躯骨骼及穴位

47. 命门　49. 厥阴俞　50. 督俞　51. 肺之俞　52. 肺俞　53. 膈俞　54. 胆俞　55. 办俞　56. 肝之俞　57. 肝俞　58. 三焦俞　59. 脾俞　60. 气海俞　61. 大肠俞　62. 关元俞　63. 小肠俞　64. 膀胱俞　88. 膊栏　165. 鬐前

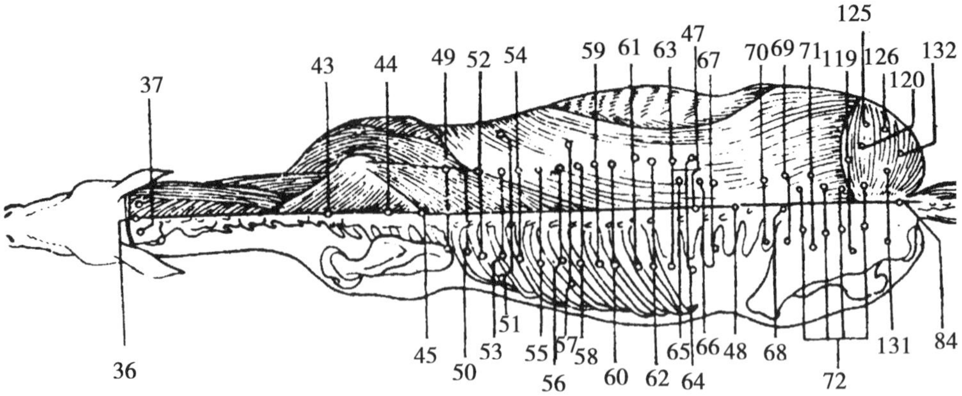

图 4－6　马的背侧穴位

36. 天门　37. 风门　43. 大椎　44. 鬐甲　45. 三川　47. 命门　48. 阳关　49. 厥阴俞　50. 督俞 51. 肺之俞　52. 肺俞　53. 膈俞　54. 胆俞　55. 胃俞　56. 肝之俞　57. 肝俞　58. 三焦俞　59. 脾俞　60. 气海俞　61. 大肠俞　62. 关元俞　67. 腰后　68. 百会　69. 肾俞　70. 肾棚　71. 肾角　72. 八窌　84. 尾根　119. 巴山　120. 路股　125. 环中　126. 环后　131. 会阳　132. 邪气

7　牛的针灸穴位

7.1　头部穴位

头部穴位共 17 个，其中天门、耳尖、山根、睛明、睛俞、太阳、通关、顺气等穴，临床上较为常用（图 4－7）。

#1. 天门

【位置】两耳根连线正中后方凹陷处，一穴。

【解剖】枕部，枕骨外结节与寰椎背侧结节之间的凹陷处。针刺项韧带索状部的起始部，有枕动脉前支及枕神经分布。

【取穴】一手指按于枕部后方，助手将牛头作上下运动，按穴的手指即可感到一活动的凹陷，此即是穴。

【操作】火针、中宽针或圆利针向后下方刺入 3 厘米；毫针刺入 3～6 厘米；或施火烙。

【主治】感冒，脑黄，癫痫，眩晕，破伤风。

#2. 龙会

【位置】两眉棱角连线的正中点处，一穴。

【解剖】额部，两外眼角连线的正中点处。其皮肤由筋膜连于骨（深部是额窦），有额动、静脉和额神经分布。

【取穴】两外眼角连线的正中点上是穴。

【操作】艾灸或烧烙。

【主治】脑黄，感冒。

#3. 耳尖（血印、罗城）

【位置】耳背面，离耳尖约 3 厘米处的并列三条血管上，每耳三穴。

【解剖】耳廓背面，距耳尖约 3 厘米处的耳大静脉的内、中，外三支。有耳大动脉和第二颈神经的耳后支分布。

【取穴】紧握耳根，使耳尖部血管显露取之。

【操作】中宽针速刺血管，出血。

【主治】中暑，感冒，中毒，腹痛，热性病。

#4. 耳根

【位置】耳根后方窝内，左右耳各一穴。

【解剖】耳根与寰椎翼前缘之间的凹陷处，针刺头前斜肌中。有耳后动、静脉和面神经的耳后支分布。

【取穴】将耳朵前后移动，在耳根后方的凹陷中取穴。

【操作】中宽针或火针向内下方刺入 1～1.5 厘米；圆利针或毫针刺入 3～6 厘米。

【主治】感冒，过劳，风湿症。

#5. 通天

【位置】两内眼角连线正中上方约 6～8 厘

米处，一穴。

【解剖】穴部皮肤借结缔组织连于骨，是额窦的背侧壁，有额动静脉和额神经分支分布。

【取穴】两内眼角连线正中上方约一掌处取之。

【操作】火针沿皮下向上平刺 2 ~ 3 厘米；火烙。

【主治】同天门穴。西藏民间兽医治疗脑包虫，于此穴附近进行开颅取虫。

#6. 山根（人中）

【位置】鼻镜部，主穴在鼻唇镜正中上沿有毛与无毛交界处，副穴在两鼻孔背角处各一穴，共三穴。

【解剖】鼻端部，主穴在鼻唇镜背侧正中，另两穴在两鼻孔背角处。针刺皮下鼻孔前开肌和鼻孔内侧开肌，有上唇动脉，鼻背侧静脉和颊上神经分布。

【取穴】先在鼻镜与鼻部皮肤交界线的中点确定中央主穴，其余两穴在两侧鼻孔背侧上缘几乎与主穴呈一等腰三角形。

【操作】中（小）宽针向后下方刺入 1 厘米，出血 3 毫针刺入 3 ~ 4.5 厘米。

【主治】中暑，感冒，咳嗽，肚痛，眩晕。

#7. 鼻中（三关、内户）

【位置】鼻镜正中，两鼻孔内侧下缘连线中点处，一穴。

【解剖】上唇背侧正中，两鼻孔之间的鼻唇镜处。皮下为鼻端开张肌，有鼻侧动、静脉及鼻背神经分布。

【取穴】两鼻孔内侧下缘连线之中点取穴。

【操作】小宽针直刺 1 厘米。

【主治】慢草，热病，唇肿，黄疸，衄血。

#8. 承浆（命牙）

【位置】下唇下缘正中，有毛与无毛交界处，一穴。

【解剖】下唇腹侧正中部，针刺口轮匝肌内，有颏动、静脉及颏神经分布。

【取穴】下唇正中，有毛与无毛交界处取穴。

【操作】中（小）宽针向后上方刺入 1 厘米，出血。

【主治】下颌肿痛，唇肿流涎，五脏积热。

#9. 睛俞（眉神、鱼腰）

【位置】上眼胞正中，眶上突（眉棱骨）下缘，左右眼各一穴。

【解剖】眶上突下缘正中，针刺眼睑与眶上突之间的深部。有眶上动脉及额神经分支。

【取穴】眶上突下缘正中取穴。

【操作】下压眼球，毫针沿眶上突下缘向内后下方刺入 3 ~ 5 厘米；或翻开眼睑，在其黏膜上以三棱针散刺，出血。

【主治】肝经风热，眼胞肿，眩晕。

#10. 睛明（睛灵、泪堂）

【位置】下眼胞眼箱骨上缘，内眼角外侧的皮肤褶上，左右眼各一穴。

【解剖】下眼睑上，两眼角内、中 1/3 交界处，针沿泪骨上缘刺入眼鞘与泪骨之间，有颧动、静脉和颧神经分支。

【取穴】下眼睑上正中偏内眼角取穴。

【操作】上压眼球，毫针向内下方刺入 3 厘米；或翻开眼睑，以三棱针散刺黏膜。

【主治】肝热传眼，睛生翳膜。

#11. 太阳

【位置】外眼角后方约 3 厘米处的颞窝中，左右侧各一穴。

【解剖】外眼角后方的颞窝中。深部有颞浅动、静脉分支，有颞深神经分支及面神经的耳、眼睑神经分布。

【取穴】颞窝中按取穴位。

【操作】毫针直刺 3 ~ 6 厘米；或用小宽针刺入 1 ~ 2 厘米，出血。

【主治】中暑，感冒，肝热传眼，睛生翳膜。

#12. 三江

【位置】内眼角前下方约 4.5 厘米处的血管上，左右侧各一穴。

【解剖】内眼角前下方的眼角静脉上；有眼角动脉和滑车神经分布。

【取穴】内眼角前下约三指处的血管上是穴。

【操作】三棱针或小宽针顺血管刺入 1 厘米，出血。

【主治】眼肿，肚痛，腹胀。

#13. 鼻俞（过梁）

【位置】鼻梁两侧，鼻孔上约 4.5 厘米处，左右侧各穴。

【解剖】鼻背两侧，鼻孔上方与鼻颌切迹之间的中点，有鼻背侧动、静脉和鼻外神经分支。

【取穴】鼻梁两侧，鼻孔上方约三指处取之。

【操作】三棱针或小宽针直刺 1.5 厘米，或从鼻的一侧透向另一侧。

【主治】感冒，肺热，鼻肿痛，咳嗽。

#14. 开关（牙关）

【位置】颊部后上方，第三对臼齿稍后上方，左右侧各一穴。

【解剖】咬肌前缘颊肌内，有面动、静脉和颊下神经分支。

【取穴】沿口角向后上方按取，在咬肌前缘的凹陷中是穴。

【操作】中宽针、圆利针或火针向上方刺入 2 ~ 3 厘米；毫针刺入 3 ~ 4.5 厘米。

【主治】歪嘴风，腮黄。

#15. 抱腮

【位置】开关穴后上方约 7.5 厘米处，最后一对臼齿间，左右侧各一穴。

【解剖】最后臼齿间，针刺咬肌内，有咬肌动、静脉和颊下神经分支。

【取穴】开关后上方最后臼齿间取穴。

【操作】中宽针、圆利针或火针稍向后方斜刺 2 ~ 3 厘米；毫针刺入 3 ~ 4.5 厘米。

【主治】同开关穴。

#16. 通关（知甘、舌底）

【位置】舌体底面两侧血管上，左右侧各一穴。

【解剖】舌底腹侧，舌系带前端两侧舌下静脉上，深部是舌肌，有舌下动，静脉和舌下神经的分支。

【取穴】将舌拉出口外，向上翻转，紧握舌体，使血管显露取之。

【操作】小宽针或三棱针刺入 1 厘米，出血。

【主治】慢草，木舌，喉黄，中暑；春秋开针洗口尚有预防疾病的作用。

#17. 顺气（嚼眼、垂津）

【位置】口内上颌嚼眼处（第一颌褶前中心之两侧），左右侧各一穴。

【解剖】硬腭前部，齿板后切齿乳头上，鼻腭管开口处，有腭大动、静脉和腭大神经分支。

【取穴】上颌前部齿龈上方的两个菱形孔内是穴。

【操作】细软柳条或榆树条去掉皮、节，徐徐插入穴内 18 ~ 30 厘米，达鼻腔内。

【主治】肚胀，感冒，睛生翳膜。

7.2 躯干及尾部穴位

躯干及尾部穴位共 37 个，其中颈脉、鬐甲、苏气、百会，脾俞、通窍、肷俞、关元俞、后海，尾本、尾尖、脱肛等穴，临床较为常用（图 4 – 8、图 4 – 9、图 4 – 10、图 4 – 11）。

#18. 风门

【位置】耳根穴下约 6 厘米处，左右侧各一穴。

【解剖】寰椎翼前缘下方与腮腺之间的凹陷处，皮下是头前斜肌，有枕动、静脉与颈神经之背克分布。

【取穴】耳根穴下方约一掌处取穴。

【操作】火针或圆利针向内下方刺入 3 厘米；毫针刺入 4.5 厘米。

【主治】感冒，风湿症，破伤风。

#19. 喉门（锁喉）

【位置】下颌骨后，喉头下，左右侧各一穴。

【解剖】第一、第二气管轮腹侧面的两侧，针刺胸前肌与胸骨甲状肌之间。有颈动、静脉及颈神经腹支分布。

【取穴】喉头正下方，第一、第二气管轮处取之。

【操作】中宽针、圆利针或火针斜向后下方刺入 3 厘米；毫针刺入 4.5 厘米。

【主治】喉肿，喉结，喉麻痹。

#20. 颈脉（大脉、鹘脉）

【位置】喉下约 6 厘米处大血管上，左右侧各一穴。

【解剖】颈静脉沟内的颈静脉上，背侧有

臂头肌，有颈动脉和颈神经腹侧支分布。

【取穴】喉头下方约一掌处的大血管上是穴。头吊起扣颈绳，使血管显露取之。

【操作】大宽针顺血管刺入 1 厘米，出血。

【主治】五脏积热，中暑，中毒，脑黄，肺风毛燥。

#21. 丹田

【位置】背中线上，第一、第二脊梁骨间隙中，一穴。

【解剖】第一、第二胸椎棘突间，棘上韧带深部、棘间肌和棘间韧带内，有肋颈动、静脉和第一胸神经背支分布。

【取穴】使颈部作上、下运动，颈椎可随颈部运动而活动，而胸椎则不活动，其棘突也较突出，从而可确定第一胸椎，然后根据穴位的自然位置取之，顺序按取鬐甲（三合）、三川（天福）、安福、苏气等穴。为便于记忆，拟下列顺口溜：一，二丹，四、五三，八、九苏，五、六川。

【操作】小宽针、圆利针或火针向前下方刺入 3 厘米，毫针刺入 6 厘米。

【主治】中暑，过劳，前肢风湿，肩痈。

#22. 鬐甲（三台）

【位置】背中线上第四、第五脊梁骨间隙中（鬐甲最高点），一穴。

【解剖】背部第四、第五胸椎棘突间，棘上韧带深部、棘间肌和棘间韧带内，有肋颈动，静脉及第四胸神经背支分布。

【取穴】详见丹田穴。

【操作】同丹田穴。

【主治】肺热咳嗽，前肢风湿，脱膊，肩部肿痛。

#23. 三川（天福、田福）

【位置】背中线上第五、第六脊梁骨间隙中，一穴。

【解剖】第五、第六胸椎棘突间，棘上韧带深部、棘间肌和棘间韧带内，有肋间动、静脉和第五胸神经背侧支分布。

【取穴】详见丹田穴。

【操作】同丹田穴。

【主治】肚痛，泄泻。

#24. 苏气

【位置】背中线上，第八、第九脊梁骨间隙中，一穴。

【解剖】第八、第九胸椎棘突间，棘上韧带深部、棘间肌与棘间韧带内，有肋间动、静脉和第八肋间神经背侧支分布。

【取穴】详见丹田穴。

【操作】同丹田穴。

【主治】肺热，咳嗽，气喘。

#25. 安福（通筋）

【位置】背中线上，第十、第十一脊梁骨间隙中，一穴。

【解剖】第十、第十一胸椎棘突间、棘上韧带深部、棘间肌与棘间韧带内，有肋间动、静脉和第十肋间神经背侧支分布。

【取穴】详见丹田穴。

【操作】同丹田穴。

【主治】肺热，腹泻，风湿。

#26. 天平（断血）

【位置】背中线上，第十三、第十四脊梁骨间隙中，一穴。

【解剖】最后胸椎与第一腰椎棘突间，棘上韧带深部、棘间肌与棘间韧带内，有肋间动、静脉及最后肋间神经背支分布。

【取穴】背中线上，背腰结合部的凹陷中取穴。

【操作】中（小）宽针、圆利针或火针直刺 3 厘米；毫针直刺 4.5 厘米。

【主治】尿闭，肠黄，阳痿，阉割后出血。

#27. 后丹田

【位置】背中线上，第一、第二腰脊骨间隙中，一穴。

【解剖】第一、第二腰椎棘突间，棘上韧带深部、棘间肌与棘间韧带内，有第一、第二腰椎动、静脉和腰神经背支分布。

【取穴】天平穴后方一个腰脊骨凹陷中。

【操作】小宽针、圆利针或火针直刺 3 厘米；毫针直刺 4.5 厘米。

【主治】慢草，腰胯痛，尿闭。

#28. 命门（肾门）

【位置】背中线上，第二、第三腰脊骨间隙中，一穴。

【解剖】第二、第三腰椎棘突间的凹陷处，棘上韧带深部、棘间肌和棘间韧带内，有第二腰椎动、静脉和腰神经背侧支分布。

【取穴】后丹田向后一个腰椎棘突间凹陷中取穴。

【操作】小宽针、圆利针或火针直刺3厘米；毫针直刺3~5厘米。

【主治】尿闭，肾痛，腰胯痛，胎衣不下，慢草。

#29. 安肾

【位置】背中线上，第三、第四腰脊骨间隙中，一穴。

【解剖】第三、第四腰椎棘突间凹陷处，棘上韧带深部、棘间肌及棘间韧带内，有第三腰椎动、静脉和腰神经背侧支分布。

【取穴】命门穴后一腰椎棘突间的凹陷中是穴。

【操作】小宽针、圆利针或火针直刺3厘米；毫针直刺3~6厘米。

【主治】同命门穴。

#30. 腰中（腰带）

【位置】腰部两侧，第三、第四腰脊骨之间旁开6厘米处，左右侧各一穴。

【解剖】第三、第四腰椎横突顶端之间，刺入背最长肌内，有腰动、静脉和第三腰神经背侧支分布。

【取穴】安肾穴旁开约6厘米处取穴。

【操作】小宽针、圆利针或火针直刺2~3厘米；毫针直刺4.5厘米；也有向脊柱方向斜刺约6~9厘米。

【主治】腰风湿，胞黄，肾痛。

#31. 百会

【位置】背中线上，腰椎与荐椎间隙中，一穴。

【解剖】第六腰椎与第一荐椎间，棘上韧带深部、棘间肌与棘间韧带内，有腰动、静脉以及最后腰神经背侧支分布，其正下方深部是脊髓。

【取穴】腰荐十字部的凹陷中是穴。

【操作】小宽针、圆利针或火针直刺3~4.5厘米；毫针直刺6~9厘米。

【主治】腰胯风湿，腰胯闪伤，肾虚，肚胀。

#32. 肾俞（左归尾、右尾归）

【位置】百会穴正旁约8厘米处的肌沟中，左右侧各一穴。

【解剖】百会穴两侧的臀中肌内，有臀前动、静脉和臀前神经的分支。

【取穴】百会穴旁开约四指处取之。

【操作】小宽针、圆利针或火针直刺3厘米；毫针直刺4.5厘米。

【主治】腰胯风湿，腰背闪伤。

#33. 带脉

【位置】肘后约10厘米处的血管上，左右侧各一穴。

【解剖】胸壁两侧，肘突后上方，胸深肌肱部上缘第七肋间处的胸外静脉上，有胸外动脉伴行，并有肩胛颈动脉和胸肌神经分支的分布。

【取穴】于胸壁肘突后方一掌处按压血管，使其显露后取穴。

【操作】中宽针沿血管刺入1厘米。

【主治】感冒，肠黄，中暑，腹痛。

#34. 六脉

【位置】倒数第一、第二、第三肋间，在髂骨翼上角水平线上，每侧三穴，共六穴。

【解剖】第十、第十一、第十二肋间，刺入背最长肌与髂肋肌之肌沟中；穴部有肋间动，静脉以及胸神经背侧支分布。

【取穴】按穴位自然位置取穴。

【操作】小宽针、圆利针或火针向内下方斜刺3厘米，毫针刺入6厘米。

【主治】便结，肚胀，积食，泄泻，慢草。

#35. 脾俞（六脉第一穴）

【位置】倒数第三肋间，髂骨翼上角水平线上，左右侧各一穴。

【解剖】倒数第三肋间，其他同六脉穴。

【取穴】同六脉穴。

【操作】同六脉穴。

【主治】同六脉穴。

#36. 关元俞

【位置】第十三、第十四脊梁骨间旁开背中线8厘米处，左右侧各一穴。

【解剖】最后肋骨后缘与第一腰椎横突顶

端之间，刺入背最长肌与髂肋肌之肌沟中，有最后肋间动、静脉和最后肋间神经分布。

【取穴】最后肋骨的后缘与第一腰椎横突顶端之间是穴。

【操作】小宽针、圆利针或火针向内下方刺入3厘米5毫针刺入4.5厘米3亦可向脊柱方向刺入6~9厘米。

【主治】慢草，便结，肚胀，积食，泄泻。

#37. 肚角（左侧关元俞）

【位置】左侧第十三、第十四脊梁骨间之间距背中线8厘米处，一穴。若在右侧同样位置再取一穴，称为关元俞。

【解剖】同关元俞穴。

【取穴】同关元俞穴。

【操作】同关元俞穴。

【主治】同关元俞穴。

#38. 食胀

【位置】胸壁左侧，倒数第二肋间，距背中线18厘米处，一穴。

【解剖】脑壁左侧，第十一肋间，髂骨结节下角水平线上，穿通胸壁及膈，刺入腹腔瘤胃背侧盲囊内，有肋间动、静脉和胸神经背侧支分布。

【取穴】按穴位自然位置取穴。

【操作】小宽针或毫针向下方刺入9厘米。

【主治】同关元俞穴。

#39. 通窍（八卦）

【位置】倒数第四、第五、第六、第七肋间，髂骨翼上角水平线上，每侧四穴，共八穴。

【解剖】胸壁两侧，第六、第七、第八、第九肋间，髂骨结节上角水平线上，刺入背最长肌与髂肋肌沟中，有肋间动、静脉和肋间神经分支。

【取穴】六脉穴前四个肋间，同位高取穴。

【操作】中（小）宽针、圆利针或火针向内下方斜刺3厘米；毫针刺入4.5~6厘米。

【主治】肺痛，咳嗽，过劳，风湿。

#40. 肷俞（饿眼）

【位置】左侧肷窝部，即肋骨后，髂骨翼前，脊梁骨下，呈三角形凹陷处，一穴。

【解剖】左腹侧髂部，瘤胃放气的部位，有旋髂深动脉，最后肋间动脉和第一、第二腰神经的浅支分布。

【取穴】瘤胃鼓气时，选其最高点刺入。

【操作】套管针向内下方刺入6~9厘米，徐徐放出气体。

【主治】急性瘤胃鼓气。

#41. 滴明

【位置】肚脐前约15厘米，距正中线旁开约12厘米凹陷处的腹壁血管上，左右侧各一穴。

【解剖】剑状软骨两侧与肋软骨之间的凹陷处，即腹皮下静脉（乳静脉）进入“乳井”之处。穴部有腹壁动、静脉及肋间神经腹支分布。

【取穴】乳井的乳静脉上是穴。

【操作】中宽针沿血管刺入2厘米，出血。

【主治】尿闭，胀气，腹下水肿。

#42. 阳明

【位置】奶头基部外侧，每个奶头一穴。

【解剖】乳头基部外侧，针刺乳腺。穴部有乳腺动、静脉和精索外神经的分支分布。

【取穴】奶头基部外侧凹陷处是穴。

【操作】中（小）宽针向内上方刺入1~2厘米。

【主治】奶黄，尿闭。

#43. 海门（天枢、云门）

【位置】肚脐两旁，离脐约3厘米处，左右侧各一穴。

【解剖】腹下脐部白线两侧，刺入胸腹皮肌深部与腹黄膜之间，达腹腔，有腹壁后动、静脉和腰神经，最后肋间神经的腹支分布。

【取穴】脐两侧约3厘米处。

【操作】中（小）宽针直向上刺1厘米，刺时应提起皮肤，或在肿胀处散刺，或插入宿水管放腹水。

【主治】尿闭，腹水3散刺治腹下浮肿。

#44. 阴俞（会阴）

【位置】在公牛阴囊后上方中心缝上，母牛在肛门下、阴门上方之间正中缝的中点处，一穴。

【解剖】公牛在阴囊缝线后上部，皮下为筋膜及阴茎缩肌，深部为球海绵体肌和尿道海绵体，有阴部外动、静脉和精索外神经分布。

母牛在肛门与阴门背侧角之间，刺向阴门外括约肌与肛门外括约肌之间的凹陷处，内有会阴的动、静脉和阴部内神经分布。

【取穴】公牛在外肾后上方中心缝上，母牛在肛门与阴门间的正中缝上中点取穴。

【操作】毫针、圆利针或火针直刺1~2厘米；毫针刺后可施艾灸。

【主治】阴囊肿胀，子宫脱。

#45. 穿黄（吊黄）

【位置】胸骨前缘正中的皮肤褶上，一穴。

【解剖】胸前部，胸正中沟两侧约1.5厘米处，经皮下穿过。穴部有肩横动脉，臂头静脉，颈神经腹支和胸廓前神经分布。

【取穴】胸前肿胀处最低部位取穴。

【操作】穿黄针，左右对穿皮肤。

【主治】胸黄。

#46. 开风

【位置】尾根穴前一节，正中一穴。

【解剖】荐部，第四、第五荐椎棘突结合部，皮下为荐髂背韧带，两侧为股二头肌的起始部，有臀后动、静脉和荐神经的背侧支分布。

【取穴】尾根穴前一节凹陷中是穴。

【操作】小宽针稍向前斜刺1厘米；毫针刺入2厘米。

【主治】中暑，尿闭，风湿症。

#47. 尾根

【位置】尾背侧正中最后接脊骨与尾骨之间的凹陷中，一穴。

【解剖】荐尾之间的凹陷处，刺入两荐尾上内侧肌之间，有臀后动、静脉和臀后神9背侧支分布。

【取穴】手摇尾巴时所动的骨节间是穴。

【操作】小宽针、圆利针或火针直刺1~2厘米；毫针直刺3厘米。

【主治】便秘，脱肛，子宫脱，热泻，热性病。

#48. 尾节

【位置】尾根穴后一节凹陷中，一穴。

【解剖】尾背侧第一、第二尾椎棘突间，皮下为尾筋膜，两侧为荐尾背内侧肌，有尾外侧背动、静脉和尾背侧纵神经分布。

【取穴】尾根穴向后一节凹陷中是穴。

【操作】拉直尾，小宽针或圆利针直刺1厘米；毫针直刺1~1.5厘米。

【主治】同尾根穴。

#49. 尾干

【位置】尾根穴后二节凹陷中，一穴。

【解剖】尾背侧第二、第三尾椎间的凹陷中，肌肉、血管和神经分布与尾节穴同。

【取穴】从尾根穴向后第二个凹陷中是穴。

【操作】同尾节穴。

【主治】血凝气滞，尿闭，五淋等。

#50. 尾本

【位置】尾下离尾根约7.5厘米处的血管上，一穴。

【解剖】尾根穴腹侧正中的尾静脉上，有尾动脉和尾腹神经分布。

【取穴】离尾根约7.5厘米处的血管上是穴。

【操作】中（小）宽针顺血管刺入1厘米，出血。

【主治】肚痛，便秘，风湿，尾痹。

#51. 尾尖（垂珠）

【位置】尾尖上，一穴。

【解剖】尾尖上，有尾外侧背动、静脉和尾背侧纵神经分布。

【取穴】尾尖部是穴。

【操作】中宽针刺入1厘米或作十字形劈开，出血。

【主治】中暑，感冒，过劳，中毒，热性病。

#52. 后海

【位置】肛门与尾根之间凹陷中，一穴。

【解剖】尾根下，肛门上方的深窝内正中处，刺入肛门外括约肌与尾肌间的皮下疏松结缔组织内，有尾动、静脉和直肠后神经分支。

【取穴】尾下肛门上的深窝内是穴。

【操作】提起尾巴，小宽针，圆利针或火针向前上方刺入3~4.5厘米；毫针刺入5~10厘米。

【主治】久痢泄泻，胃肠热结，脱肛。

#53. 肛脱

【位置】肛门两侧旁开2厘米处，左右侧

各一穴。

【解剖】穴部皮下为疏松结缔组织，有阴部神经、直肠后神经和阴部内动、静脉分支。

【取穴】肛门两侧旁开 2 厘米处是穴。

【操作】圆利针或毫针向前下方斜刺 3～5 厘米；或电针、水针。

【主治】直肠脱。

#54. 阴脱

【位置】阴唇两侧，阴唇上下联合的中点旁开 2 厘米处，左右侧各一穴。

【解剖】穴部皮下为疏松结缔组织，针体经过阴门缩肌、前庭缩肌，周围有阴部神经和阴部动、静脉分支。

【取穴】阴唇两侧中点旁开 2 厘米处是穴。

【操作】毫针向前下方斜刺 4～8 厘米；或电针、水针。

【主治】阴道脱，子宫脱。

7.3　前肢穴位

共有 19 个，其中膊尖、膊栏、抢风、肩井、胸堂、夹气、前缠腕、涌泉、前蹄头等穴，临床较为常用。

#55. 轩堂（前通膊）

【位置】三台穴两旁，距背中线 3～6 厘米，左右侧各一穴。

【解剖】鬐甲两侧，肩胛软骨上缘正中，刺入斜方肌和菱形肌的深部与棘横筋膜之间的肌间隙内，有颈神经背支及肋颈动，静脉的分支。

【取穴】肩胛软骨上缘正中向两侧内下方按取之。

【操作】中宽针、圆利针或火针向内下方刺入肩胛软骨内侧 9 厘米；毫针刺入 10～15 厘米。

【主治】失膊，夹气。

#56. 膊尖（雁翅、云头）

【位置】肩胛软骨前角下凹陷中，左右侧各一穴。

【解剖】肩胛骨前角与肩胛软骨连接处的凹陷中，刺入斜方肌和菱形肌的深部，达菱形肌与颈下锯肌的肌间隙内，有颈横动、静脉和颈神经背支及胸神经分支分布。

【取穴】沿肩胛软骨向下方触摸，转弯处有一凹陷即是穴。

【操作】中宽针、圆利针或火针刺入肩胛软骨内侧 6 厘米；毫针刺入 10～15 厘米。

【主治】失膊，前肢风湿。

#57. 肺门

【位置】膊尖穴斜前下方约 8 厘米处，左右侧各一穴。

【解剖】肩胛前缘中部，沿胸深肌肩胛部前缘，刺入颈下锯内，有颈外动、静脉，颈神经背侧支及肩胛背侧神经分支。

【取穴】沿膊尖穴向下触摸，在肩胛前缘颈中部凹陷处是穴。

【操作】小宽针、圆利针或火针向内后下方刺入 3 厘米；毫针刺入 6～9 厘米。

【主治】失膊，咳嗽，肺气痛，前肢风湿。

#58. 膊栏（滋元、爬壁）

【位置】肩胛软骨后角下凹陷中，左右侧各一穴。

【解剖】肩胛骨后角与肩胛软骨相接处的凹陷中，穿过背阔肌和下锯肌，达肋骨外肌，有肋间动、静脉，胸神经背侧支及胸背神经通过。

【取穴】沿肩胛软骨向后下方触摸，转弯处有一凹陷是穴。

【操作】小宽针、圆利针或火针斜向前内下方刺入 3 厘米，毫针刺入 6～9 厘米。

【主治】同膊尖穴。

#59. 肺攀

【位置】肩胛骨后缘与肺门穴同位高，左右侧各一穴。

【解剖】肩胛骨后缘上、中 1/3 的交界处，针刺三角肌后缘的臂三头肌长肌内。有肩胛下动、静脉肌支，肋间神经外侧支和桡神经的分支。

【取穴】肩胛骨后缘与肺门同位高取穴。

【操作】小宽针、圆利针或火针斜向前内下方刺入 3 厘米；毫针刺入 4.5 厘米。

【主治】同肺门穴。

#60. 肩井（中膊、撞膀）

【位置】臂骨节外上缘之凹陷中，左右侧各一穴。

【解剖】肩关节上部，臂头大结节上方凹陷处，刺入岗上肌与岗下肌的肌间隙内，有肩峰动、静脉与肩胛上神经的分支。

【取穴】由肩胛骨向下摸至肱骨外结节上方的凹陷中是穴。

【操作】小宽针、圆利针或火针向内下方刺入3～4.5厘米；毫针刺入6～9厘米。

【主治】失膊，肩膊痛，前肢风湿，肩膊麻木。

#61. 抢风（中腕）

【位置】臂骨节后下方约8厘米处的凹陷中，左右侧各一穴。

［解剖）臂骨三角肌隆起背侧的凹陷中，即三角肌深部、小圆肌后缘、臂三头肌长头与外头所形成的方孔中，有桡神经和臂深动脉、肩胛下动脉及其伴行的静脉分支。

【取穴】从肩井穴向后下方摸有一较深的凹陷是穴。

【操作】小宽针、圆利针或火针直刺3～4.5厘米；毫针刺入6厘米。

【主治】同肩井穴。

#62. 冲天

【位置】抢风穴后斜上方约6厘米处的肌沟中，左右侧各一穴。

【解剖】肩胛骨后缘中部，刺入三角肌深部，臂三头肌之长头内，有胸背动、静脉和桡神经分支。

【取穴】抢风穴后斜上方约6厘米处的肌沟中取之。

【操作】小宽针、圆利针或火针直刺3～4.5厘米；毫针直刺6厘米。

【主治】同肩井穴。

#63. 肩外髃

【位置】肩关节后缘的凹陷中，左右侧各一穴。

【解剖】臂骨大结节嵴中部的后上方，刺入三角肌深部，达小圆肌后缘。有旋臂后动、静脉和腋神经分支。

【取穴】沿肩关节后缘触摸到凹陷处是穴。

【操作】同冲天穴。

【主治】同肩井穴。

#64. 肘俞（下腕、肘前）

【位置】肘头直前方凹陷处，左右侧各一穴。

【解剖】臂骨外上髁～亡部的凹陷处，即尺骨肘突前缘，臂三头肌长头与外头之间的肌间隙内，有臂深动、静脉及桡神经的分支。

【取穴】从肘头直前方摸至尺骨与肱骨外上髁之间的凹陷中是穴。

【操作】小宽针、圆利针或火针向内下方刺入3厘米；毫针直刺4.5厘米。

【主治】肘头肿胀，风湿，闪伤，肩膊麻木。

#65. 胸堂

【位置】胸骨两旁，腋窝前方血管上，左右侧各一穴。

【解剖】胸外侧沟下部臂头静脉上，有肩横动脉与肌皮神经皮支分布。

【取穴】头高抬，双腿站齐，在胸前血管怒张处取之。

【操作】中宽针沿血管刺入1厘米，出血。

【主治】胸膊痛，失膊，中暑，心肺积热。

#66. 夹气

【位置】腋窝内，左右侧各一穴。

【解剖】前肢与躯干相接处的腋窝内，刺入肩胛下肌与下锯肌间的疏松结缔组织内，有臂动、静脉与臂神经丛分支分布。

【取穴】患肢提起，向外扳开，在胸壁与前肢之间的前方凹陷中。

【操作】大宽针向上刺破皮肤，再以夹气针向同侧抢风穴方向刺入10～15厘米，出针后将患肢前后摇动数次即可。注意严密消毒。

【主治】里夹气。

#67. 腕后（曲尺、追风）

【位置】前肢腕骨（膝盖骨）节后面正中，左右侧各一穴。

【解剖】腕关节后方，副腕骨与指浅屈肌之间的凹陷中，穿通腕掌侧横韧带，达指浅屈肌与副腕骨深部，有骨间总动脉的掌侧支和尺神经分布。

【取穴】腕关节正后方筋前骨后凹陷处是穴。

【操作】中（小）宽针直刺1.5～2.5厘米；用毫针由外向对侧穿透。

【主治】腕部肿痛，前肢风湿，闪伤。

#68. 膝眼（跪膝）

【位置】膝盖外下缘的凹陷中，左右侧各一穴。

【解剖】腕关节背侧，腕桡侧伸肌腱与指总伸肌腱之间的陷沟中，刺入关节囊内。有腕背动、静脉网及桡神经分布。

【取穴】屈曲腕关节，在其外下方触按取穴。

【操作】中（小）宽针向后刺入1厘米。

【主治】腕部肿痛。

#69. 前缠腕（前寸子）

【位置】前肢悬蹄上方约3厘米处凹陷中，每肢内外各一穴。

【解剖】掌部，球节上方，指深屈肌腱与骨间中肌之间凹陷处。深部是滑液囊，有指动脉和指神经伴行。

【取穴】沿悬蹄向上，在屈肌腱与骨间中肌之间按取；或提起患肢，使球节部弯曲显现出凹陷处是穴。

【操作】中（小）宽针顺血管刺入1.5厘米，出血。

【主治】蹄黄，寸腕肿痛，扭伤。

#70. 涌泉

【位置】前肢蹄叉前缘正中稍上方，左右侧各一穴。

【解剖】第三、第四指的第一指节骨中部背侧面。皮下有蹄冠动、静脉和指背侧神经分布。

【取穴】蹄叉正上方约三指处凹陷中，是穴。

【操作】中宽针顺血管刺入1～1.5厘米，出血。

【主治】蹄肿，扭伤寸子，中暑，感冒。

#71. 前蹄头（前八子、答子）

【位置】前蹄叉上缘两侧，有毛与无毛交界处，每肢内外各一穴。

【解剖】第三、第四指的蹄匣上缘正中，刺入蹄冠静脉丛，有蹄冠动脉和指背侧神经分布。

【取穴】蹄匣上缘正中有毛与无毛交界处取穴。

【操作】中宽针直刺1厘米，出血。

【主治】蹄黄，扭伤，便结，肚痛，感冒。

#72. 前灯盏（背风）

【位置】前肢两悬蹄后下方正中的凹陷中，每肢内外各一穴。

【解剖】第三、第四指上部之间与近籽骨下凹陷中，针刺指忱，深部为指深屈腱。有指枕动脉和指掌侧神经分布。

【取穴】悬蹄后下方正中凹陷处是穴。

【操作】中宽针向前下方刺入1厘米，出血。

【主治】蹄黄，中暑。

#73. 前蹄门（交泥）

【位置】蹄球后缘，每肢内外各一穴。

【解剖】蹄后，指枕上部，刺入指枕的结缔组织内。有指枕动脉与指掌侧神经分布。

【取穴】轻按蹄踵上缘，指下感有血管处是穴。

【操作】中宽针向前上方刺入1厘米，出血。

【主治】蹄黄，过劳。

7.4　后肢穴位

共17个，其中肾堂、掠草、阳陵、后三里、后缠腕、滴水等穴，临床较为常用。

#74. 气门

【位置】髂骨翼后方，荐椎两侧约9厘米处，左右侧各一穴。

【解剖】髋结节和坐骨结节的内侧连线与股二头肌前缘相交部的肌沟中，刺入臀中肌内。深部有坐骨神经和臀前动、静脉分支。

【取穴】自髋结节与尾根穴作一连线，再以坐骨结节与百会穴作一连线，两线相交处是穴。

【操作】毫针直刺6厘米；火针或圆利针直刺3厘米。

【主治】后肢风湿症，不孕症。

#75. 居髎

【位置】髂骨翼后下方的凹陷中，左右侧各一穴。

【解剖】髋结节后下方，臀肌下缘的凹陷处，针刺股阔筋膜张肌深部，达股四头肌，有髂股动、静脉和股神经分支。

【取穴】髋结节后下方按取之。

【操作】圆利针或火针直刺 3～4.5 厘米；毫针直刺 6 厘米。

【主治】腰胯痛，腰胯风湿，后肢麻木，不孕症。

#76. 大胯

【位置】股骨大转子正上方 9～12 厘米凹陷处，左右侧各一穴。

【解剖】股骨大转子正上方 9～12 厘米凹陷处，刺入臀中肌内，有臀前和髂腰动，静脉以及坐骨神经分支。

【取穴】沿股骨大转子正上方按取凹陷处是穴。

【操作】火针或圆利针直刺 3～4.5 厘米；毫针直刺 6 厘米。

【主治】后肢风湿，闪伤腰胯，后肢麻木。

#77. 小胯

【位置】股骨大转子正下方约 6 厘米处凹陷中，左右侧各一穴。

【解剖】大转子正下方约 6 厘米处，刺入股二头肌内，有股深和股后动，静脉以及坐骨神经分支。

【取穴】沿大转子正下方按取凹陷处即是穴。

【操作】同大胯穴。

【主治】同大胯穴。

#78. 大转

【位置】股骨大转子正前方 6 厘米处的凹陷中，左右侧各一穴。

【解剖】股骨大转子正前方 6 厘米处凹陷中，刺入股阔筋膜张肌、臀中肌与股四头肌直头之间隙内，有臀前旋髂深动脉，髂腰动、静脉以及臀前神经的分支。

【取穴】沿大转子正前方按取凹陷处是穴。

【操作】同大胯穴。

【主治】同大胯穴。

#79. 邪气（黄金）

【位置】股骨大转子后上方 9 厘米处的肌沟中，左右侧各一穴。

【解剖】坐骨弓上万，股二头肌与半腱肌间之肌沟中，有臀后动、静脉和臀后神经分支分布。

【取穴】自股骨大转子与坐骨结节作一连线，与股二头肌沟相交处是穴。

【操作】火针或圆利针直刺 3～4.5 厘米，毫针直刺 6 厘米。

【主治】后肢风湿，胯部肿痛。

#80. 汗沟

【位置】邪气穴下方约 7 厘米处肌沟中，左右侧各一穴。

【解剖】股部，大转子后下方、股二头肌与半腱肌之间的肌沟中，有股深动、静脉及胫神经近侧支的分布。

【取穴】在邪气下方约四指处的股二头肌与半腱肌之间的肌沟中。

【操作】同邪气穴。

【主治】同邪气穴。

#81. 肾堂

【位置】后肢内侧，大腿褶下方 9 厘米处的血管上，左右侧各一穴。

【解剖】股内侧上部皮下的隐静脉上，有隐动脉及隐神经伴行。

【取穴】自膝盖骨上方作一水平线与隐静脉相交处是穴。

【操作】中宽针顺血管刺入 1 厘米，出血。

【主治】外肾黄，后肢闪伤。

#82. 掠草（梳子骨）

【位置】膝盖骨前下缘稍偏外方的凹陷中，左右侧各一穴。

【解剖】膝关节背侧面，膝外、中直韧带之间的凹陷处，刺入关节囊前面的脂肪内。有膝上及腘动、静脉以及胫神经关节支分布。

【取穴】先摸到膝盖骨，在其下缘稍偏外方的凹陷中是穴。

【操作】火针或圆利针向后上方刺入 3～4.5 厘米。

【主治】掠草痛，后肢风湿。

#83. 阳陵（后通膊）

【位置】膝盖骨后方约 12 厘米处的凹陷中，左右侧各一穴。

【解剖】膝关节后，胫骨外髁后上缘的凹陷处，刺入股二头肌前、中支肌间隙内。有股后动、静脉和股后皮神经及胫神经分支。

【取穴】大转子垂直线与膝关节水平线交点处是穴。

【操作】火针或圆利针直刺 3 厘米；毫针直刺 4.5 ~6 厘米。

【主治】掠草痛，后肢风湿，后肢麻木。

#84. 后三里

【位置】掠草穴下缘斜外下方 9 厘米处肌沟中，左右侧各一穴。

【解剖】小腿上部外侧，腓骨小头下部腓骨长肌与趾外侧伸肌之间肌沟中，有胫前动、静脉与腓神经分支分布。

【取穴】摸取腓骨小头，在其下方肌沟中按取穴位。

【操作】毫针向内后下方刺入 6 ~ 7.5 厘米。

【主治】脾胃虚弱，后肢风湿，后肢麻木。

#85. 曲池（承山）

【位置】第二跗骨前外方的血管上，左右侧各一穴。

【解剖】跗关节背侧，中横韧带下，趾长伸肌腱外侧凹陷处的趾外侧静脉上，有跗外侧动脉、胫浅神经分布。

【取穴】跗关节屈曲，在曲纹稍外方，小腿外侧皮下静脉上。

【操作】中宽针顺血管刺入 1 厘米，出血。

【主治】合子肿痛，后肢风湿。

#86. 后缠腕

【位置】后肢悬蹄上方约 3 厘米处凹陷中，每肢内外各一穴。

【解剖】蹠部，球节上方，趾深屈肌腱与骨间中肌之间凹陷处。深部是滑液囊，有趾动脉和趾神经伴行。

【取穴】沿悬蹄向上在屈肌腱与骨间中肌之间按取，或提起患肢，使球节部弯曲，显现出凹陷取之。

【操作】中（小）宽针顺血管刺入 1.5 厘米，出血。

【主治】蹄黄，寸腕肿痛，扭伤。

#87. 滴水

【位置】后肢蹄叉前缘正中稍上方，左右侧各一穴。

【解剖】第三、第四趾的第一趾节骨中部背侧面。皮下有蹄冠动、静脉和指背侧神经分布。

【取穴】蹄叉正上方约三指处凹陷中是穴。

【操作】中宽针顺血管刺入 1 ~ 1.5 厘米，出血。

【主治】蹄肿，扭伤寸子，中暑，感冒。

#88. 后蹄头（笞子）

【位置】后蹄叉上缘两侧，有毛与无毛交界处，左右侧内外各一穴。

【解剖】第三、第四趾的蹄匣上缘正中，刺入蹄冠静脉丛，有蹄冠动脉和趾背侧神经分支。

【取穴】蹄匣上缘正中有毛与无毛交界处按取穴位。

【操作】中宽针直刺 1 厘米，出血。

【主治】蹄黄，扭伤，便结，肚痛，感冒。

#89. 后灯盏

【位置】后肢两悬蹄后下方正中的凹陷中，左右侧内外各一穴。

【解剖】第三、第四趾上部之间与近籽骨下的凹陷中，针刺趾枕，深部为趾深屈腱。有趾枕动脉和趾跖侧神经分支。

【取穴】悬蹄后下方正中处凹陷中是穴。

【操作】中宽针向前上方刺 1 厘米，出血。

【主治】蹄黄，中暑。

#90. 后蹄门

【位置】蹄球后缘，左右侧内外各一穴。

【解剖】蹄后，趾枕上部，刺入趾枕的结缔组织内。有枕动脉与趾跖侧神经分支。

【取穴】轻按蹄踵上缘，指下感有血管处是穴。

【操作】中宽针向前下方刺入 1 厘米，出血。

【主治】蹄黄，过劳。

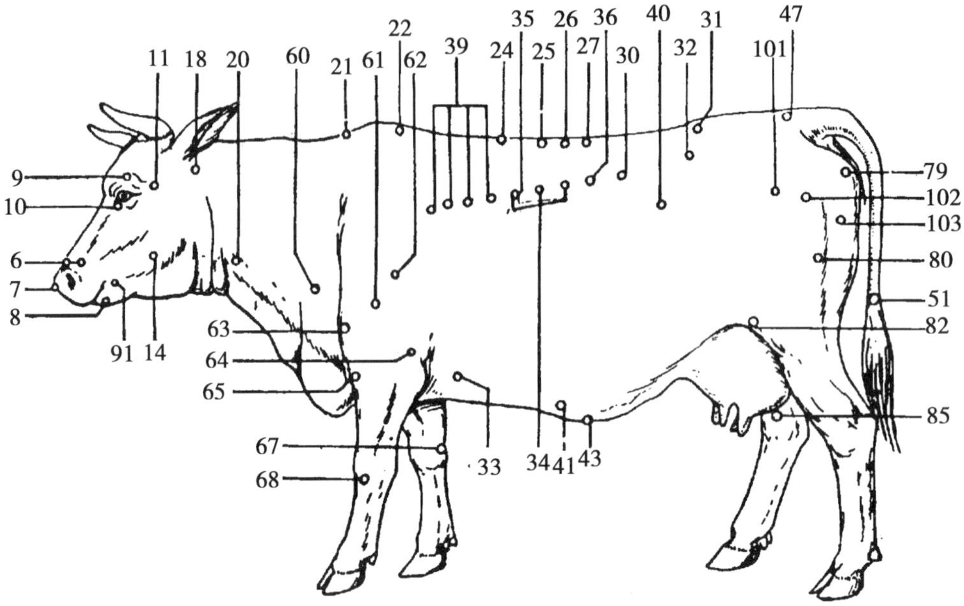

图 4－7　牛体表穴位

5. 山根　7. 鼻中　8. 承浆　9. 睛俞　10. 睛明　11. 太阳　14. 开关　18. 风门　20. 颈脉　21. 丹田　22. 鬐甲　24. 苏气　25. 安福　26. 天平　27. 后丹田　30. 腰中　31. 百会　32. 肾俞　33. 带脉　34. 六脉　35. 脾俞　36. 关元俞　39. 通窍　40. 膁俞　41. 滴明　43. 海门　47. 尾根　51. 尾尖　60. 肩井　61. 抢风　62. 冲天　63. 肩外颙　64. 肘俞　65. 胸堂　67. 腕后　68. 膝眼　79. 邪气　80. 仰瓦　82. 掠草　85. 曲池　91. 锁口　101. 环中　102. 环后　103. 汗沟

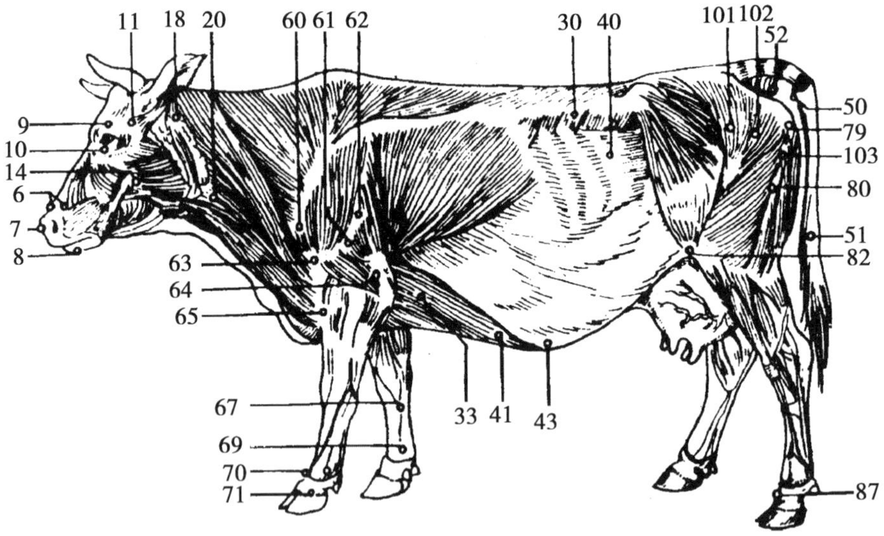

图 4－8　牛的肌肉及穴位

6. 山根　7. 鼻中　8. 承浆　9. 睛俞　10. 睛明　11. 太阳　14. 开关　18. 风门　20. 颈脉　30. 腰中　33. 脉　40. 膁俞　41. 滴明　43. 海门　50. 尾本　51. 尾尖　52. 后海　60. 肩井　61. 抢风　62. 冲天　63. 肩外颙　64. 肘俞　65. 胸堂　腕后　69. 前缠腕　70. 涌泉　71. 前蹄头　79. 邪气　80. 仰瓦　82. 掠草　87. 滴水　101. 环中　102. 环后　103. 汗沟

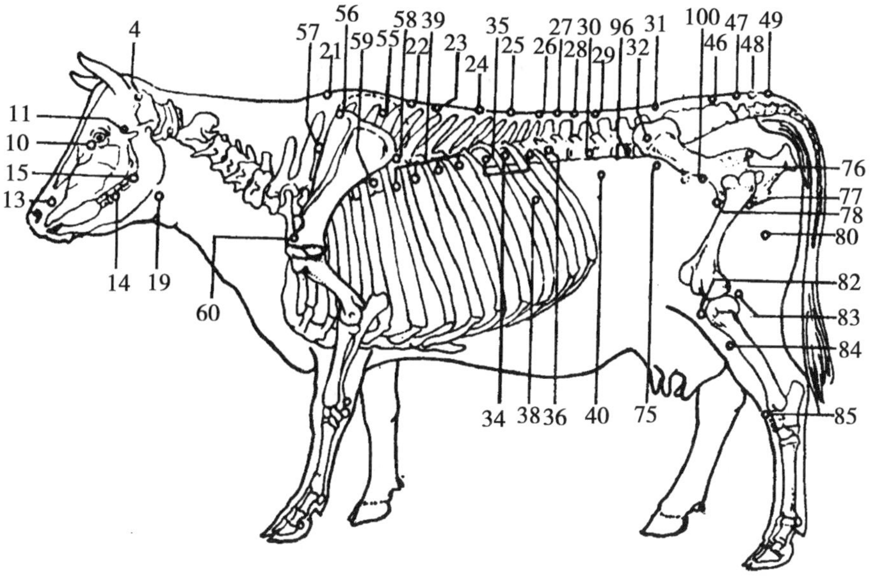

图 4－9　牛的骨骼及穴位

4. 耳根　10. 睛明　11. 太阳　13. 鼻俞　14. 开关　15. 抢腮　19. 喉门　21. 丹田　22. 鬐甲　23. 三川　24. 苏气　25. 安福　26. 天平　27. 后丹田　28. 命门　29. 安肾　30. 腰中　31. 百会　32. 肾俞　34. 六脉　35. 脾俞　36. 左关元俞（肝角）　38. 食胀　39. 通窍　40. 膁俞　46. 开风　47. 尾根　60. 肩井　75. 居髎　76. 大胯　77 小胯　78. 大转　80. 仰瓦　82. 掠草　83. 阳陵　84. 后三里　85. 曲池　96. 肾棚　100. 环跳

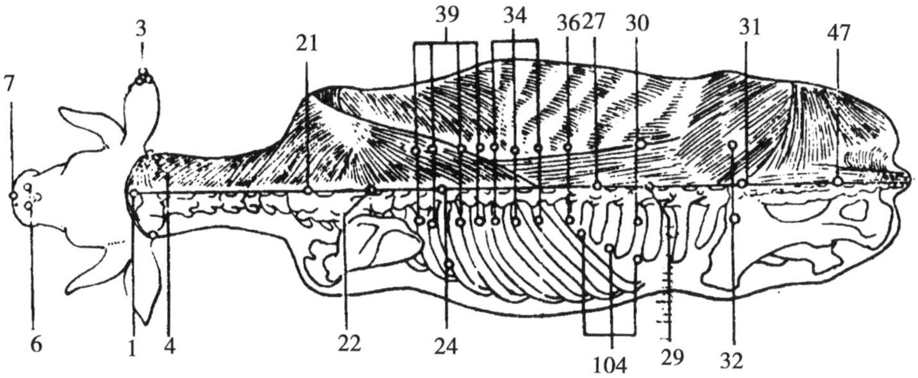

图 4－10　牛背侧穴位

1. 天门　3. 耳尖　4. 耳根　6. 山根　7. 鼻中　21. 丹田　22. 鬐甲　24. 苏气　27. 后丹田　29. 安肾　30. 腰中　31. 百会　32. 肾俞　34. 六脉　36. 关元俞　39. 通窍　47. 尾根　104. 腰旁三穴

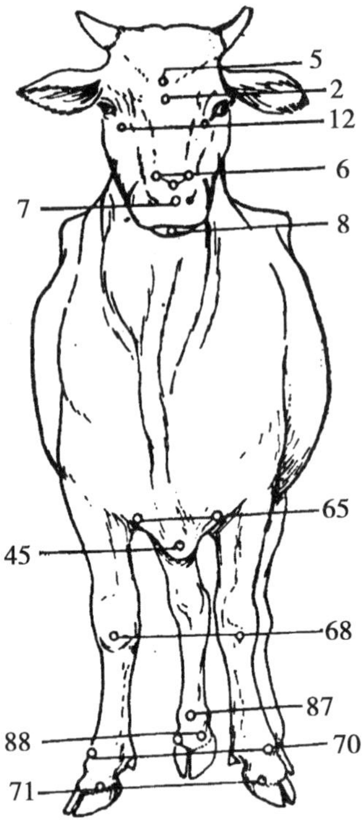

图4－11　牛体前面穴位

2. 龙会　5. 通天　6. 山根　7. 鼻中　8. 承浆　12. 三江　45. 穿黄　65. 胸堂　68. 膝眼　70. 涌泉　71. 前蹄头　87. 滴水　88. 后蹄头

8　狗的针灸穴位

参见图4－12、图4－13、图4－14。

#1. 人中

【穴位】上唇正中沟上1/3处，一穴。

【解剖】有口轮匝肌，眶下神经上唇支厘眼神经的筛神经支。

【操作】直刺0.5厘米。

【主治】中风、中暑，支气管炎。

#2. 山根

【穴位】鼻背上方，有毛无毛交界处，一穴。

【解剖】有鼻唇提肌，眶下神经鼻外支，鼻背侧动静脉。

【操作】刺0.2～0.5厘米，出血。

【主治】中风、中暑、鼻窦炎、感冒、犬瘟热初期。

#3. 三江

【穴位】内眼角下的眼角静脉处，左右眼角一穴。

【解剖】有鼻唇提肌颧骨肌、眼角动静脉。

【操作】刺0.2～0.5厘米，出血。

【主治】便秘、腹痛、眼结膜炎。

#4. 承泣

【穴位】眶下缘中部上方，左右侧各一穴。

【解剖】睑结膜与球结膜间的结膜穹窿，深层为眼球下直肌、眼球缩肌、动眼神经和外展神经。

【操作】把眼球向上推，沿眼眶直刺约2～4厘米。

【主治】急慢性结膜炎、视神经萎缩、视网膜炎、白内障。

#5. 睛明

【穴位】内眼角上下眼睑交界处，左右眼各一穴。

【解剖】有眼角动静脉、滑车下神经。

【操作】直刺，把眼球向外推，直刺入0.2～0.5厘米。

【主治】眼结膜炎、角膜炎、瞬膜肿胀。

#6. 翳风

【穴位】耳基部与下颌关节后上方间的凹陷中，左右侧各一穴。

【解剖】有腮耳肌、腮腺、面神经穿出处，颞浅动静脉。

【操作】直刺 1 ~3 厘米。

【主治】颜面神经麻痹、聋症。

#7. 上关

【穴位】咬肌后上方，与颧弓背侧间的凹陷中，左右侧各一穴。

【解剖】皮下为颧肌，前缘为咬肌后缘为腮腺，深层为颞肌有耳睑神经颧支，颞浅动静脉分布。

【操作】直刺 3 厘米，或艾灸。

【主治】颜面神经麻痹、聋症。

#8. 下关

【穴位】咬肌后缘中部，下颌关节下方与下颌骨角间的凹陷中（与上关相对），左右侧各一穴。

【解剖】有腮腺，面神经颊背神经及颊腹神经。

【操作】直刺 3 厘米，或艾灸。

【主治】颜面神经麻痹、聋症。

#9. 耳尖

【穴位】耳廓尖端外面脉管上，左右耳各一穴。

【解剖】耳后静脉。

【操作】耳后静脉点刺，出血。

【主治】中风、中暑、疝痛、痉挛、眼结膜炎。

#10. 天门

【穴位】头顶部枕骨后缘正中，一穴。

【解剖】有颈耳浅肌，颈耳深肌、臂头腹、枕神经。

【操作】直刺 1 ~3 厘米，或艾灸。

【主治】癫痫发作、热症、脑炎、四肢抽搐、惊厥。

#11. 颈脉

【穴位】颈外静脉前 1/3 处，一穴。

【解剖】上界为臂头肌，下界为胸头肌。

【操作】刺 0.5 ~1 厘米，出血。

【主治】肺炎、中毒、中暑。

#12. 大椎

【穴位】第七颈椎与第一胸椎棘突间，一穴。

【解剖】有第八颈神经背支。

【操作】直刺 1 ~3 厘米，或艾灸。

【主治】发热，神经痛、风湿痛、支气管炎、癫痫、创伤性炎症、并有降温、强壮和防病作用。

#13. 陶道

【穴位】第一、第二胸椎棘突间 一穴。

【解剖】有第一胸神经背支。

【操作】斜向前下方刺入 0.5 ~1 厘米，或艾灸。

【主治】神经痛、肩扭伤、前肢扭伤、癫痫、发热。

#14. 身柱

【穴位】第三、第四胸椎棘突间，一穴。

【解剖】有第三胸神经背支。

【操作】斜向前下方刺入 1 ~1.5 厘米，或艾灸。

【主治】肺炎、支气管炎、犬瘟热、肩部扭伤和神经痛。

#15. 灵台

【穴位】第六、第七胸椎棘突间，一穴。

【解剖】有第六胸神经背支。

【操作】同身柱穴。

【主治】肝炎、肺炎、支气管炎，胃痛。

#16. 中枢

【穴位】第十、第十一胸椎棘突间，一穴。

【解剖】有第十胸神经背支。

【操作】直刺 0.5 ~1 厘米，或艾灸。

【主治】胃炎、食欲缺乏。

#17. 脊中

【穴位】第十一、第十二胸椎棘突间，一穴。

【解剖】有第十一胸神经背支。

【操作】同中枢穴。

【主治】消化不良，肠炎、食欲缺乏、肝炎、腹泻。

#18. 悬枢

【穴位】第十三胸椎与第一腰椎棘突间，一穴。

【解剖】有第十三胸神经背支。

【操作】直刺0.5~1厘米，或艾灸。

【主治】风湿病，腰部扭伤，消化不良，肠炎、腹泻。

#19. 命门

【穴位】第二、第三腰椎棘突间，一穴。

【解剖】有第二腰神经背支。

【操作】同悬枢穴。

【主治】必尿器障碍，食欲缺乏、风湿病、腰部扭伤、慢性肠炎、激素失调，阳痿，肾炎。

#20. 阳关

【穴位】第四、第五腰椎棘突司，一穴。

【解剖】有第四腰神经背支。

【操作】同悬枢穴。

【主治】性机能减退，子宫内膜炎、卵巢囊肿、子宫卵巢萎缩，情期延长，风湿病，腰部扭伤。

#21. 关后

【穴位】第五、第六腰椎棘突同，一穴。

【解剖】第五腰神经背支。

【操作】直刺0.5~1厘米，或艾灸。

【主治】子宫内膜炎，卵巢囊肿、膀胱炎、大肠麻痹、便秘。

#22. 百会

【穴位】第七腰椎棘突与荐骨间，一穴。

【解剖】第七腰神经背支。

【操作】同关后穴。

【主治】各种神经错乱、坐骨神经痛，后躯瘫痪，直肠脱。

#23. 尾根

【穴位】最后荐椎与第一尾椎棘突间，一穴。

【解剖】第三荐神经背支

【操作】同尾干穴。

【主治】同尾干穴。

#24. 尾节

【穴位】第二、第三荐椎棘突间，一穴。

【解剖】第二荐神经背支。

【操作】同尾干穴。

【主治】同尾干穴。

#25. 尾干

【穴位】第一、第二荐椎棘突间，一穴。

【解剖】第一荐神经背支。

【操作】直刺0.3~0.5厘米，或艾灸。

【主治】后躯瘫痪、尾麻痹、脱肛、便秘或腹泻。

#26. 尾尖

【穴位】尾末端，一穴。

【解剖】尾神经。

【操作】直刺，针从末端插入0.5~1厘米。

【主治】中风、中暑、胃肠炎。

#27. 二眼

【穴位】第一、第二背荐孔处，每侧各两穴。

【解剖】有阔筋膜胀肌、臀肌、臀前动静脉臀后神经。

【操作】直刺1~1.5厘米，或艾灸。

【主治】后躯瘫痪、神经痛子宫疾病。

#28. 尾本

【穴位】尾根腹侧正中，一穴。

【解剖】有尾肌、荐尾付肌、尾动中脉。

【操作】刺0.5~1厘米，出血

【主治】腹痛、尾神经麻痹、腰部风湿

#29. 后海（交巢）

【穴位】尾根与肛门间的凹陷中，一穴。

【解剖】肛门括约肌、尾肌、直肠神经。

【操作】直刺1~1.5厘米。

【主治】腹泻、直肠麻痹，括约肌麻痹、阳痿。

#30. 肺俞

【穴位】肩关节与髋关节连线与第3肋骨后缘交点的肋间隙左右侧各一穴。

【解剖】有背阔肌、肋间肌、肋间血管及肋间可神经。

【操作】沿着肋间斜刺入1~2厘米或艾灸。

【主治】肺炎，支气管炎。

#32. 心俞

【穴位】肘突内侧第四、第五肋骨与肋软骨连接处的间隙内，左右侧各一穴。

【解剖】有胸肌、肋间肌、肋间血管及肋间神经。

【操作】同肺俞穴

【主治】精神紧张、心脏病、癫痫。

#35. 肝俞

【穴位】肩关节与髋关节连线与第九肋间交点处，左右各一穴。

【解剖】腹外斜肌、肋间肌、肋间血管及肋间神经。

【操作】同肺俞穴。

【主治】肝炎、黄疸、眼病、神经痛。

#37. 脾俞

【穴位】背正中线左外侧约 10 重米第十三肋后缘处，左侧一穴。

【解剖】腹外斜肌、腹内斜肌，腹横机，肋间神经及血管。

【操作】直刺或斜刺，沿肋骨后缘刺入 1 ~ 2 厘米，或艾灸。

【主治】消化不良、慢性腹泻、食欲缺乏、呕吐、贫血。

#38. 胃俞

【穴位】肩关节与髋关节连线与倒数第二第三肋间交点处，左右侧各一穴。

【解剖】同肝俞穴。

【操作】同肺俞穴。

【主治】胃炎、胃扩张、消化不良、食欲缺乏肠炎。

#39. 三焦俞

【穴位】髂肋肌沟中与第一腰椎横突末端相对，左右侧各一穴。

【解剖】内侧为最背长肌，外侧为髂肋肌，第一腰神经及血管。

【操作】同脾俞穴。

【主治】同脾俞穴。

#40. 肾俞

【穴位】髂肋肌沟中与第二腰椎横突末端相对，左右侧各一穴。

【解剖】内侧为背最长肌，外侧为髂肋肌，第二腰神经及血管。

【操作】直刺 0. 5 ~ 1 厘米，或艾灸。

【主治】肾炎、泌尿器机能障碍、多尿症，性腺机能减退，性激素失调、不孕症，阳痿、腰部风湿和扭伤。

#41. 气海俞

【穴位】髂肋肌沟中与第三腰椎横突末端相对，左右侧各一穴。

【解剖】内侧为背最长肌，外侧为髂肋肌，第三腰神经及血管。

【操作】同肾俞穴。

【主治】便秘，气胀。

#42. 大肠俞

【穴位】髂肋肌沟中与第四腰椎横突末端相对，左占侧各一穴。

【解剖】内侧为背最长肌，外侧为髂肋肌，第四腰神经血管。

【操作】同肾俞穴。

【主治】消化不良、肠炎、便秘。

#43. 关元俞

【穴位】髂肋肌沟中与第五腰椎横突末端相对，左右侧各一穴。

【解剖】内侧为背最长肌，外侧为髂肋肌，第五腰神经及血管。

【操作】同肾俞穴。

【主治】消化不良，便秘、泄泻。

#44. 小肠俞

【穴位】在髂肋肌沟中，与第六腰椎横突末端相对，左右侧各一穴。

【解剖】内侧为背最长肌、外侧为髂肋肌、第六腰神经及血管。

【操作】同肺俞穴。

【主治】肠炎、肠痉挛，腰骶痛。

#45. 膀胱俞

【穴位】距第六、第七腰椎横突腹侧约 10 厘米处，左右侧各一穴。

【解剖】腹外斜肌、腹内斜肌，腹横直肌，腰血管，腰神经，腹壁后动脉。

【操作】直刺入 0. 5 ~ 1 厘米。

【主治】膀胱炎、血尿、膀胱痉挛、尿液潴留，腰骶痛。

#46. 胰俞

【穴位】肾俞穴腹侧约 3 厘米处，左右侧各一穴。

【解剖】腹外斜肌，腹内斜肌，腹横肌，腰神经及血管。

【操作】直刺入 0. 5 ~ 1 厘米。

【主治】胰腺炎、消化不良、慢性腹泻、多尿症。

#47. 卵巢俞

【穴位】距第四腰椎横突末端约 3 厘米处即第四、第五腰椎横突间，左右侧各一穴眶第五，第六腰椎横突。

【解剖】同胰俞穴。

【操作】直刺入 1.5 ~3 厘米。

【主治】性腺机能减退，卵巢激素机能不全、卵巢机能减退，卵巢炎、卵巢囊肿。

#48. 子宫俞

【穴位】距第五、第六腰椎横突间约 3 厘米处，左右侧各一穴。

【解剖】同胰俞穴。

【操作】直刺入 1.5 ~3 厘米。

【主治】子宫囊肿、子宫内膜炎、子宫机能障碍、腰部风湿、子宫炎。

#51. 膏肓俞

【穴位】肩胛骨后角内侧，左右侧各一穴。

【解剖】背阔肌、下据肌、肩胛下肌、肩胛下神经、胸背神经、肩胛下动脉。

【操作】斜刺进针 2 ~4 厘米，或艾灸。

【主治】神经痛、肩部扭伤、肩胛神经麻痹、肩风湿、肺炎、支气管炎、贫血、久病体弱。

#52. 胸膛

【穴位】胸前两外侧臂三头肌与臂头肌间的臂头静脉上，左右测各一穴。

【解剖】有臂头肌、胸深浅肌、臂神经及臂头静脉。

【操作】刺 0.5 ~1 厘米，出血。

【主治】中暑，肩肘扭伤和神经痛，风湿症。

#53. 肩井

【穴位】肩峰前下方的凹陷中，左右侧各一穴。

【解剖】臂头肌，三角肌，岗上肌，肌皮神经。

【操作】直刺入 1 ~3 厘米。

【主治】前肢和肩部神经痛或麻痹、肩部刳 I 伤，肱神经和岗上肌麻痹。

#54. 肩外颙

【穴位】肩峰后下方的凹陷中，左右侧各一穴。

【解剖】三角肌、岗下肌、三头肌，肩胛上神经、桡神经。

【操作】直刺入 2 ~4 厘米，或艾灸。

【主治】前肢和肩部神经痛、麻痹或扭伤、岗上肌或臂肱神经麻痹。

#55. 抢风

【穴位】肩外俞与肘俞间连线的上寺与中古交界处，左右侧各一穴。

【解剖】三角肌、臂三头肌、桡神经。

【操作】直刺入 2 ~4 厘米。

【主治】一般感觉麻；忙、前肢神经障碍，扭伤。

#56. 郗上

【穴位】肩外俞与肘俞间连线的下士处肘俞穴前上方，左右侧各一穴。

【解剖】臂三头肌，桡神经。

【操作】同抢风穴。

【主治】前肢扭伤、神经痛或神经麻痹、臂肱和桡骨神经麻痹。

#57. 肘俞

【穴位】臂骨外上髁与肘突间的凹陷中，左右侧各一穴。

【解剖】臂三头肌、肘肌，桡神经。

【操作】同抢风穴。

【主治】关节炎、前肢及肘 部神经痛或神经麻痹。

#58. 四渎

【穴位】臂骨外上髁与桡骨外髁间前方的凹陷中，左右侧各一穴。

【解剖】腕桡侧伸肌，指外侧伸肌，桡神经。

【操作】直刺 2 ~4 厘米，或艾灸。

【主治】前肢扭伤、神经痛或神经麻痹、臂肱和桡神经麻痹。

#59. 前三里

【穴位】前臂外侧上寺处，腕外屈肌与第五指伸肌司，左右侧各一穴。

【解剖】前为第五指伸肌，后为腕外侧屈肌桡神经。

【操作】同四渎穴。

【主治】桡、尺神经麻痹，前肢神经痛或风湿痛。

#60. 外关

【穴位】前臂骨外侧下奇处桡骨与尺骨的间隙中，左右侧各一穴。

【解剖】前为第五指伸肌，后为腕外侧屈肌桡神经。

【操作】直刺入1~2厘米，或艾灸。

【主治】桡、尺神经麻痹、前肢神经痛或风湿痛、便秘、乳汁分泌不足。

#61. 内关

【穴位】前臂骨内侧与外关相对的前臂骨间隙中，左右侧各一穴。

【解剖】前为桡骨、后为腕桡屈肌与指深屈肌、正中神经及血管。

【操作】直刺入1~2厘米，或艾灸。

【主治】胸支神经障碍、胃肠痉挛、急腹痛、心脏病，中风。

#62. 阳辅

【穴位】前臂远端正中，阳池穴上放厘米处，左右侧各一穴。

【解剖】前臂头静脉，桡浅神经。

【操作】直刺入0.5~1厘米，或艾灸。

【主治】前胸神经紊乱，腕腱扭伤，桡神经麻痹。

#63. 阳池

【穴位】腕关节背侧，桡一中间腕骨与桡骨远端连接处的凹陷中，左右侧各一穴。

【解剖】同阳辅穴。

【操作】同阳辅穴。

【主治】指、趾扭伤，前肢神经痛或神经麻痹感冒、腕关节炎。

#64. 腕骨

【穴位】尺骨远端和付腕骨间的凹陷中，左右侧各一穴。

【解剖】尺神经。

【操作】针从前臂内侧直刺入0.5~1厘米，或艾欠。

【主治】胃炎、腕、肘及指关节炎。

#65. 膝脉

【穴位】第一腕掌关节内侧下方第一、第二掌骨间的掌心浅静脉上，左右侧各一穴。

【解剖】掌心浅静脉。

【操作】刺入0.5~1厘米，出血。

【主治】球、腕关节肿痛、屈腱炎。

#66. 涌滴（前涌泉、后滴水）

【穴位】第三、第四掌（跖）骨间的掌（跖）背侧的静脉上，前后肢每足一穴。

【解剖】掌（跖）背侧静脉，第三指（趾）背侧总神经。

【操作】同膝脉穴。

【主治】指、趾扭伤、中暑、腹痛、风湿、感冒。

#68. 环跳

【穴位】股骨大转子后上方，左右肢各一穴。

【解剖】臀浅肌、臀深肌、股二头肌、坐骨神经、臀后动脉。

【操作】直刺入2~4厘米，或艾灸。

【主治】后躯麻痹，骨盆支神经麻痹和神经痛坐骨神经痛、股骨神经麻痹。

#69. 膝上

【穴位】股骨上缘外侧约0.5厘米处，左右肢各一穴。

【解剖】股四头肌、股神经。

【操作】直刺入0.5~1厘米。

【主治】后肢骨盆神经紊乱，神经节炎。

#70. 膝凹

【穴位】股骨与胫骨外髁之间的凹陷中，左右肢各一穴。

【解剖】股二头肌、腓肠肌、腓神经。

【操作】同膝上穴。

【主治】同膝上穴。

#71. 膝下

【穴位】股骨与胫骨隆起间，膝外与膝中直韧带间，左右侧各一穴。

【解剖】膝外、膝中直韧带，深部为膝关节囊。

【操作】直刺0.5~1厘米，或艾灸。

【主治】扭伤、神经痛、腺关节炎。

#72. 后三里

【穴位】小腿外侧上寺处，胫腓骨间隙中距腓骨头暇侧约5厘米处，左右侧各穴。

【解剖】胫前肌，趾长伸肌（前），趾深屈肌（后）、腓神经。

【操作】直刺入1~5厘米，或艾灸。

【主治】后躯麻痹、骨盆支神经痛或麻痹、胃炎、肠症挛和急腹痛、关节炎、发热、消化不良，并有防病保健和强壮作用。

#73. 解溪

【穴位】径骨内侧与胫、跗骨间的凹陷中左右侧各一穴。

【解剖】有腓浅神经，深部为关节囊。

【操作】直刺入 0.5 厘米，或艾灸。

【主治】扭伤、后肢麻痹或神经疾患。

#75. 后跟

【穴位】跟骨与腓骨远端间凹陷中，左右侧各一穴。

【解剖】跖神经。

【操作】同解溪穴。

【主治】同解溪穴。

#76. 肾堂

【穴位】股内侧隐静脉上，左右侧各一穴。

【解剖】有腓肠肌、股薄肌、腓浅神经、隐神经、隐静脉。

【操作】刺入 0.5 ~1 厘米，出血。

【主治】髋关节炎、扭伤、神经痛。

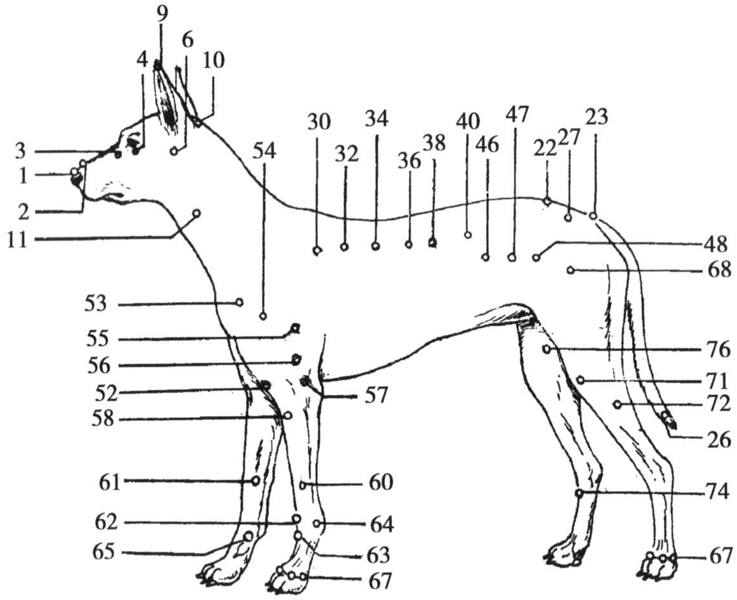

图 4-12　犬体表穴位

1. 人中　2. 山根　3. 三江　4. 承泣　6. 翳风　9. 耳尖　10. 天门　11. 颈脉　22. 百会　23. 尾根　26. 尾尖　27. 二眼　30. 肺俞　32. 心俞　34. 膈俞　36. 胆俞　38. 胃俞　40. 肾俞　46. 胰俞　47. 卵巢俞　48. 子宫俞　52. 胸堂　53. 肩井　54. 肩颙　55. 抢风　56. 郗上　57. 付俞　58. 四渎　60. 外关　61. 内关　62. 阳辅　63. 阳池　64. 腕骨　65. 膝脉　67. 六缝　68. 环跳　71. 膝下　72. 后三里　74. 中付　76. 肾堂

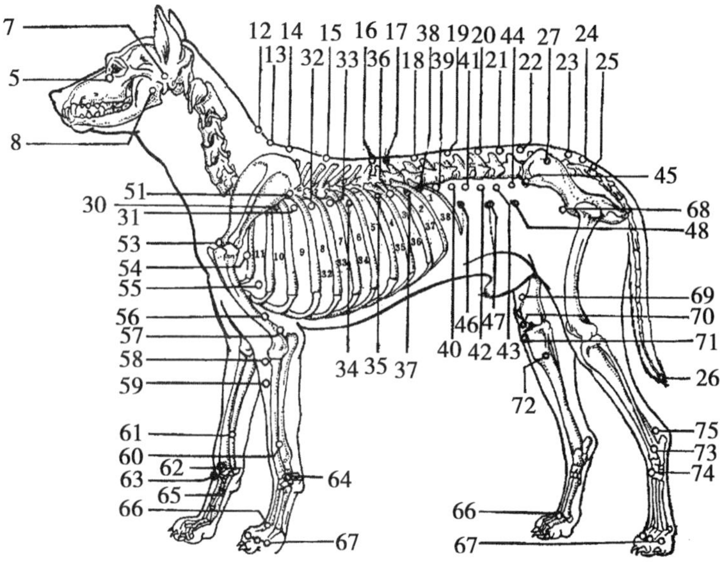

图 4－13　犬的骨胳及穴位

5. 睛明　7. 上关　8. 下关　12. 大椎　13. 陶道　14. 身柱　15. 灵台　16. 中枢　17. 脊中　18. 悬枢　19. 命门　20. 阳关　21. 关后　22. 百会　23. 尾根　24. 尾节　25. 尾干　26. 尾尖　27. 二眼　30. 肺俞　31. 厥阴俞　32. 心俞　33. 督俞　34. 膈俞　35. 肝俞　36. 胆俞　37. 脾俞　38. 胃俞　39. 三焦俞　40. 肾俞　41. 气海俞　42. 大肠俞　43. 关元俞　44. 小肠俞　45. 膀胱俞　46. 胰俞　47. 卵巢俞　48. 子宫俞　51. 膏肓俞　53. 肩井　54. 肩外颙　55. 抢风　56. 郗上　57. 肘俞　58. 四渎　59. 前三里　60. 外关　61. 内关　62. 阳辅　63. 阳池　64. 腕骨　65. 膝脉　66. 涌滴　67. 六缝　68. 环跳　69. 膝上　70. 膝凹　71. 膝下　72. 后三里　73. 解溪　74. 中付　75. 后跟

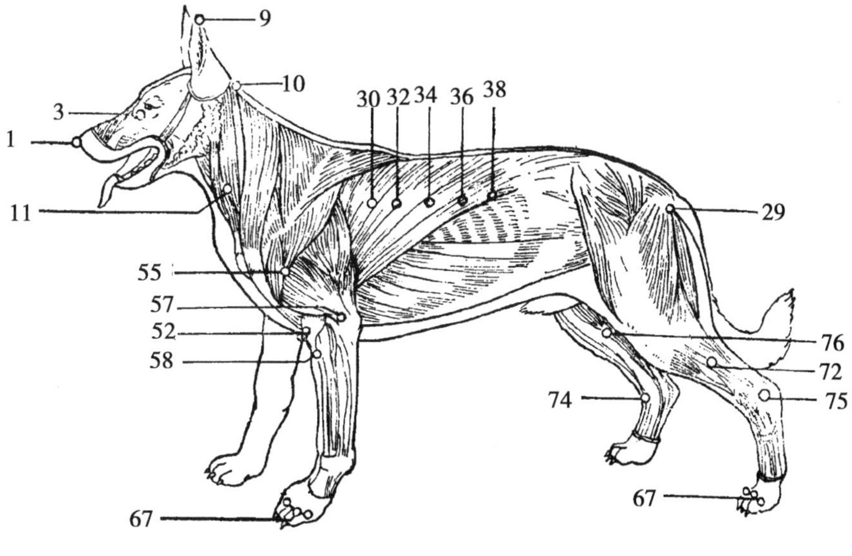

图 4－14　犬的肌肉的及穴位

1. 人中　3. 三江　9. 耳尖　10. 天门　11. 颈脉　29. 后海　30. 肺俞　32. 心俞　34. 膈俞　36. 胆俞　38. 胃俞　52. 胸堂　55. 抢风　57. 肘俞　58. 四渎　67. 六缝　72. 后三里　74. 中付　75. 后跟　76. 肾堂

第五讲

中兽医内科学

中兽医内科学是运用中兽医学理论研究内科疾病辨证论治规律的一门临床学科。中兽医内科疾病主要可以分为外感病和内伤病两大类。一般说来，外感病主要指伤寒、风温、暑温、湿温等急性病，主要根据六经、卫气营血和三焦辨证方法遣治；内伤病包括脏腑经络诸病，主要以脏腑、气血津液、经络等辨证方法论治。

1　中兽医内科疾病的特点

1.1　病因以内伤为主

内科疾病的病因大都以内伤为主。如劳倦过度，可引起水肿、消渴、阳痿、遗精等；水草失宜，常致胃痛、呕吐、呃逆、噎膈之病。

在内科疾病的病因中，还有一类因脏腑功能失调而产生的内风、内寒、内湿、内燥、内火等内生之邪，因其临床表现与外感六淫之邪有相似之处，但究其原因，不是外来之邪，而是由内而生，故又称为“内生五邪”。如“内风”的表现主要是肢体麻木、四肢抽搐等，常见于中风等病证；“内寒”的表现是畏寒肢冷、可视黏膜苍白、泛吐清水、脘腹冷痛等，常见于胃痛、腹痛等证；“内燥”的主要表现是口咽干燥、干咳无痰、皮肤干涩、大便干结等，常见于咳嗽、便秘等证；“内火”主要表现是目赤、易怒、口干等，常见于血证、郁证、中风等；“内湿”的表现主要是呕吐、纳呆腹胀、大便溏薄等，常见于泄泻等病证。

在内科疾病的病因中还有痰饮、瘀血等病邪，也都是因为脏腑功能失调而产生的病理产物。这一类病理产物形成之后，也可成为致病的“病邪”，进一步引起多种病理变化，形成多种病证。如痰饮伏肺，可引起咳嗽、哮喘等病证。瘀血内停可致胃痛、腹痛等多种病证。

1.2　病机多寒热虚实错杂

内科疾病由于大多病程较长，既有寒热等病邪内盛的一面，又有脏腑气血阴阳亏虚的一面，所以内科疾病的病机大多寒热虚实错杂。

内科疾病虽多，但究其病理性质不外虚实两类：凡气虚、血虚、阴虚、阳虚，以及气血两虚，阴阳俱损等正气不足之候，皆属虚证；气滞、血瘀、水停、湿热、痰饮等病理产物，其病理过程大多属实证。

寒证与热证，是用以辨明病证的属性。一般而言，内伤疾病中的寒证与热证，多属脏腑功能失调、阴阳失于平衡所致，故其寒热常与其他虚实之象并见，较少单独出现。

2　中兽医内科疾病的诊治步骤

中兽医内科疾病的诊治步骤，一般可归纳为四诊识病、辨性定位、求因审机、确立治法和遣方用药等几个方面。

2.1　四诊识病

诊治疾病，首先是通过“望、闻、问、切”四诊，对病畜做周密的观察和全面的了解。既要了解病畜的病史和临床表现，又要了解病畜的外在环境对疾病发生、发展的可能影响。将检查所得，进行分析归纳，运用从外测内、见证推病、以常衡变的方法来判断患者的病情，以此作为辨证、立法、处方用药的依据。

四诊是辨证论治的第一步，也是最重要的一个环节。

四诊搜集的资料是否准确，是否切合病情，与辨证正确与否有着密切的关系。因此，在进行四诊时，不但要做到全面系统，还要做到重点突出，详而有要，简而不漏，使四诊资料更好地为辨证提供必要依据。

2.2　辨性定位

辨性，就是辨别病证的性质。疾病发生的根本在于邪正斗争引起的阴阳失调，故其病性无非是阴阳的偏盛偏衰。阳盛则热，阴盛则寒，故病性具体表现在寒热属性上。而虚实是邪正双方消长盛衰的反映，也是构成病变性质的一个重要方面。寒热虚实是一切病变中最基本的性质，各种疾病都离不开这四个方面。由于基本病变是虚实寒热，所以治疗的总原则，就是补虚、泻实、清热、温寒。辨清病变性质的目的，在于对病证有一个基本的认识，治疗上有一个总的原则，因此辨识病证性质是辨证中的一项重要内容。

定位，指判定病变部位。一般包括：表里定位，这在外感病辨证时很重要；脏腑、经络定位，多用于杂病；气血定位，通常杂病要区分气分病、血分病。这些定位方法或简或繁，各有其适用范围，有时需结合应用。

2.3　求因审机

求因就是审证求因。它是辨证的进一步深化，是根据病畜一系列具体证候，包括病畜临床症状、四诊和某些化验检查结果，以及发病的时间、季节、病畜所处的环境、气候等，加以综合分析，求得疾病的症结所在，为临床治疗提供确切的依据。

2.4　确立治法

确立治法，简称“立法”，是根据辨证的结果确立相应的治疗原则和方法。如肝火犯肺的咳嗽，采用清肝肃肺法。立法必须与病机相吻合。

2.5　选方遣药

选方是根据辨证立法的结果选用适当的治疗方剂。方剂是因法而立、针对证候而设，具有固定的药物组成与配伍，有其一定的适用范围。因此，要选择好恰当的方剂，必须熟悉方剂的组成、方义和药物配伍关系及其适用范围。方剂是前人临床经验的总结，也是历代医家在有关学术理论指导下，在对某些病证长期实践的基础上，所创造的各种治疗方法的具体应用。临证时应该重视它，学习运用它，并在前人的基础上不断发展和创新。在临床上要防止有法无方、杂药凑合的弊病；也有不拘成方，随证检药而法度井然者，在临床实践中，都必须不断总结和提高。选方必须在治疗法则的指导下进行，所选方剂应符合其病机和治疗法则。

遣药是在选定方剂的基础上，随证加减药物。由于病证的复杂多变，很难有与具体的病情完全吻合、可直接应用的成方。所以，应根据病证的兼挟情况，针对具体病情加减药物。这是对方剂的灵活应用，使之更能贴切病情。

总之，上述各个诊治疾病的步骤包含了辨证论治的全过程，即“理、法、方、药”各个方面。“理”就是病因病机，“法”即治疗法则，“方”即所选方剂，“药”即所用药物。理、法、方、药必须一致，才能达到良好的治疗效果。以上四诊识病、辨性定位、求因审机的内容，都属于辨证的范围，是辨证论治中的“理”；确立治法、选方遣药，则是论治的具体体现。这样，便构成了辨证论治的理、法、方、药的统一。只是为了叙述方便和利于学习、掌握，才分为若干具体的步骤加以表述，在临床应用时，并不是绝对按这样的顺序，有时需相互运用。例如，诊察是搜集临床资料的阶段，是辨证论治的前提，但在诊察过程中，实际上涉及辨性定位、求因审机，彼此之间又有着紧密不可分割的联系。所以，在临床上不必拘泥于这种程式或先后次序，可以根据具体病情和自己掌握的熟练程度灵活运用。

3　中兽医内科疾病的治疗原则

治疗原则又称治则，治疗法则。它是在中兽医基本理论和辨证论治精神指导下制定的，对疾病治疗的立法、处方、用药等具有指导意义。

3.1 正治反治

正治也称逆治，是最常用的治法。即指疾病的征象与本质相一致时所采取的治疗原则。寒者温之，热者寒之，虚者补之，实者泻之，均为正治法。

反治也称从治，是在特殊情况下所采取的治疗法则。即指疾病的某些征象与病变的本质不相一致时，采取顺从这些假象而治的法则。但实际上还是逆其疾病本质而治疗。如寒因寒用，热因热用，塞因塞用，通因通用，均为反治。再如真热假寒证，用寒凉的白虎汤治疗，均是反治的具体运用。

3.2 扶正祛邪

扶正即补法，用于虚证；祛邪即泻法，用于实证。疾病的过程，在某种意义上可以说是正气与邪气相争的过程，邪胜于正则病进，正胜于邪则病退。因此，扶正祛邪就是改变邪正双方力量的对比，使之有利于疾病向痊愈转化。

用于扶正的补法有益气、养血、滋阴、助阳等；用于祛邪的泻法有发表、攻下、渗湿、利水、消导、化瘀等。扶正与祛邪，两者又是相辅相成的，扶正有助于抗御病邪，而祛邪则有利于保存正气和恢复正气。

在一般情况下，扶正适用于正虚邪不盛的病证，而祛邪适用于邪实而正虚不显的病证。扶正祛邪并举，适用于正虚邪实的病证。在具体应用时，也应分清以正虚为主，还是以邪实为主。以正虚较急重者，应以扶正为主，兼顾祛邪；以邪实较急重者，则以祛邪为主，兼顾扶正。若正虚邪实以正虚为主，正气过于虚弱不耐攻伐，倘兼以祛邪反而更伤其正，则应先扶正后祛邪；若邪实而正不甚虚，或虽邪实正虚，倘兼以扶正反会助邪，则应先祛邪后扶正。总之，应以扶正不留邪，祛邪不伤正为原则。

3.3 脏腑补泻

由于畜体是有机的整体，脏腑之间在生理上相互联系，在病理上相互影响，一脏有病往往影响到他脏，而他脏的情况有了改变，也会反过来影响原发病的脏腑。临床上常应用脏腑之间的生克表里关系，作为补泻治法的原则。这些原则可概括为虚则补其母，实则泻其子；壮水制阳，益火消阴；泻表安里，开里通表，清里润表三个方面。

3.4 异法方宜

异法方宜治则，是指疾病的治疗不能固守一法，对不同的个体、不同的时间、不同的地域等应采取有针对性的、不同的治疗方法，方为适宜。

4 中兽医内科疾病的常用治法

4.1 汗法

汗法又称发汗法，是通过开泄腠理，以发汗的形式逐邪外出的一种治法。本法具有解表、退热、消肿、祛风湿、透斑疹等作用，广泛适用于邪遏肌表，宜从汗泄的病证。如外感表证或某些急性传染病的初期；具有表证的水肿、痹证及麻疹将透未透阶段。运用汗法时不宜过汗、妄汗，注意个体差异，掌握用药的峻缓。

4.2 吐法

吐法又称涌吐法，是运用具有催吐作用的方药，或以物探导，通过引发呕吐反应，促使有害病邪排出体外的治法。本法主要包括峻吐法、缓吐法、探吐法、洗胃法，适用于痰涎、宿食、毒物滞留在咽喉、胸膈、胃脘等病证。如癫狂、昏迷实证、误食毒物、过食草料等。对年老体弱、病危或有肺出血、上消化道出血病史者，以及妊娠和产后者，应禁用本法。

4.3 下法

下法又称攻下法，是运用有泻下或润下作用的药物，以通导大便，消除积滞，荡涤实热，攻逐水饮的治法。本法有寒下、温下、润下及逐水之别，具有通便、下积、泻火、逐水等作用，广泛适用于有各种里实证的病畜，如燥屎秘结、食积虫积、蓄血、痰饮等阻于肠胃者；内热壅盛、痢疾初期、食物或药物中毒等病证者。使用下法应把握时机，权衡攻下之峻缓，以邪去为度，不宜过量。

4.4　和法

和法是和解少阳、扶正达邪、协调内脏功能的一种治法。和法的内容丰富，应用广泛，如外感疾病宜采用和解表里、和解营卫，内伤杂病则主要采用调和肝脾、调和胆胃以及调和胃肠诸法。运用和法时应掌握适应证，不可认为和法药性平稳而滥用。

4.5　温法

温法又称温里法，是使用温性或热性的药物，以祛除寒邪，补益阳气的一种治法。本法具有回阳救逆、温中散寒、温经通阳等作用，广泛应用于因寒邪所致的呕逆泄泻、肢冷畏寒、拘挛疼痛的里寒病证。素体阴虚、咽喉干燥者，或有吐血、衄血等出血病史者，应慎用本法。

4.6　清法

清法又称清热法，是采用寒凉泄热的药物和措施，以消除热证的治法。本法具有清气分热、清营凉血、清热解毒、清脏腑热、保津存阴等作用，广泛应用于因温热病邪所致的发热、脏腑功能亢进、烦躁、出血斑疹、疮疡等病证。运用清法时注意辨别寒热真假、虚火、实火，审证而清，因畜而清，以防伤正。

4.7　消法

消法又称消导或消散法，即使用具有消导、消散、软坚、散结等功效的方药，治疗有形积滞的一种治法。本法应用广泛，适用于饮食停滞、瘕积肿块、痰核瘰疬、结石疮痈等病证。消法用药多为克伐之剂，对气血虚弱、脾肾虚寒者应慎用。

4.8　补法

补法又称补益法，是用具有补益作用的药物，补益畜体阴阳气血的不足，或补益某一脏腑之虚损的治法。本法有补气、补血、补阴、补阳等作用，广泛适用于先天不足，或后天失调，或久病、重病后证见气、血、阴、阳、津液等不足之虚证，或应用攻伐之剂而须固护正气者。运用补法时应兼顾气血，调补阴阳，五脏分补，不可妄补、滥补。

5　疾病治疗

5.1　感冒

【概述】

感冒主要是由病毒引起的上呼吸道感染性疾病，分为普通感冒与流行性感冒两种。临床主要表现为鼻塞，流涕，喷嚏，咳嗽，恶寒发热。普通感冒主要由鼻病毒所引起，以鼻咽部证候为主，一般病情较轻微；流行性感冒是由流感病毒引起，全身中毒证候明显，易引起大流行。本病发病率高，具有反复感染的特点，一年四季均可发生，尤以冬春多见。

根据本病的临床表现，属于中兽医学“伤风”、“时行感冒”等范畴。其病因主要为外感风邪疫毒。主要病机为邪侵肺卫，卫表不和，一般以肺卫表实证居多。

【诊断要点】

（1）有与感冒病畜接触史，起病多急。

（2）鼻塞流涕，喷嚏，咳嗽，恶寒发热，无汗或少汗。

（3）高热或伴呕吐、腹泻、流涕，咳嗽较甚，又有流行趋势者，应考虑患流行性感冒的可能。

（4）实验室检查可见血液白细胞总数正常或偏低，中性粒细胞减少，淋巴细胞相对增多。若由细菌感染引起的，白细胞总数可以增高。

【辨证论治】

本病辨证主要是辨其表寒和表热。恶寒重，发热轻，无汗者，多属表寒；恶寒轻，发热重，有汗，咽喉红肿者，多属表热。临床主要分为风寒证、风热证、表寒里热证和暑湿证。治疗当采取解表达邪的原则，风寒治以辛温发汗，风热治以辛凉清解，表寒里热治以散寒清热，暑湿挟感者又当清暑祛湿解表。

5.1.1　风寒证

【主证】恶寒重，发热轻，无汗，鼻塞声重，流涕清稀，喉痒咳嗽，痰吐稀薄，舌苔薄白，脉浮或浮紧。

【治法】辛温解表。

【方药】荆防败毒散加减：荆芥30克，防风30克，前胡25克，杏仁30克，枳壳25克，川芎21克，紫苏叶21克，桂枝21克，甘草21克，生姜21克。

【加减】恶寒重，无汗等表寒症状重者，加麻黄21克，桔梗24；挟湿者，加白芷30克，羌活30克；鼻塞流清涕者，加苍耳子21克，辛夷18克；咳痰无力，舌苔淡白，脉浮无力者，加党参30克，茯苓30克，葛根30克。

5.1.2 风热证

【主证】发热重，微恶风寒，汗出不畅，咽干咽痛或咽喉乳蛾红肿疼痛，口渴欲饮，鼻塞流浊涕，咳嗽痰黄，舌尖红，苔薄黄，脉浮数。

【治法】辛凉解表。

【方药】银翘散合葱豉桔梗汤加减：金银花35克，连翘35克，淡豆豉30克，薄荷（后下）15克，牛蒡子30克，杏仁30克，桔梗24克，甘草15克。

【加减】咳嗽痰多者，加浙贝母21克，瓜蒌皮25克；咽痛红肿者，加玄参20克，土牛膝25克，一枝黄花35克；高热口渴甚者，加知母30克，天花粉35克；时行热毒主要证候明显者，加大青叶40克，板蓝根40克，草河车30克；秋季兼燥邪伤畜，伴有咳呛痰少，口咽鼻唇干燥，苔薄少津者，加南沙参33克，天花粉35克；兼见干咳痰少，舌红少苔，脉细数者，加玉竹30克，白薇30克。

5.1.3 表寒里热证

【主证】发热恶寒，无汗口渴，鼻塞声重，咽痛，咳嗽气急，痰黄黏稠，尿黄便秘，舌边尖红，苔黄白相兼，脉浮数。

【治法】疏风宣肺，散寒清热。

【方药】麻杏石甘汤加减：麻黄35克，杏仁40克，防风30克，荆芥40克，黄芩40克，栀子40克，连翘30克，生石膏（先煎50克，桔梗24克，薄荷（后下）25克，生甘草15克.

【加减】外感较甚，恶寒，加紫苏叶30克，桔梗28克；里热较甚，咽喉红肿疼痛者，加板蓝根40克，土牛膝35克；大便秘结不通者，加制大黄25～30克；如风寒化热入里，症见高热不退，口渴，舌苔黄少津，舌质红绛，脉数者，宜用水牛角（先煎）60克，生地黄45克，竹叶30克，连翘35克，金银花35克，玄参40克，黄连30克，黄芩300克，石菖蒲40克，甘草15克。

5.1.4 暑湿证

【主证】身热，微恶风，汗少或汗出热不解，咳嗽痰黏，鼻流浊涕，渴不多饮，胸闷泛恶，小便短赤，舌苔薄黄而腻，脉濡数。

【治法】清暑祛湿解表。

【方药】新加香薷饮加减：香薷36克，金银花30克，鲜扁豆花30克，连翘40克，鲜荷叶40克，厚朴28克，芦根35克，生甘草15克。

加减暑热偏盛，加黄连25克，青蒿26克；湿困卫表，加豆卷30克，佩兰30克；里湿偏重，加苍术40克，白蔻仁（后下）35克；小便短赤，加六一散（包煎）30克，茯苓45克。

5.2 咳嗽

【概述】

咳嗽是肺系疾病的主要证候之一。一般认为，有声无痰为咳，有痰无声为嗽，但临床多为痰声并具，难以截然分开，故一般通称咳嗽。咳嗽既是独立性的病症，又是多种疾病的一个常见症状。究其成因，不外外感、内伤两大类。外感咳嗽为六淫外邪侵袭于肺；内伤咳嗽为脏腑功能失调，内邪干肺。不论邪从外入，或自内而发，均可引起肺失宣肃，肺气上逆作咳。

西医学的上呼吸道感染、急慢性支气管炎、支气管扩张、肺炎、肺结核等疾病均可表现以咳嗽为主症。

【诊断要点】

（1）咳逆作声，或伴咽痒咳痰。

（2）外感咳嗽，起病急，可伴有恶寒发热等外感表证。内伤咳嗽，多反复发作，病程较长，伴有其他脏腑功能失调的证候。

（3）两肺听诊可闻及呼吸音增粗，或伴有干湿性啰音。

（4）实验室检查急性期血白细胞总数和中

性粒细胞增高。

（5）肺部X线摄片检查，肺纹理正常或增多、增粗等。

【辨证论治】

本病证临床分为外感咳嗽与内伤咳嗽两大类。前者主要包括风寒袭肺证、风热犯肺证、温燥伤肺证、凉燥伤肺证；后者主要包括痰湿蕴肺证、痰热郁肺证、肝火犯肺证、肺阴亏耗证、肺气亏虚证。治疗应分清邪正虚实。外感咳嗽，治以祛邪利肺；内伤咳嗽治以祛邪扶正，标本兼顾。

5.2.1　外感咳嗽

5.2.1.1　风寒袭肺证

【主证】咳嗽声重，咳痰稀薄色白，气急咽痒，鼻塞流涕，恶寒发热，无汗，舌苔薄白，脉浮紧。

【治法】疏风散寒，宣肺止咳。

【方药】止嗽散合三拗汤加减：麻黄30克，杏仁40克，荆芥40克，紫菀30克，百部30克，紫苏叶30克，白前30克，桔梗24克，甘草15克。

【加减】风寒表证重者，加防风30克，羌活30克；外寒内热者，去白前、紫菀，加生石膏（先煎）40克，桑白皮45克，黄芩30克；咳嗽较重者，加金沸草30克。

5.2.1.2　风热犯肺证

【主证】咳嗽气粗，咯痰色白或黄，喉燥咽痛，或有发热，微恶风寒、鼻流黄涕，口渴，舌尖红，苔薄黄，脉浮数。

【治法】疏风清热，肃肺化痰。

【方药】桑菊饮加减：桑叶40克，菊花40克，杏仁40克，桔梗20克，芦根45克，连翘35克，薄荷（后下）24克，甘草15克。

【加减】咳嗽重者，加浙贝母30克，枇杷叶（包煎）30克，前胡30克；发热口渴明显者，加黄芩30克，知母40克，瓜蒌皮30克；咽痛声嗄者，加射干30克，赤芍药35克；口干少津，舌质红者，加南沙参40克，天花粉40克；夏季挟暑者，加六一散（包煎）30克，鲜荷叶40克。

5.2.1.3　温燥伤肺证

【主证】干咳少痰，咯痰不爽，咽喉干痛，鼻燥咽干，口干，舌尖红，苔薄黄少津，脉细数。

【治法】疏风清肺，润燥止咳。

【方药】桑杏汤加减：桑叶30克，杏仁30克，南沙参40克，浙贝母40克，前胡30克，栀子35克，淡豆豉24克。

【加减】燥热明显者，加石膏（先煎）60克，知母30克；咽痛明显者，加玄参40克，马勃30克；鼻衄或痰中挟血者，加白茅根45克，生地黄40克；津伤较甚者，加麦门冬30克，玉竹40克。

5.2.1.4　凉燥伤肺证

【主证】咳嗽，痰少或无痰，喉痒，咽干唇燥，恶寒发热，无汗，舌苔薄白而干，脉浮紧。

【治法】疏散风寒，润肺止咳。

【方药】杏苏散加减：杏仁40克，紫苏叶40克，桔梗30克，白前40克，百部40克，紫菀30克，款冬花30克，陈皮35克，甘草15克。

【加减】恶寒甚者，加荆芥40克，防风40克。

5.2.2　内伤咳嗽

5.2.2.1　痰湿蕴肺证

【主证】咳嗽痰多，咳声重沉，晨起为甚，痰黏腻或稠厚成块，色白量多，呕恶食少，体倦，舌白腻，脉濡滑。

【治法】健脾燥湿，化痰止咳。

【方药】二陈汤合三子养亲汤加减：陈皮40克，法半夏30克，苍术35克，厚朴35克，茯苓45克，紫苏子30克，白芥子30克，莱菔子40克，薏苡仁40克。

【加减】寒痰较重，痰黏白如沫，怕冷者，加干姜30克，细辛25克；久病脾虚，神倦者，加党参40克，白术40克；痰湿化热，痰转黄者，加黄芩40克，瓜蒌皮40克。

5.2.2.2　痰热郁肺证

【主证】咳嗽气粗，痰多质稠，色黄难咯，或有热腥味或吐血痰，胸肋胀满，咳时引痛，面赤或有身热，口干欲饮，舌质红，苔黄腻，脉滑数。

【治法】清热化痰，肃肺止咳。

【方药】清金化痰汤加减：桑白皮45克，黄芩40克，栀子40克，浙贝母30克，知母30克，桔梗30克，全瓜蒌35克，橘红26克，甘草15克。

【加减】痰黄如脓或腥臭者，加鱼腥草50克，金荞麦60克，冬瓜子45克；胸满，咳逆痰涌，便秘者，加葶苈子30克，生大黄（后下）25克；津伤口渴甚者，加南沙参40克，麦门冬30克，天花粉40克。

5.2.2.3　肝火犯肺证

【主证】咳呛气逆阵作，咳时胁痛，面红目赤，咽喉干燥，心烦口苦，常感痰滞咽喉，咯之难出，量少质黏，舌苔薄黄少津，脉弦数。

【治法】清肺平肝，顺气降火。

【方药】加减泻白散合黛蛤散加减：桑白皮45克，地骨皮45克，青黛（另冲）30克，海蛤壳30克，黄芩30克，知母30克，天花粉40克，甘草15克。

【加减】火热较盛，咳嗽频作者，加牡丹皮40克，栀子30克。

5.2.2.4　肺气亏虚证

【主证】病久咳声低微，咳而伴喘，咯痰稀薄色白，食少，气短胸闷，神倦乏力，舌淡嫩，苔白，脉细弱。

【治法】补益肺气，化痰止咳。

【方药】补肺汤加减：黄芪45克，党参40克，紫菀30克，桑白皮30克，五味子46克，法半夏30克，茯苓40克，白术40克，陈皮35克，甘草15克。

【加减】痰多呈白沫者，加干姜30克，细辛30克，吴茱萸35克；咳逆气短者加诃子30克，补骨脂40克；畏寒，肢冷者，加肉桂（后下）30克，制附子30克。

5.2.2.5　肺阴亏耗证

【主证】干燥，咳声短促，痰少黏白，或痰中带血，咽干口燥，舌质红、少苔，脉细数。

【治法】滋阴润肺，止咳化痰。

【方药】沙参麦冬汤加减：南沙参45克，麦门冬42克，玉竹42克，百合32克，天花粉42克，川贝母30克，杏仁30克，桑叶30克，甘草15克。

【加减】咳而气促者，加五味子40克，诃子40克；盗汗量多者，加乌梅46克，浮小麦45克；咯吐黄痰者，加知母40克，黄芩40克；痰中带血者，加牡丹皮40克，栀子40克，藕节40克。

5.3　便秘

【概述】

便秘指大便秘结不通，排便时间延长，或欲解大便而艰涩不畅的一种病证。

本病主要因饮食不当、素体阳盛及病后体虚所致。病机总属肠腑传导失常。病在大肠，与脾胃肝肾相关。病理性质有虚实之分。因肠胃积热、气机郁滞者属实，因阳气不足、阴血亏虚者属虚。

本病可见于西医学中的功能性便秘、肠神经官能症及药物所致的便秘。

【诊断要点】

（1）排便次数减少，排便周期延长；或粪质坚硬，便下困难，或排出无力，出而不畅。

（2）常兼有腹胀、腹痛、纳呆、口臭、排便带血以及汗出气短等兼杂证。

（3）发病常与外感寒热、饮食、脏腑失调、坐卧少动、年老体弱等因素有关。起病缓慢，多表现为慢性病变过程。

（4）实验室检查如纤维结肠镜等有关检查，常有助于部分便秘的诊断。

【辨证论治】

便秘一证有虚有实，实证包括肠胃积热、肠道气滞、阴寒积滞证，虚证包括脾气虚弱、血虚津少、脾肾阳虚证。治疗以通下为原则，但决不可单纯用泻下药，需结合其病变性质而施以相应的治法。实证常用清热泻火、理气导滞、散寒温中法，虚证常用补益脾气、养血润肠、温阳开秘法。

5.3.1　肠胃积热证

【主证】大便干结，腹胀腹痛，口干口臭，小便短赤，舌质红，苔黄燥，脉滑数。

【治法】泻热导滞，润肠通便。

【方药】麻子仁丸加减：火麻仁35克，白芍药30克，枳实30克，厚朴30克，大黄（后下30克，杏仁30克，白蜜100毫升。

【加减】津液已伤，可加生地黄30克，玄

参40克，麦门冬40克；易怒目赤者，加牡丹皮210克，栀子30克；热甚便血者，加槐花30克，地榆30克。

5.3.2　肠道气滞证

【主证】大便干结，或不甚干结，欲便不得出，或便而不爽，肠鸣矢气，腹中胀痛，胸胁满闷，嗳气频作，食少纳呆，舌苔薄腻，脉弦。

【治法】顺气导滞。

【方药】六磨汤加减：槟榔30克，乌药35克，广木香30克，枳实30克，沉香粉（冲服）33克，大黄（后下）40克，郁李仁30克。

【加减】气郁化火，加黄芩30克，栀子30克，龙胆草30克；气逆呕吐者，加制半夏20克，旋覆花（包煎）20克，代赭石（先煎）50克；跌仆损伤，便秘不通，属气滞血瘀者，加桃仁40克，红花30克，赤芍药30克。

5.3.3　阴寒积滞证

【主证】大便艰涩，腹痛拘急，胀满拒按，胁下偏痛，肢体不温，呃逆呕吐，舌苔白腻，脉弦紧。

【治法】温里散寒，通便止痛。

【方药】大黄附子汤加减：熟附子30克，生大黄（后下）30克，细辛20克，枳实30克，厚朴30克，木香30克，干姜25克，茴香20克。

【加减】腹部冷痛者，加肉桂30克，乌药25克。

5.3.4　脾气虚弱证

【主证】粪质并不干硬，虽有便意，但便难排出，汗出气短，便后乏力，神疲肢倦，舌质淡，苔白，脉弱。

【治法】补气润肠。

【方药】黄芪汤加减：炙黄芪45克，党参40克，陈皮30克，火麻仁40克，当归30克，白蜜100毫升。

【加减】气虚较甚者，加白术40克；气虚下陷脱肛者，加升麻30克，柴胡30克，桔梗25克；大便燥结难下者，加郁李仁40克，杏仁40克，肉苁蓉30克；爪甲淡白者，加生何首乌30克，生地黄40克；喘促甚者，加蛤蚧粉（冲）20克；纳谷不佳者，加炒麦芽45克。

5.3.5　血虚津少证

【主证】大便干结，面色无华，气短，口唇色淡，舌质淡，苔白，脉细。

【治法】养血润燥。

【方药】《尊生》润肠丸加减：制大黄40克，当归42克，生地黄30克，火麻仁50克，桃仁40克，枳壳30克，生何首乌40克，柏子仁40克。

【加减】血虚内热者，可加知母30克，胡黄连35克；阴血已复，大便仍干燥者，加郁李仁40克，松子仁30克。

5.3.6　脾肾阳虚证

【主证】大便干或不干，排出困难，小便清长，四肢不温，腹中冷痛，得热则减，腰膝冷痛，舌质淡，苔白，脉沉迟。

【治法】温阳通便。

【方药】济川煎加减：当归42克，怀牛膝30克，肉苁蓉45克，升麻30克，枳壳30克，干姜35克，炙附片（先煎）30克，肉桂30克。

【加减】夜尿多，加金樱子35克，乌药30克，山药45克；腹痛明显者，加木香30克，延胡索30克。

5.4　泌尿系感染

【概述】

泌尿系感染是常见的感染性疾病，指病原体在尿中生长繁殖并侵犯泌尿道黏膜或组织而引起的炎症。临床分为上泌尿道感染（输尿管炎和肾盂肾炎）和下泌尿道感染（膀胱炎和尿道炎）。下泌尿道感染可单独存在，上泌尿道感染则多伴发下泌尿道炎性症状，临床上不易严格区分。肾盂肾炎又分为急性和慢性两期，大都由下尿道感染引起。慢性肾盂肾炎是导致慢性肾功能不全的一个重要原因。

本病属于中兽医学“淋证”、“癃闭”、“腰痛”等病证范畴。主要因下阴不洁，秽浊之邪从下入侵，热蕴膀胱，由腑及脏；饮食不节，过食肥甘辛辣，脾失健运，酿湿生热，湿热下注；年老体衰，肾气亏虚，膀胱气化无权。病位在肾和膀胱，主要病机是湿热蕴结下焦，膀胱气化失调。病久不愈，湿热耗伤正气，可致脾肾亏虚。

【诊断要点】

（1）本病急性期主要表现为尿频、尿急、尿痛，腰痛或向阴部下传的腹痛，常伴寒战、发热、头痛、乏力、食欲不振、恶心等全身症状。慢性期患者平日也常有尿频、尿急、尿痛、腰痛等不适症状。慢性期急性发作时，全身症状可与急性期一样剧烈。

（2）体征主要有脊肋点（腰大肌外缘与十二肋交叉点）压痛，肾区叩击痛阳性。

（3）尿常规检查可见脓尿，高倍镜下每视野白细胞数常在5个以上，并常出现白细胞管型。尿细菌培养（清洁中段尿培养），菌落计数 $>10^5$ 个/毫升。

【辨证论治】

本病临床辨证主要是辨虚实。实证系湿热蕴结下焦，膀胱气化不利所致，病程较短，小便涩痛不利，苔黄舌红，脉实数；虚证系脾肾两虚，膀胱气化无权，病程长，小便频急，痛涩不甚，苔薄舌淡，脉细数。实证治予清热利湿，虚证治当培补脾肾。

5.4.1 膀胱湿热证

【主证】小便频数，淋沥不爽，尿色黄赤，灼热刺痛，急迫不舒，恶心呕吐，大便秘结，苔薄黄，脉濡数。

【治法】利湿通淋，清热解毒。

【方药】八正散加减：车前子（包煎）42克，瞿麦32克，木通30克，滑石（包煎）45克，萹蓄32克，生大黄（后下）30克，甘草梢25克，灯心草30克。

【加减】若伴有泌尿系结石者，加石韦35克，鸡内金30克，金钱草60克，郁金40克；若发热重，加金银花45克，黄芩42克，黄柏40克；若舌苔厚腻，湿热困重者，加薏苡仁40克，栀子40克。

5.4.2 热伤血络证

【主证】小便热涩刺痛，尿色深红，甚则夹有血块，甚则尿道挛急疼痛，腰痛拒按，高热恶寒，舌质红，苔黄厚燥，脉滑数。

【治法】清热利湿，凉血止血。

【方药】小蓟饮子加减：生地黄45克，小蓟35克，滑石（包煎）42克，木通30克，蒲黄（包煎）30克，淡竹叶30克，当归30克，栀子40克，炙甘草21克。

【加减】若热邪较重，高热恶寒，加金银花40克，连翘45克；大便秘结者，加生大黄（后下）40克；热邪伤阴，阴虚火旺，加知母40克，黄柏40克，女贞子42克，旱莲草42克。

5.4.3 脾肾两虚证

【主证】发病日久，缠绵难愈，时轻时重，遇劳及遇寒加重，小便赤涩不甚，淋沥不已，夜尿次数增多，舌淡，脉虚数。

【治法】健脾益肾。

【方药】无比山药丸加减：山药40克，茯苓42克，泽泻32克，熟地黄45克，山茱萸42克，巴戟天42克，菟丝子42克，杜仲32克，牛膝32克，五味子45克，肉苁蓉30克。

【加减】若肾阴不足，湿热留恋，咽干唇燥，尿频而短，小便涩痛者，加知母40克，黄柏40克；若因劳倦病情加重，小便点滴而出，精神疲惫，加党参40克，黄芪42克，白术40克，升麻36克，柴胡36克。若湿热未尽者，加萹蓄35克，瞿麦35克。

5.5 急性卡他性结膜炎

【概述】

急性卡他性结膜炎是常见的细菌感染性眼病。临床主要表现为眼部结膜明显充血，脓性或黏液性分泌物，有自愈趋势。本病传染性强，常在温暖季节流行。

本病的致病菌最常见的有结膜炎杆菌（Koch—Weeks）、肺炎双球菌、葡萄球菌和流行性感冒杆菌等。

根据本病的临床表现，当属于中兽医学“暴风客热”范畴。外因多为风热之邪，突从外袭，内因为内热阳盛。主要病机为风热内外相合，上攻白睛，以致猝然发病。

【诊断要点】

临床表现

（1）轻者患眼瘙痒不适和异物感，重者羞明灼热，眼睑沉重。也可因为分泌物过多而视物不清，但拭去分泌物后即可恢复视力。

（2）轻者仅睑结膜及穹窿部结膜充血，重

者还有明显的球结膜充血，甚至结膜水肿，眼睑肿胀，结膜囊内大量黏液或脓性分泌物。部分患者球结膜有点片状出血。

（3）本病多为双侧型，双眼同时或先后发病。轻症通常3～4日内发展到最高峰，8～14日消退。肺炎双球菌引起者，8～10日到达极限，而后立即好转。Koch—Weeks杆菌引起者较重，约需2～4周方可痊愈。

实验室检查

结膜囊分泌物培养多可找到Koch—Weeks杆菌、肺炎双球菌、葡萄球菌或流行性感冒杆菌。这些细菌在发病3～4日内繁殖旺盛，晚期或临床用药之后即不易找到。

【辨证论治】

本病总属肺经风热为患，临床主要分为风重于热证、热重于风证、风热并重证。治疗以宣肺疏风，清热泻火为原则，并辨清其风重、热重或风热并重，风重以祛风为主，热重以清热为主，风热并重宜祛风清热。

5.5.1　风重于热证

【主证】胞睑微肿，白睛红赤轻，痒涩兼作，羞明多泪，恶风发热。舌苔薄白或微黄，脉浮数等。

【治法】疏风解表，兼以清热。

【方药】羌活胜风汤加减。羌活40克，防风40克，独活40克，荆芥30克，柴胡30克，白术40克，薄荷（后下）30克，白芷30克，川芎30克，枳壳30克，黄芩40克，桔梗25克，前胡25克，甘草21克。

【加减】风邪不盛者，去羌活、独活；热象较明显者，加金银花40克，连翘40克，桑白皮40克，菊花30克。

5.5.2　热重于风证

【主证】白睛红赤肿胀较甚，眵多粘结，热泪如汤，胞睑红肿，烦躁口渴，溺黄便秘。舌红苔黄，脉数有力。

【治法】清热泻火，兼以疏风。

【方药】泻肺饮加减。石膏（先煎）40克，黄芩20克，桑白皮20克，连翘20克，山栀子20克，赤芍药10克，枳壳20克，木通20克，荆芥30克，防风30克，白芷30克，羌活20克，甘草15克。

【加减】热毒炽盛者，去羌活、白芷，加金银花30克，蒲公英35克，鱼腥草35克；大便秘结甚者，加大黄35克，芒硝36克。

5.5.3　风热并重证

【主证】胞睑、白睛红赤肿胀，痛痒交作，灼热羞明，热泪如汤，眵多粘结，恶寒发热，便秘溲黄，口渴思饮。舌红苔黄，脉数有力。

【治法】外散风邪，内泻火热。

【方药】防风通圣散加减。防风30克，荆芥30克，桔梗26克，薄荷（后下）26克，连翘30克，山栀子30克，黄芩30克，滑石40克，川芎30克，当归30克，白芍药30克，大黄26克，甘草15克。

外治法

（1）点眼药。①黄连西瓜霜眼药水：每日点眼3～4次。适用于本病轻症者。②千里光眼药水：每日点眼3～4次。适用于本病轻症者。③熊胆眼药水：每日点眼3～4次。适用于本病各证。

（2）熏洗法。用新鲜野菊花40克，蒲公英60克，车前草60克，紫花地丁60克煎水熏洗患眼。每日1～2次。适用于本病重症者。

5.6　肠梗阻

【概述】

肠梗阻是指由于肠管的扭转、痉挛、麻痹或索带粘连等引起肠腔内容物不能顺利通过肠道，其临床特点是腹痛、腹胀、呕吐、肛门停止排气排便，并具有病因复杂、病情多变、发展迅速，并发症多等特点。中兽医认为本病属“关格”、“肠结”、“腹胀”范畴，多由于饮食不节，脾运失健，湿热内蕴，瘀血内停或寒邪凝滞、燥屎内结、蛔虫聚团致肠腑气机不畅，气血瘀滞，通降功能失常而发为本病。

【诊断要点】

（1）腹痛、腹胀、呕吐、肛门停止排气排便。

（2）机械性肠梗阻的腹痛多为阵发性绞痛，若发展为绞窄性肠梗阻则呈持续性绞痛。梗阻位置较高，则呕吐频繁，呕吐物为食物或胃液；梗阻位置较低，则腹胀明显，呕吐出现较晚且次数少，呕吐物多为粪样物。梗阻后期可出现唇干舌燥、眼窝内陷、皮肤弹性消失、尿少或无、面色苍白、四肢厥冷等休克症状。

（3）单纯性肠梗阻腹部检查可有压痛，或游走性包块，但一般无反跳痛和肌紧张，病情进一步发展，腹部压痛可逐渐加重，伴有固定性包块，反跳痛和肌紧张，并可见到肠型和蠕动波，听诊肠鸣音亢进，有气过水声或金属声，如出现麻痹性肠梗阻，听诊肠鸣音则减弱或消失。

（4）实验室检查可见患者外周血白细胞总数及中性粒细胞比例均明显升高；梗阻时间较长合并脱水时血红蛋白和红细胞压积可明显升高；合并酸中毒时血生化检查可见二氧化碳结合力明显下降；X 线摄片检查是判断肠梗阻的可靠方法，病变早期腹部透视可见小肠肠腔内积气，梗阻 4～6 小时后，腹部平片（立、卧位）可发现梗阻以上部位肠腔内有多个阶梯状气液平面，麻痹性肠梗阻则可见整个肠腔明显积气。

【辨证论治】

按照本节概述的分析，辨证论治，明确饮食、湿热、瘀血、内结、蛔虫等主因。

5.6.1 肠腑气滞证

【主证】相当于西医学早期肠梗阻和粘连性肠梗阻。症见腹中胀痛阵作，且胀重于痛，腹部包块时有时无，叩之如鼓，压痛轻，伴恶心呕吐、无肛门排气排便等症，舌淡红苔薄白，脉弦。

【治法】行气散结，通里攻下。

【方药】大承气汤加减。生大黄（后下）36 克，厚朴 27 克，枳实 30 克，黄芩 36 克，乌药30 克，青皮27 克，木香30 克，莱菔子36 克，芒硝（冲服）30 克，生甘草 18 克。

【加减】腹痛较甚者，加延胡索 36 克，川楝子 30 克；恶心呕吐较甚者，加制半夏 24 克，姜竹茹 30 克。

5.6.2 肠腑瘀结证

【主证】相当于各型肠梗阻已有不同程度的血运障碍或肿瘤性肠梗阻。症见腹部疼痛剧烈，痛重于胀，痛有定处，腹部拒按，有时呕吐咖啡样物，舌黯红或有瘀斑，苔黄腻，脉涩。

【治法】化瘀行气，通里攻下。

【方药】桃核承气汤加减。桃仁 30 克，当归 36 克，赤芍药 30 克，丹参 45 克，红花 18 克，生大黄（后下）36 克，厚朴 30 克，枳实 30 克，青皮 27 克，木香 30 克，莱菔子 36 克，芒硝（冲服）30 克。

【加减】腹胀较甚者，加香附 36 克，川楝子 30 克 ，伴有明显肿块、质地坚硬者，加炙乳香 18 克，炙没药 18 克，三棱 30 克。

5.6.3 虫结阻滞证

【主证】本证多有蛔虫病史。症见腹痛阵作，攻窜不定，腹部有时可扪及蛔虫团块，肛门排气时有时无，舌淡苔薄白，脉弦。

【治法】驱蛔行气，通腑攻下。

【方药】驱蛔承气汤加减。槟榔 30 克，乌梅 36 克，川楝子 30 克，生大黄（后下）30 克，厚朴 27 克，枳实 30 克，青皮 18 克，木香 30 克，莱菔子 30 克，芒硝（冲服）30 克。

【加减】伴有食积者，加神曲 36 克，炒麦芽 36 克；伴有呕吐者，加姜竹茹 30 克。

5.6.4 肠腑燥结证

【主证】本证多见于年老体弱、有习惯性便秘病史者。症见大便干结难解，渐出现腹胀、腹痛，左下腹触及粪块，伴不思纳谷等症，舌干少津苔厚腻，脉细数。

【治法】滋阴润肠，通里攻下。

【方药】增液承气汤加减。生地黄 45 克，玄参 36 克，麦门冬 45 克，生大黄（后下）30 克，厚朴 27 克，枳实 30 克，青皮 18 克，木香 30 克，莱菔子 36 克，芒硝 30 克（冲服），生甘草 18 克。

【加减】年老体虚者，加党参 30 克，当归 30 克；口渴欲饮者，加石斛 45 克，沙参 30 克。

5.6.5 水结湿阻证

【主证】本证相当于各型肠梗阻合并有肠腔积液者。症见腹痛阵阵加剧，肠鸣辘辘有声，腹胀拒按，伴恶心呕吐、口渴不欲饮、肛门无排气排便、小便短少等症，舌质红苔白腻，脉滑数。

【治法】行气逐水，通腑攻下。

【方药】甘遂通结汤加减。甘遂末（冲服）3 克，桃仁 30 克，炒枳壳 30 克，赤芍药 30

克，厚朴30克，生大黄（后下）30克，木香30克，牛膝30克 。

【加减】腹胀较甚者，加香附30克，川楝子30克；呕吐明显者，加姜半夏30克。

5.7　急性乳腺炎

【概述】

急性乳腺炎是指细菌侵入乳房所引起的乳腺急性化脓性疾病。致病菌多为金黄色葡萄球菌或链球菌。本病好发于产后未满月的产畜，其临床特点是高热恶寒、乳房内结块、局部红、肿、热、痛，排乳不畅，甚则破溃流脓。中兽医认为本病属“乳痈”范畴，发于哺乳期的称为“外吹乳痈”，发于怀孕期的称为“内吹乳痈”。本病多由病畜产后饮食不节，阳明积热，致乳络阻滞，乳汁郁结而化热成脓所致。另外本病也可由产畜乳头畸形，乳汁排泄不畅，或乳头破碎、毒邪外侵，肉腐成脓而成。

【诊断要点】

（1）本病好发于产后3～4周的哺乳期母畜，且多为初产畜。

（2）根据本病典型临床表现可将之分为三期。

初期：乳房肿胀疼痛，局部皮肤不红或微红，乳汁排泄不畅，乳房结块可有或无。

中期：乳房可扪及大小不等的结块，压痛明显，继则乳房结块渐增，红肿灼热，局部跳痛，波动应指。

溃后：乳痈破溃，脓水稠厚，脓去热泄，肉芽鲜活，疮面愈合迅速。少数疮面脓水淋漓，肉芽不鲜，经久难愈，甚者可发生袋脓现象。

（3）本病初中期可伴有明显的全身症状，如恶寒发热、小便黄赤、大便秘结等症，溃后正气虚弱者可伴乏力、精神疲劳等症。

（4）本病初中期外周血白细胞总数及中性粒细胞比例均明显增高。

【辨证论治】

按病情分析，明确饮食、湿热、郁结或病原菌等主因。

5.7.1　淤乳表热证

【主证】乳房胀痛，皮色不红或微红，乳汁排泄不畅，乳房内可扪及结块。伴恶寒发热、大便干结等症，舌淡红苔薄白，脉浮数。

【治法】疏风清热，通乳散结。

【方药】瓜蒌牛蒡子汤加减。全瓜蒌（打碎）45克，牛蒡子45克，金银花45克，连翘45克，柴胡30克，陈皮20克，黄芩30克，赤芍药30克，路路通30克，王不留行30克，荆芥30克，防风30克，生甘草15克。

【加减】哺乳期乳汁壅滞者，加鹿角霜27克，漏芦30克；乳房结块明显者，加当归尾30克，橘核60克；偏于气郁者，加川楝子30克，枳壳45克；产后恶露未尽者，加川芎30克，益母草60克 。

5.7.2　肝胃郁热证

【主证】乳房红肿热痛，结块表面肌肤灼热，触痛明显，甚则结块变软，波动应指，伴全身壮热烦躁、口渴欲饮、小便黄赤、大便干结等症。舌红苔黄，脉滑数。

【治法】清热解毒，消肿散结。

【方药】仙方活命饮加减。金银花30克，连翘45克，皂角刺60克，炮穿山甲（先煎）30克，当归36克，牡丹皮30克，山栀子45克，赤芍药30克，蒲公英90克，全瓜蒌（打碎）45克，生甘草15克。

【加减】偏于热甚，体温较高者，加生地黄30克，生石膏（先煎）60克；疼痛较甚者，加川楝子30克，炙乳香18克，炙没药18克；产畜断乳后乳汁壅胀者，加生麦芽90克，生山楂90克。

5.7.3　瘀热郁滞证

【主证】多见于急性乳腺炎治疗过程中大剂量使用抗生素而形成僵块，持续不消者。症见乳房结块僵硬，排乳不畅，疼痛隐隐，舌质紫黯苔薄黄，脉弦。

【治法】化瘀行气，通乳散结。

【方药】桃红四物汤加减。桃仁30克，红花30克，当归30克，赤芍药30克，丹参30克，川芎30克，乳香18克，没药18克，皂角刺60克，陈皮18克，路路通30克。

【加减】乳房肿块质地偏硬者，加三棱30克，莪术30克，橘核60克；乳汁量少或不通者，加通草20克，王不留行45克。

5.7.4 气血两虚证

【主证】溃后脓液稀薄，肿痛虽减，但疮面肉芽色淡红，且愈合不易，伴面色少华、神疲乏力等症，舌淡苔薄，脉细。

【治法】补益气血，生肌收口。

【方药】十全大补汤加减。当归30克，生黄芪60克，党参30克，白术30克，茯苓30克，熟地黄30克，太子参30克，麦门冬30克，红花20克，赤芍药30克，川芎30克，丹参30克，炙甘草20克。

【加减】余邪未尽者，加金银花30克，连翘45克；舌苔腻者，加陈皮20克，砂仁（后下）20克；胃纳不佳者，加焦山楂30克，神曲60克。

5.8 不孕症

不孕症是临床常见疾病，且有逐年上升趋势。在畜体生殖活动过程中，由综合性的、多方面的因素影响了受孕的任何一个环节，都可导致不孕症。

5.8.1 原发性不孕症

肾主生殖而藏精气，当肾气盛，精血充沛，天癸至，两精相搏，则可受孕。反之，由于某些因素影响了上述任何一个环节，都会导致不孕。因此，不孕与肾的关系最为密切。肾虚可导致肝、脾、心等脏腑功能失调，出现肝肾不足、脾肾两虚、心肾不交等证，并影响冲任气血紊乱，则胞宫难以摄精成孕。

【诊断要点】

（1）母畜检查判断有无性器发育异常等。

（2）阴道脱落细胞涂片检查，表现单纯雌激素作用或雌激素功能低下则提示无排卵功能；宫颈黏液检查，以宫颈评分监测排卵。

（3）激素水平测定：用以分析卵巢、垂体及下丘脑功能有无异常而影响排卵。

（4）B超检查可了解子宫、卵巢有无器质性病变及监测卵泡发育与排卵。

（5）子宫输卵管造影术可了解官腔病变及输卵管通畅程度。

（6）腹腔镜、宫腔镜检查则可直观了解盆腔脏器及官腔。

【辨证论治】

原发性不孕症的根本在于肾，但肝、脾、气血的影响也是非常重要的。临证应当注意辨别。

5.8.1.1 肾气亏损证

【主证】屡配久不孕，小便清长，大便溏薄，产科检查子宫偏小，舌淡苔白，脉沉细或沉迟。

【治法】温肾填精，补益冲任。

【方药】毓麟珠加减：党参45克，茯苓30克，山药45克，熟地黄30克，山茱萸30克，当归30克，川芎20克，续断30克，菟丝子30克，杜仲30克，鹿角胶（霜）45克，紫河车30克。

【加减】若小腹清冷，加小茴香20克，紫石英30克，淫羊藿30克；大便溏薄，加炮姜20克，煨木香20克，炒白扁豆30克；小便频多，加益智仁30克，桑螵蛸30克；形体消瘦，加牡丹皮30克，龟板（先煎）30克，白薇20克，知母30克。

5.8.1.2 肝肾不足证

【主证】久配不孕，形体消瘦，舌质红，苔少，脉弦或细弦。

【治法】补益肝肾，填精养阴。

【方药】二至地黄汤加减：女贞子45克，旱莲草45克，山药45克，熟地黄30克，山茱萸30克，当归30克，白芍药30克，茯苓30克，续断30克，枸杞子30克，紫河车30克，甘草15克。

5.8.1.3 气血虚弱证

【主证】多年不孕，肌肤不泽，形体虚弱，可伴子宫发育不良，舌淡苔白，脉细弱。

【治法】补益气血，固养胞脉。

【方药】八珍汤加减：黄芪45克，党参36克，白术30克，茯苓30克，山药60克，当归30克，川芎20克，白芍药30克，熟地黄30克，女贞子36克，阿胶36克，香附30克。

【加减】畏寒肢冷加附子25克，补骨脂10克。

5.8.1.4 心肝气郁证

【主证】多年屡配不孕，舌质黯红，苔薄白，脉弦或细弦。

【治法】疏肝解郁，调和气血。

【方药】开郁种玉汤加减：当归36克，赤芍药36克，白芍药36克，白术30克，茯苓30克，香附30克，青皮20克，柴胡20克，郁金30克，川楝子20克，延胡索30克，丹参30克，牛膝30克。

【加减】若乳房胀痛结块加橘叶30克，橘核30克，全瓜蒌36克，路路通30克。

5.8.2　继发性不孕症

曾有过妊娠（包括分娩与流产）史，以后不再受孕者，称为继发性不孕。继发性不孕的发生常因产后摄生不慎，邪入胞宫，与血相搏结，以致瘀血或湿热下注，胞脉受阻，任脉不通，两精不能抟合而不孕。或因肝气郁结，疏泄失常，气血不和，冲任瘀滞而致不孕。或因素体肥胖，以致痰湿内生，冲任胞脉闭塞，而致不能摄精成孕。相当于西医输卵管炎症造成的梗阻性不孕、子宫内膜异位症的不孕、免疫性不孕等。

【诊断要点】

（1）曾有过妊娠，以后未再受孕，常伴有乳胀或胀痛，腹部有包块，时有低热等异常表现。

（2）子宫输卵管造影术可了解官腔病变及输卵管通畅程度；腹腔镜、宫腔镜检查则可直观盆腔脏器及宫腔内有无生殖系统炎症、肿块等。

（3）诊断性刮宫及子宫内膜组织学检查，既可了解宫颈及宫腔的病变，又可了解黄体功能状态。若为增生期内膜则视为无排卵，若为分泌期内膜提示有排卵；若为分泌早期或分泌不足表示虽有排卵但黄体功能欠佳。

（4）B超检查可了解子宫、卵巢有无器质性病变及监测卵泡发育与排卵。

（5）性激素水平测定：以分析卵巢、垂体及下丘脑功能有无异常而影响排卵。

【辨证论治】

继发性不孕症可追溯其妊娠或流产史，或有可致不孕的其他病史。常有本虚标实，或虚实夹杂之证，临证需根据相关因素仔细辨析。

5.8.2.1　气滞血瘀证

【主证】曾受孕，因流产、早产及产科手术后多年不孕，小腹胀痛拒按，舌紫黯，或有瘀点，脉弦涩。

【治法】活血化瘀，疏通胞脉。

【方药】少腹逐瘀汤加减：当归36克，赤芍药30克，川芎20克，桃仁30克，红花30克，川牛膝30克，五灵脂36克，香附30克，乌药30克，枳壳30克，丹参45克，延胡索36克。

【加减】若腹痛甚加制没药20克，制乳香20克，葛根27克；小腹冷痛加艾叶30克，肉桂20克，小茴香30克；时有腹痛低热者，加败酱草45克，牡丹皮30克，蒲公英30克，地骨皮30克；如血瘀较重，身体健壮者，可用朴硝荡胞汤治之。

5.8.2.2　痰湿内阻证

【主证】曾因流产而后多年不孕，形体较肥，倦怠乏力，舌淡胖，苔白腻，脉滑。

【治法】燥湿化痰，健脾理气。

【方药】启宫丸加减：半夏30克，苍术30克，白术30克，茯苓30克，陈皮30克，神曲30克，石菖蒲27克，厚朴18克，香附30克，川芎18克，远志30克，海藻45克。

【加减】若纳呆加薏苡仁15克，佩兰10克。

5.8.2.3　湿热下注证

【主证】产科手术后多年不孕，舌红，苔黄腻，脉弦数。

【治法】清化湿热，调畅冲任。

【方药】四妙丸合红藤败酱散加减：黄柏30克，苍术30克，牛膝30克，薏苡仁45克，红藤45克，败酱草36克，茯苓30克，香附30克，延胡索30克，路路通36克，天仙藤45克，枳壳30克。

【加减】腹胀痛，加制没药20克，制乳香20克；苔黄腻加胆星20克，瓜蒌皮30克；湿热偏于热者，用龙胆泻肝汤治之。

5.9　卵巢囊肿

卵巢囊肿是产科常见肿瘤，有良性和恶性之分。多由脏腑失调，气血不和，因新产伤于风冷，寒湿内侵，经产余瘀阻滞，又致肾阳不振；或气机阻滞，痰饮夹瘀内留所致。

【诊断要点】

（1）临证以下腹部肿块，或伴有腹胀、腹痛、压迫症状、疼痛等为主要症状。

（2）产科检查见子宫旁肿块，边界清楚，或可活动。

（3）细胞学检查、细针穿刺活检、B超、放射学诊断、腹腔镜检查、肿瘤标志物可辅助检查良、恶性肿瘤。

（4）良性卵巢肿瘤应与卵巢瘤样病变、输卵管卵巢囊肿、子宫肌瘤、妊娠子宫、腹水等相鉴别；恶性卵巢肿瘤须与子宫内膜异位症、盆腔结缔组织炎、结核性腹膜炎、生殖道以外的肿瘤、转移性卵巢等肿瘤相鉴别。

【辨证论治】

本病临床主要分为气滞血瘀证、痰湿凝结证及湿热郁毒证。治疗以软坚消癥为原则，根据病情分别采用行气、化痰、清热利湿法。

5.9.1 气滞血瘀证

【主证】下腹有囊性肿块，巨大囊肿可见腹胀腹痛，口干不欲饮，唇燥，二便不畅，舌紫暗，脉弦细。

【治法】行气活血，软坚消症。

【方药】七制香附丸合血府逐瘀汤加减：苍术36克，白术36克，当归27克，赤芍药30克，桃仁30克，琥珀粉（吞）1.5克，木香30克，山楂30克，生鸡内金18克，炒枳壳15克。

【加减】大便秘结，加大黄（后下）18克；舌红苔少，加干地黄27克，炙龟板（先煎）45克。

5.9.2 痰湿凝结证

【主证】形体较肥，舌苔白腻，脉弦滑。

【治法】化痰行气，软坚消症。

【方药】海藻玉壶汤：海藻36克，昆布36克，夏枯草36克，石菖蒲27克，胆星27克，生牡蛎（先煎）90克，苍术27克，陈皮18克，莪术27克，三棱27克，桃仁30克，赤芍药30克，焦山楂30克，焦六曲30克，肉桂（后下）10克。

【加减】偏寒者，加制附片27克，白芥子27克，肉桂改为15克。

5.9.3 湿热郁毒证

【主证】小腹部肿块，腹胀或痛或满，或不规则阴道出血，甚至伴有腹水，大便干燥，尿黄灼热，口干不欲饮，舌黯红，苔厚腻，脉弦滑或滑数。

【治法】清热利湿，解毒散结。

【方药】清热利湿解毒汤：半枝莲90克，龙葵90克，白花蛇舌草90克，川楝子30克，车前草90克，土茯苓90克，瞿麦45克，败酱草90克，鳖甲90克，大腹皮30克。

【加减】若热毒盛者加龙胆草45克，苦参45克，蒲公英45克；若腹水多者加水红花子30克，油葫芦30克。

第六讲

中兽医药学现代研究

1　不孕症

不孕症是指导致家畜不能正常发情、配种或受孕的一类病症。家畜特别是奶牛，引起不孕症的最常见病症是子宫内膜炎和卵巢机能障碍（主要包括卵巢静止、持久黄体、卵巢囊肿等）。

1.1　子宫内膜炎

子宫内膜炎属于中兽医的带证。临床主要表现为持续或间断的从外阴排出白色、黄白色脓汁，或带有少量脓汁的黏液，或稀薄浑浊的黏液，也就是白带。母牛表现不发情或发情周期异常，也有少数发情正常。

【病因】

现代兽医学认为是由病原微生物感染所致。常见病原菌有葡萄球菌、链球菌、化脓棒状杆菌、大肠杆菌等。

中兽医学认为是气滞血瘀、湿热下注所致。

【治疗】

现代兽医学以抗菌、消炎、促进子宫收缩为治疗原则。临床上主要使用抗生素治疗。

中兽医学以活血化瘀、清热燥湿、缩宫排脓、祛瘀生新为治则。例如“清宫液”系列药就是典型药剂。

清宫液

组成：益母草、桃仁、蒲黄、郁金、连翘、败酱草和少量抗菌药。

性状：棕褐色液体，气味芳香。

适应证：治疗各种类型子宫内膜炎。

用法：子宫灌注，隔日一次，每次 100 毫升，4 次为一疗程。

疗效：治愈率 85% 以上。

药理作用：现代药理学研究表明，“清宫液”具有抑菌杀菌、抗炎、促进子宫收缩、加速损伤组织的恢复、调节卵巢机能的作用。

清宫液 2 号

组成：丹参、连翘、忍冬藤、千里光、红花等。

性状：红褐色混悬液。

适应证：治疗各种类型子宫内膜炎。

用法：子宫灌注，用时充分震摇，隔日一次，每次 100 毫升，4 次为一疗程。

疗效：治愈率 85% 以上。

药理作用：现代药理学研究表明，“清宫液 2 号”具有抑菌、抗炎、促进子宫收缩、加速损伤组织的恢复。

清宫液 3 号

性状：紫红色油悬剂。

适应证：治疗各种类型子宫内膜炎，更适合用于慢性子宫内膜炎的治疗。

用法：子宫灌注，用时充分震摇，每 4 日一次，每次 25 毫升，3 次为一疗程。

疗效：治愈率 83% 以上。

药理作用：现代药理学研究表明，“清宫液 3 号”具有抑菌、抗炎、促进子宫收缩、加速损伤组织的恢复。

1.2　卵巢机能障碍

卵巢机能障碍是家畜正常卵巢周期性活动丧失，在临床上最常见的是母牛表现不发情。

【病因】

现代兽医学认为是由于饲养管理不当，致

使内分泌失调而发病。

中兽医学认为是由于饲养管理不当，致使母牛气血亏虚、脾肾虚弱而不孕。

【治疗】

现代兽医学采取激素疗法。要求准确的诊断和适当的剂量。但在实际应用过程中，经常会出现因诊断不确切或激素用量难以准确把握而产生严重的副作用。

中兽医学采取益气养血、补肾壮阳、催情助孕的方法。没有副作用。代表药剂有“催情助孕液”，单味药有羊洪膻。

催情助孕液

组成：当归、黄芪、益母草、菟丝子、淫羊藿、阳起石。

性状：棕红色透明液体。

适应证：母牛不发情。

用法：子宫灌注，隔日1次，每次100毫升，4次为1疗程。

疗效：催情率87.23%。

羊洪膻　具有养血活血、益气健脾、补肾壮阳的作用。用法，羊洪膻全草500～600克粉末，开水烫成稀糊状，候温灌服或混于饲料中饲喂，隔天一剂，5剂为一疗程。可取得良好效果。

2　乳房炎

乳房炎中兽医称乳痈。

【病因】

现代兽医学认为是由于病原菌感染而引起的乳房炎症。

中兽医学认为是由外邪或内毒蕴结、乳络不畅、气血瘀滞所致。

【治疗】

现代兽医学采用抗生素疗法，可以乳房注射、肌内注射或静脉注射给药。通过药敏试验选择敏感抗生素可以提高疗效，治疗期间乳中有抗生素残留。

中兽医学以清热解毒、散瘀消肿、行气通络为主要治则进行选药治疗。传统用药途径主要是内服，必要时配合局部外用；新型中药制剂也可采用乳房注射、肌内注射和静脉注射。下面介绍2种有效的制剂。

乳源康

组成：甘肃丹参、连翘、黄连、金银花。

性状：红色透明液体。

适应证：奶牛乳房炎。

用法：乳房灌注，每个发病乳区每次50毫升，每日1次，连续使用3～5天。

乳康2号合剂

组成：蒲公英、黄芩、地肤子、金银花、王不留行。

性状：深棕色液体，味微苦。

适应证：奶牛临床型乳房炎。

用法：内服，1.5毫升/kg体重，每日1次，重症或慢性乳房炎患牛每天服用2次，5天为一疗程。

3　胎衣不下

母牛分娩后12小时胎衣仍没有排出称作胎衣不下或胎盘滞留。

【病因】

现代兽医学认为因难产、流产、胎儿过大等多种原因引起母牛子宫收缩无力导致胎衣不下。

中兽医学认为因饲养管理不善使得母牛气血运行不畅，胞宫活动力减弱，不能使胎衣正常排出。

【治疗】

现代兽医学使用兴奋子宫的药物。常用的有催产素、麦角新碱、雌激素、新斯的明等。治疗效果很差。

中兽医学以是活血化瘀、行气利水为主要治则，另外根据辨证，或清热或温经或补气血。

《牛医经鉴》所载治疗胎衣不下方为：“滑石、桂枝、枳壳、桃仁、元明粉、木通、三棱、归尾、大黄、厚朴、莪术，用长流水煎，香油和服”。该方的治疗原则为，活血祛瘀、行气逐水。

临床上使用极为普遍的生化汤：“当归、川芎、桃仁、炮姜、甘草”，诸药协同，温经散寒、活血化瘀。在临床应用中，常常随症加减，如加党参、黄芪、白术等以补其气虚；加五灵脂、香附等以行其气滞；加红花、益母草、生蒲黄、丹参等以增强活血化瘀作用；加肉桂、艾叶以增强温经作用；加大戟以逐水湿。

山楂、红糖可预防牛胎衣不下。

益母草、红糖可预防牛胎衣不下。

4　中药预防禽霍乱的试验研究

在本试验中进行了两次中药预防禽霍乱的强毒攻击试验，现将结果简报如下：

试验一：第一次强毒攻击试验

材料和方法

（1）鸡源。兰州市榆中陈仓养殖加工厂的洛斯蛋鸡，463 天，经临检、剖检均无异常。除用鸡瘟Ⅱ系、马立克苗预防外，未用其他菌（疫）苗。

（2）种毒。中国兽药监察所 84.6.4 冻干强毒 $C_{48}{}^{-1}$ 移入脑心汤培养液培养 24 小时，取此菌液 0.1 毫升，鸡胸肌处注射复壮，待复壮鸡（18 小时）死后剖检（肝表面有灰白色针尖大散在的坏死点），取肝脾涂血平皿分菌培养，24 小时后将平皿菌落移斜面保存（留为攻毒种菌）。另于脑心汤中培养，再取此培养液稀释计数，摸索筛选最佳培养基和被检菌液最适合的存放温度，进行 3 次检测结果之后证实。取样初检计数后的菌液在 8℃存放一夜其死亡率分别为 16.6%、20%、30%。

（3）药物及制法。将刺五加、土骨蛇、黄芪、贯仲、獭肉、板蓝根、山豆根、大白、枝子等 14 味中药，冷水浸泡 30 分钟后，武火至沸，再用文火熬 20 分钟，滤后，再加水武火至沸，文火熬 25 分钟，再滤其液，合并两次煎液，使药液中含生药 1 克/10 毫升，然后存入 4℃冰箱备用。或粉碎为细末，填服。

（4）菌苗。制苗菌液由中国兽药监察所提供的 731（哈兽研）、B26（广西兽研所）培养灭活苗（均系弱毒），佐剂用中监所的白油，每 120 毫升菌液加佐剂 240 毫升，按 8 000转/分钟离心搅拌 5 分钟后，再加白油佐剂 120 毫升，离心搅拌 5 分钟。抽检取样培养阴性，每鸡胸肌注射 0.6 毫升，注后有轻微反应（精神委顿、减食），次日恢复。

（5）分组。分 5 组。

①疫苗组：9 只，于攻毒前 24 天，每只胸肌注射白油苗 0.6 毫升。

②粉剂组：10 只，将中药粉碎于攻毒前 22 天，按每日 1 克/只的量添加于饲料中直至攻毒为止。

③汤剂 A 组：10 只，于攻毒前 10 天，按每日 15 毫升/只，由口灌服。直至攻毒后第 3 天为止。

④汤剂 B 组：10 只，于攻毒前 2 天按每 15 毫升/只，经口灌服，至攻毒为止，共 3 次。

⑤空白对照组：10 只，不作任何处理，仅用于强毒对照。

（6）饲养管理。笼架喂养，饲料购自兰州市饲料公司的混合料，令自由采食。

（7）毒力测定。根据 10^{-8} 的计数结果（20.5 个菌/毫升）注射测毒。见表 6－1。

表 6－1　第一次强毒攻击试验毒力测定

注射毒液量（毫升）	0.2	0.4	0.6	0.8
初检内含菌数（个）	4	8	12	16
复检内含菌数（个）	3.2	6.4	9.6	12.8
注射鸡数（只）	4	4	4	2
鸡死亡比例（%）	4/4	4/4	4/4	2/2

其最小致死量应为 3.2 个细菌。

测毒鸡于次日上午开始按菌量大小依次死亡，注射后 20～24 小时内死亡 4 只（0.6、0.8、0.8、0.6），其余 10 只均于 25～28 小时内死亡。

（8）强毒攻击。9 月 13 日 10～11 时，将 12 日初检的菌液（32 个/毫升），按每只鸡 0.1 毫升的菌量分别于各鸡的胸肌处进行肌注，同时用脑心汤血琼脂平皿复检攻毒的真实菌数，14 日复检结果证明，每只鸡攻毒的真实菌数为：0.1 毫升中含 2.12 个细菌。

结果

（1）临床体征。攻毒于 10 小时内未见异常反应，11～12 小时内，粉剂组、对照组内相继出现精神不振、食欲下降、眼闭、活动度降低，粉剂组反应比对照组更为明显，随之死亡开始。

（2）死亡时间。20～28 小时内粉剂组死亡 4 只，对照组死亡 3 只；28～36 小时内粉剂组死亡 2 只，对照组死亡 4 只，疫苗组死亡 2 只；36～48 小时，对照组死亡 1 只，疫苗组死亡 1 只；48～72 小时内，对照组死亡 2 只，疫苗组死亡 1 只，72～108 小时内，疫苗组死亡 1 只。

（3）预防效果。见表 6－2。

表 6－2　第一次强毒攻击试验预防效果

组别＼类别	N	死亡数	死亡率（%）	存活数	保护率（%）	其他
粉剂组	10	6	60	4	40	攻毒前 22 天，每日灌药至攻毒前 1 天停药
疫苗组	9	5	55.6	4	44.4	攻毒前 24 天注苗
对照组	10	10	100	0	0	不作处理
汤剂 A 组	10	0	0	10	100	攻毒前 10 天，连灌 至攻毒后第 3 天停药
汤剂 B 组	10	0	0	10	1 00	攻毒前 2 天，灌至攻毒为止

结语

（1）该试验的汤剂组效果最好，保护率为 100%，但这只能说明有免疫治疗作用，不能证明其免疫期，或者说是一种预防性治疗。

（2）粉剂组抗禽霍乱有一定的作用，剖检肝脾肿大，而有慢性药蓄积中毒的可能性。

（3）汤剂组抗禽霍乱有良效，其服药次数可减至 3 次为宜。

（4）攻毒致死鸡只，最短时间为 23 小时（粉剂组），最长为 96 小时（连续灌药的汤剂 A 组）。

（5）存活鸡只有坚强免疫力，经 15 天观察，一直未见异常。

试验二：第二次强毒试验

材料和方法

（1）鸡源。兰州市五星坪鸡场伊沙褐蛋鸡，无注射禽霍乱疫苗史。

（2）种毒。中国兽药监察所 $C_{48}{}^{-1}$强毒。

（3）中药及制法。将前药分别组成方 1、方 2，粉碎过 40 目筛待用。方 1 组是在方 2 组的基础上去癞肉、土骨蛇等，加入金银花、连翘等药组成。

（4）疫苗。中国兽药监察所提供。

（5）分组及攻毒时间。

① 预防性治疗：

方 1 组：6 只，从 5 月 22～26 日喂药，每天 5 克/只，26 日停药，29 日攻毒（间隔 3 天）。

方 2 组：5 只，喂药攻毒时间同上。

②预防及免疫期：5 月 29 日攻毒。

1 个月组（方一）：11 只，5 月 4～7 日，每天喂药 5 克/只。

2 个月组（方二）：11 只，4 月 7～10 日，每天喂药 5 克/只。

3 个月组（方二）：11 只，3 月 5～8 日，每天喂药 5 克/只。

③疫苗对照组：10 只，5 月 4 日皮下注射疫苗 1 毫升（距攻毒时间 25 天）。

④强毒对照组：10 只，不作任何处理，只作攻毒对照。

（6）饲养管理。同第二批试验。

（7）测毒。10^{-8}的种毒菌液，放于 12℃的冰箱中，其存活率为 90%（95.48＋98.44＋76.90%），将 30 个菌/毫升的初检菌液进行鸡胸肌注射，分设以下 3 组：

0.1 毫升，2 只，死亡鸡数 0/2

0.2 毫升，2 只，死亡鸡数 1/2（36～40 小时间）

0.3 毫升，2 只，死亡鸡数 2/2（24～36 小时间）

将此测毒菌液进行复检，其结果为 8 个/毫升，则实际攻毒菌液中的菌数为 8×0.3 毫升菌液：2.4 个菌，故其攻毒致死量为 2.4 个菌。

（8）攻强毒。5 月 29 日将 8 个菌/毫升的

10^{-8}的初检菌液，按90%打折后，于10时30分至11时35分，每只鸡胸肌注入0.1毫升，其内应含细菌数为8个/毫升×0.4×90%；2.88个菌，将此菌液进行复试，实际注菌为2.8个。

结果见表6-3。

表6-3　第二次强毒试验试验结果

组别	N	死亡数（只）	相对保护率（%）	绝对保护率（%）
对照组	10	9	10	10
方1组	6	2	66.6	56.6
方2组	5	0	100	90
1个月组（方1）	11	3	72.3	62.3
2个月组（方2）	11	0	100	90.9
3个月组（方2）	11	1	90.9	80.9
疫苗组	10	4	60	50

5　中草药有效成分及复方制剂研究应用

5.1　抗蜂螨中草药有效成分的提取分离及其药效研究

5.1.1　蜜蜂螨病的危害及防治现状

蜂螨病（bee acariasis）是蜜蜂的常见侵袭性疾病之一，其种类主要是大蜂螨（*Varroa jacobsoni*）与小蜂螨（*Tropilaelaps clareae*）两种。被大蜂螨寄生的蜜蜂表现发育不良，体质衰弱，采集力下降，寿命缩短，蜂群逐渐减弱。受小蜂螨危害的蜂群常常使得幼虫未化成蛹就死去，羽化出房时翅膀残缺不全。螨害严重的蜂群可引起成年蜂大批死亡，甚至全群死亡，给养蜂生产造成极大的危害。在世界范围内广泛流行，温带地区，未防治的蜂群大蜂螨侵染引起的死亡率接近100%。蜂螨的寄生率高，寄生密度大。全世界主要经济蜂种的意大利蜜蜂100%感染大蜂螨，根据世界各地的气候和生态条件，意大利蜜蜂的蜂群每年要用药物防治1~3次，才能保证蜂群正常发展。工蜂房中的自然寄生率平均6%，最高达11%，母螨寄生密度1.05个螨/巢房；雄蜂房的螨寄生率平均15%，最高达26%，寄生密度1.47个螨/巢房。雄蜂封盖房内寄生螨率达48%左右，而工蜂封盖房内为8.9%。大蜂螨的寄生率和寄生密度与蜂群群势成负相关，4月下旬以后寄生率达15%~20%；夏季，寄生率则保持相对稳定，6月中旬~8月中旬，蜂螨寄生率均在10%上下；到了秋季，蜂群群势下降，寄生率上升，蜂螨的寄生率达49%；直到初冬，蜂王停产，蜂螨也停止繁殖，以成螨的形态在蜂体上越冬。由此可见，蜂螨病不但流行范围广，而且一年四季都发生。

在中国，将蜜蜂螨病归为一类疾病，可见蜂螨的危害性之大。目前使用的杀螨药多以化学药品为主，一方面大多已产生耐药性，另一方面化学药物在蜂蜜及王浆等蜂产品中残留，部分药物若用药剂量不适也可导致蜜蜂中毒和脱子，这不仅严重影响了我国蜂产品的出口，阻碍养蜂业的发展，还危害食用者的健康。因此，研制低毒、低残留、不易产生耐药性的药物显得尤为重要。利用中草药的纯天然、多功能、毒副作用小和不易产生耐药性就有避免这些不足的可能。

5.1.2　从中草药中提取和筛选抗螨有效成分具有广阔前景

随着现代制剂技术和中药药理学的发展，从纯天然药物中提取和分离有效成分，经过筛选和大量的临床实践来研制新药，成为当代中药学研究的一个亮点。对蜜蜂螨病的防治应以寻求“高效、安全、经济、无公害”的杀螨制剂为主，选择来自天然物的生理活性物质，研制新型杀螨剂显得极为重要。

一般来说，中草药有效成分用于食品动物疾病的防治可具有以下优势和特点：

- 毒性小；
- 残留少；
- 不易产生耐药性；
- 与传统中兽药不同，可具有明确的作用机理或作用靶点。

5.1.3 研究内容

本研究对对叶百部、橘皮、薄荷、博落回的有效成分进行了提取，并对前3种中草药的有效成分进行分离和结构鉴定。把4种中草药的有效成分粗提物与氧化苦参碱等用于杀灭大、小蜂螨的活性研究，进行初步筛选。并对博落回与百部提取物进行了体外杀螨试验。为临床研制杀螨新制剂及其药理学研究奠定了基础。

百部的主要成分以生物碱为主，至今已分离并鉴定了35种。其药理作用有以下几种。

（1）抗菌作用。体外实验表明，百部煎剂及乙醇浸剂对肺炎球菌、乙型溶血性链球菌、脑膜炎球菌、金黄色葡萄球菌等多种细菌有抑制作用。

（2）对呼吸系统的作用。百部生物碱具有松弛支气管平滑肌，抑制平滑肌痉挛的作用，并能降低动物呼吸中枢的兴奋性，抑制咳嗽反射。

（3）抗病毒作用。百部煎剂有抗亚洲甲型流感病毒和新城疫病毒的作用。

（4）杀虫作用。百部对蛔虫、蛲虫、绦虫、血吸虫、螨、蜱、蚊蝇幼虫、头虱、衣虱及臭虫等有驱杀作用。

（5）百部的毒性作用与其他药理作用。对叶百部碱小鼠静脉注射的LD_{50}为62.0毫克/千克，口服的LD_{50}为1 079.4毫克/千克。此外，百部还有镇静、镇痛作用，可抑制烟酸痉挛反应。

对叶百部、博落回的提取采用乙醇冷浸渍法，用柱层析法结合薄层层析法并联合应用重结晶法对对叶百部、橘皮、薄荷的有效成分进行分离。通过测定熔点、紫外、核磁共振氢谱和质谱，参考有关文献对单体进行结构鉴定。共鉴定了3个化合物，分别为对叶百部碱、薄荷醇、百里酚。

初步筛选试验表明，25%薄荷醇、不同浓度的百部碱提取物（0.2%、0.1%、0.05%）与博落回（1%、2%、4%）提取物对大蜂螨与小蜂螨都有杀灭活性，并且对蜜蜂无伤害。对照成分草酸、乳酸与苦参碱对蜂螨有一定疗效，但效果不明显；此外，25%橘皮提取物百里酚对蜂螨有杀灭作用，而50%橘皮提取物百里酚却对蜜蜂有伤害。

药效试验表明，三种中草药提取物对大蜂螨与小蜂螨都有杀灭活性，且对蜜蜂安全（$P>0.05$），其中25%薄荷提取物对蜂螨的杀灭作用最明显。结合初步筛选试验确定了薄荷提取物的最佳用药浓度，其值为25%，百部与博落回的用药浓度分别在0.05%和1%以上，其剂量与其他提取物剂量和用药浓度有待进一步研究。

体外杀螨试验表明，3种不同浓度的百部提取物和博落回提取物对大、小蜂螨都有效。其中0.05%百部碱提取物对小蜂螨的杀灭活性与15%乳酸相当，而对大蜂螨的杀灭活性明显优于15%乳酸。

6 丹参制剂的药理学和临床应用研究

6.1 前言

乳房炎是危害奶牛业的主要疾病之一。据报道，美国1 100万头泌乳奶牛中有约50%患有乳房炎，日本平均乳房炎患病率为45.1%；我国奶牛乳房炎发病率更高，孙福先（1996）、萧乾庆（1997）、杨章平（1998）等报道奶牛隐性乳房炎头数阳性率分别达85.7%、51.3%、54.1%。在奶牛乳房炎的治疗上，主要还是以抗生素为主，然而由于西药抗生素治疗急性炎症期易转化成慢性期，乳房出现硬块，乳腺增生，导致奶汁停止，不易治愈，甚至奶牛被淘汰，并且大量抗生素药物的残留及耐药性等给饮用鲜奶的人体造成危害。因此，常用的抗生素类药物由于残留将逐渐被淘汰，而选用新的无残留药物在治疗奶牛乳房炎显得越来越重要。用中草药防治奶牛乳房炎则可以避免以上问题。在我国，虽然在用中药治疗奶牛乳房炎取得了很好的成绩，但由于药物制剂和质量的原因未能在临床中得到广泛的应用。

传统的治疗乳房炎的中草药制剂是选用或配制方剂，再用水或酒精提取，加工成制剂。缺点在于未能除去一些大分子物质，使得制剂不稳定；制剂成分复杂，难以质量控制。所以，选用合适的中药，组方和制剂是中草药制剂治疗乳房炎的关键技术。

丹参为唇形科植物丹参（Salvia miltiorrhiza Bunge）的干燥根及根茎。丹参味苦，性微寒，入心、肝经，具有活血化瘀，活络通痹，养心安神，解毒凉血，排脓生肌，消肿止痛、疗疮止痢等功效，是我国常用中药。近代对丹参的研究主要是从丹参的化学成分和药理作用等两方面进行。

对丹参化学成分的研究大致可分为两个阶段。第一阶段为20世纪30年代初期至40年代，主要由中尾万山和泷浦洁等人对丹参脂溶性成分进行了研究。第二阶段为20世纪60年代末至今，国外学者主要是Kakisawa等人的工作。我国学者对丹参的化学成分的研究主要起始于20世纪70年代后期。当时随着中医中药研究的深入开展和丹参制剂的实验和临床研究，以及现代结构仪器的引进，使得丹参成分的研究工作有了迅速的发展。丹参中80%的成分均为20世纪70年代以后报道，且大多数成分均为我国学者所发现。丹参的主要成分可分为水溶性成分和脂溶性成分两大类，水溶性成分主要以酚性酸类化合物为主，脂溶性的主要以二萜醌类化合物为主。有些化合物由于获得的量甚微，未能作结构和药理作用的研究。目前丹参丹参中已阐明结构的有50余种，其中二萜醌类化合物占一半以上。对丹参的成分研究虽取得了较大的进展，但据丹参临床实践和药理研究的结果，现已分得的成分还不能完全说明的丹参全部生物活性，而且，丹参的化学成分还有待于作进一步的探讨和研究。

丹参药理的研究随着中西医的结合和中医中药研究的深入开展，以及现代结构仪器的引进和丹参成分研究取得的进展取得了显著的成果。主要有以下几个方面。

（1）对心血管系统有着广泛的药理作用，可以降低心肌的耗氧量，缩小心肌的梗死面积；抗动脉硬化，扩张冠状动脉防止动脉狭窄，改善微循环障碍等。

（2）对消化系统，丹参可保护受损的肝细胞促进肝细胞的再生和抗纤维化作用，为临床治疗慢性肝炎提供了依据；丹参有保护胃粘膜完整性并增强其防御功能的作用。

（3）丹参具有杀伤和诱导分化肿瘤细胞的作用。

（4）天然抗氧化的作用，丹参的水溶性成分和脂溶性成分都含有较强活性的消除自由基和抗脂质过氧化物质。

（5）抗菌消炎作用，丹参脂溶性成分总丹参酮在体外试验中证明对金黄色葡萄球菌及其耐药菌株链球菌、分枝杆菌、蜡样芽孢杆菌、人结核杆菌、都有明显抑杀作用。其中抗菌主要成分为二萜丹参醌类化合物，甘西鼠尾草中的总丹参酮含量最高，抗菌作用最强。

本研究选用丹参为主要药物，利用中药化学的方法提取、分离和鉴定丹参等中草药的活性成分，用获得活性成分和其他几味中药具有抗菌、提高机体免疫力和活血化瘀的活性成分合理组方，制成注射剂，通过对乳房炎致病菌的药敏、制剂稳定性、安全和药效试验，对其进行药理研究。试验室药理药效稳定后，将其用于临床，并作临床疗效验证试验，最终确定组方和制剂。将其大量应用于奶牛乳房炎的治疗，必将促进奶业事业的发展，增加出口创汇能力，具有巨大的经济社会效益；同时为新兽药的研制提供新的思路。

6.2　研究内容

采用二氯甲烷冷浸法从丹参中提取其脂溶性有效成分，用薄层层析（TLC）、柱层层析（LSC）和结晶相结合的方法对提取的总丹参酮进行分离和纯化，其单体化合物的纯度均在97%以上。通过和硫酸的特殊显色反应，同时测定熔点、红外（IR）、紫外（UV）、核磁共振谱（^{1}HNMR）等数据并结合文献以鉴定提纯的丹参单体化合物分子结构，最后确定提纯的化合物分别为丹参酮 IIA（tanshinone IIA）、丹参酮 I（tanshinone I）、隐丹参酮（cryptotanshinone）和二氢丹参酮 I（dihyrotanshinone I）。

用上述提纯的总丹参酮和其他中药的有效成分进行组方研究，开发出乳源康注射剂。经系列实验研究证实其制剂稳定性好，在 100℃ 恒温水浴加热 15 小时，仍然清澈透明，颜色保持不变。药理试验表明，乳源康注射液能对多种乳房炎致病菌有较强的抑杀作用；兔子点眼试验表明无局部刺激作用，对照组和实验组间无显著性差异；口服时，乳源康对小白鼠的 $LD_{50}=36.80$ 克/千克体重，LD_{50}95% 可信限为 32.12 ~ 41.47 克/千克体重；腹腔注射时 $LD_{50}=13.85$ 克/千克体重，LD_{50}95% 可信限为 12.76 ~ 14.94 克/千克体重；对二甲苯所致小白鼠耳廓肿胀试验表明，乳源康注射液与氢化可的松的消炎效果相同，均与空白对照组差异显著（$P<0.05$）。

对 122 例临床乳房炎乳区进行了乳源康的疗效观察试验，以青链霉素和环丙沙星治疗为对照，总有效率分别为 94.7%、93.7%、93.2%，总痊愈率分别为 81.7%、70.8%、82.2%。乳源康临床应用安全，效果显著，降温消肿较快。

本试验将不同的中草药活性成分分别提取并组方，开发成中草药注射剂，用于奶牛乳房炎临床，为首次报道。

7　六茜素说明书

六茜素系中草药六茜草的有效成分，中医临床应用证明对呼吸道、消化道泌尿道及创口感染均有良好的治疗效果，特别是对细菌性泄泻和菌痢有特效。

自 1991 年以来，我们人工合成了六茜素。通过药理药效试验证明，六茜素是一种低毒（小白鼠肌内注射 LD_{50}：230.73 毫克/千克体重；静脉注射 LD_{50}：227 毫克/千克体重）广谱的抗菌药，如对鸡白痢沙门氏菌、禽霍乱巴氏杆菌、大肠杆菌、金色葡萄球菌、伤寒杆菌、绿脓杆菌、无乳链球菌、停乳链球菌及猪丹毒杆菌等有较强的抑制作用，对真菌也有一定的作用。由于本药属于中草药天然成分，与其他药物相比具有毒性低、不产生耐药性等优点，在兽医临床上有较高的应用价值。

【性状】本品为白色结晶性粉末。

【功能与主治】广谱抗菌药物，具有疗效好、毒性低、不产生耐药性等特点。主要用于治疗雏鸡白痢、奶牛乳房炎、犊牛腹泻和仔猪黄白痢等病。

【用法用量】①雏鸡白痢：与饲料混匀饲喂。剂量为每千克饲料添加 0.5 ~ 1 克六茜素，饲喂 5 ~ 7 天。

②奶牛乳房炎：乳房局部注射，药液现配现用。剂量为 400 毫克/天，3 天为一疗程。

③仔猪黄白痢：口服或肌内注射，80 ~ 120 毫克/天，2 ~ 3 次/日，连用 7 天。

【贮藏】密封，干燥处保存。

8　茜草素注射液说明书

茜草素是在六茜素研制成功的基础上，利用 Topliss、Free-Wilson 模型和 Hansch 分析法对该类化合物的抗菌活性进行了 QSAR 研究。3 种方法较好的配合而筛选出的一种抗菌活性较强的化合物。并经大量的临床试验表明，茜草素对奶牛乳房炎、子宫内膜炎和仔猪水肿病的临床效果明显强于六茜素。

茜草素在治疗奶牛乳房炎和子宫内膜炎时不仅具有疗效好、毒性低、无残留、消肿、回乳快等特点，而且可避免因大量滥用抗生素而引起的菌株耐药性问题，深受广大养殖户的青睐和认同，目前已在全国许多地区广泛使用，获得了良好的治疗效果。在治疗仔猪水肿病时，茜草素能有效控制该病，且价格低廉，治愈一头只需花 10 ~ 14 元，及易推广，并已成为许多省、市、地区治疗仔猪水肿病的首选药。

【性状】本品为白色结晶性粉末。

【功能与主治】茜草素主要治疗奶牛乳房炎、子宫内膜炎和仔猪水肿病等。具有有抗菌疗效好、毒性低、不产生耐药性、无残留、消肿、回乳快以及价格低廉等特点。

【用法用量】

①乳房炎：溶于灭菌水后用乳导管导入乳池，每个乳区 200～400 毫克/次，病情严重者可加至 600～800 毫克/次，2 次/日，3 天为一疗程。

②子宫内膜炎：溶于灭菌水后直接导入子宫，800～1 200 毫克/次，肌内注射同量，1 次/日，8 天为一疗程。

③仔猪水肿病：溶于灭菌水后肌内注射，100～200 毫克/次（4～8 毫克/千克体重），2～3 次/日。

【规格】400 毫克/瓶，50 瓶/盒，20 盒/箱。

【贮藏】密封，干燥处保存。

9　催乳速添加剂

催乳速是以葛根为主要成分的数味中药复方，在动物体内具有明显的生物活性。能显著地促进动物乳汁分泌，改善繁殖性能，提高免疫力，并且具有剂量小，见效快，毒性低等优点。通过试验证明：对低产奶牛添加“催乳速”后，增奶效果明显提高 5%～10% 以上，并且不会对奶牛机体及其性能产生任何不良影响。

【主要成分】葛根等数味中药制剂。

【主治】主要用于增加低产奶牛的产奶量，治疗牛、羊、马等母畜气血不足所致缺乳症。具有疏通经络，活血化瘀，调整阴阳，促进母畜下乳之功效。

【使用方法】牛、马每日添加 50 克，猪、羊每日添加 30 克，3～5 日为一添加期。拌于精料或其他饲料中饲喂，干湿均可。

【储藏】3 年。

【规格】100 克/袋或 300 克/袋。

10　金丝桃素络合物说明书

金丝桃素是贯叶连翘中最具活性成分之一，近年来在欧、美等国家和地区被用于抗抑郁症、艾滋病以及甲型、乙型肝炎等其他病毒性疾病。国外试验证明，金丝桃素在体外和体内能有效地抑制 HIV 和其他反转录病毒（如 HSV-1、HSV-2、副流感病毒、牛痘病毒等）。

中国农业科学院兰州畜牧与兽药研究所成功地从贯叶连翘中提取、分离出金丝桃素，完成了结构鉴定及药理学、毒理学试验。并进行了较系统的体外、体内抗禽流感和口蹄疫病毒试验。结果表明，金丝桃素对 H_5N_1 和 H_9N_1 亚型的禽流感病毒和 O 型口蹄疫有很强的杀灭率，对感染的畜禽有较高治愈率。

金丝桃素蛋白络合物是旨在提高金丝桃素稳定性，而将金丝桃素与功能蛋白有效结合而制成的。

【性状】棕黑褐色粉末状，无臭，味微苦。

【用途】用于治疗或预防禽流感、口蹄疫病毒或其他 RNA 病毒引起的病毒性疾病。

【用量】预防时每千克饲料中混入本品 0.5 克，拌匀，自由采食。治疗时每千克饲料混入本品 25 克自由采食，连用 3～5 天。

【储藏】在密封、低温、闭光环境中保存。

11　中草药

中草药是中药和草药的总称。通常，中药是指中医常用的药物，草药是指民间所应用的药物。但是，随着草药的不断发掘、研究和推广应用，一些疗效较好的草药也已逐渐被中（兽）医所应用。这样，中药和草药的区分将愈加困难，所以，我们把中药和草药统称为中草药。

中草药根据其自然属性可分为以下几类。

植物药：如当归、黄芪、大黄、双花等。

动物药：如鹿茸、龙骨、海马等。

矿物药：如石膏、明矾、硼砂等。

据文献资料报道，我国现有药用植物 12 694种，其中植物药11 020种，动物药1 590种，矿物药 84 种，而实际作为商品使用的中药材只有1 000余种。

11.1　中草药研究的内容

11.1.1　来源及分类

按来源分：野生、人工栽培（或饲养）；按药物功能分类：如解毒药、清热药、理气药、活血化淤药等；按药用部分分类：如根类、叶类、花类、皮类、种子类、根茎类、全草类。按有效成分分类：如含生物碱类的中草药、含黄酮类中草药、含挥发油的中草药、含苷类中草药等。

11.1.2 采集、加工、炮制

采集　中草药所含成分是药物具有防病治病功能的物质基础。而有效成分的质和量与中药材的采收季节、时间和方法有着十分密切的关系。中药适时采集是确保药材质量的重要环节之一，也是影响药物性能和疗效好坏的重要因素。

在不同的生长发育阶段，植物中化学成分的积累是不相同的，同时受气候、产地、土壤等多种因素的影响，甚至有很大区别。首先，植物生长年限的长短与药物中所含化学成分的质量有密切关系。如甘草酸是甘草中的主要成分，生长三四年的甘草中含量较之一年者高出1倍。还有人参中总皂苷的含量以6~7年采收者最高。其次，植物在生长过程中，随月份的变化，有效成分的含量也各不相同。又如在甘草不同生长发育阶段，进行甘草甜素的含量测定，结果开花前期甘草甜素的含量最高为10%，故甘草应在开花前期采收为宜。

中草药加工　主要是清除杂质，及时干燥，以免霉烂，使其便于保存和应用。

中草药炮制　是根据中医临床用药理论和药物配制的需要所进行的进一步加工工艺。其与药效有着密切的关系。实践证明，通过炮制能消除或降低药物的毒性或副作用，改变药性或提高疗效，便于粉碎加工及贮藏等。例如元胡，功能是活血、理气、止痛，但醋制元胡的活血、理气、止痛作用明显提高。

常用的炮制方法有以下几种。

炒：有清炒、土炒、麸炒等，目的在于除去水分，使组织疏松，利于成分的浸出和粉碎，且可杀死植物中的酶便于贮藏。

炙：将药材与液体辅料共炒，使辅料渗入药物内部。常用的有蜜炙、盐炙、醋炙、酒炙、油炙等。

煅：将药材高温处理，使药料松脆纯净，便于粉碎及浸出有效成分。

蒸：利用水蒸气加热的方法。常用的有清蒸、醋蒸、酒蒸等。

煮：将药材与辅料共煮。常用的为醋煮。

水制：有漂、泡、烂、水飞等数法。

11.1.3 鉴定

常用的中药鉴定方法有6种：原植物鉴定（基原鉴定）、性状鉴定、显微鉴定、理化鉴定、生物鉴定、质量分析。

11.1.4 化学成分

中草药所含成分很复杂，通常含有糖、氨基酸、蛋白质、脂肪、有机酸、挥发油、生物碱、苷类、鞣质、无机盐、色素、维生素等。每一种中草药都可能含有多种成分，在这些成分中，有一部分具有明显生物活性并起医疗作用的，常称为有效成分，如生物碱、苷类、挥发油、氨基酸等。如黄连中的黄连素、麻黄中的麻黄碱、黄芩中的黄芩苷等。而另一些曾认为无效的成分，如多糖、蛋白质、鞣质等因发现了它们具有生物活性而成为有效成分。如茯苓多糖、当归多糖、枸杞多糖等。

11.1.5 药理作用

我国广大医药人员在中医理论的指导下，结合现代科学的理论、方法和手段，密切结合临床，深入系统地开展了中药药理研究。

目前，主要是对中药复方、单味药进行研究，以阐明中医方药的药理作用及临床应用；另一方面，应用现代药物学的方法，对从中药中提取分离的生物活性物质进行研究，以说明中药的作用机理，加深对中医药理论的科学认识，并进行结构鉴定，以至合成创新，开发新药。抗疟药—青蒿素的研制就是一个很好的例子。

各类中药的药理研究如下。

活血化淤药：当归、阿胶、何首乌等。

清热解毒药：连翘、板蓝根、鱼腥草等。

清热凉血药：紫草、白头翁等。

清热燥湿药：苦参、黄连、黄芩等。

补益药：党参、黄芪、枸杞子等。

11.1.6 临床应用

中草药的性能　是指每味药物本身所固有的性质和功能。

四气五味　四气是指药物的寒、热、温、凉四种不同的性质。五味即酸、苦、甘、辛、咸五种不同的滋味。

升降沉浮　升降沉浮是指药物作用于机体的四种趋向。

归经　主要是指药物作用于机体某一脏腑或经络而发生选择作用的一种归纳方法。也就是说凡某药能治某经之病症，该药即归某经。

药物的相互关系　古代医家把各种药物的相互关系概括为：相须、相使、相畏、相杀、相恶、相反药物的用量 中草药用量的安全范围一般比较大，但某些有毒的药物仍须十分注意，不能过量。

11.2　中药及中药复方

11.2.1　中药及中药图谱

在此，我们重点介绍几味。

11.2.2　中药复方制剂

近年来，中草药防治畜禽疾病已显示出极好的前景，引起国内外同行的浓厚兴趣。例如，奶牛乳房炎、奶牛繁殖病、奶牛肠道疾病、猪痢疾、猪呼吸道疾病以及禽类的各种疾病均应用中药复方制剂进行临床治疗取得了很好的效果。中国农业科学院中兽医研究所研制的中药复方制剂“敌球灵”主要用于球虫病的预防和治疗。该产品经全国20余个省市，数百万余羽鸡的推广应用，统计数据显示，预防效果98%以上，治愈率93%～98%；其次复方制剂“禽瘟王”通过大面积临床应用，不仅对禽（鸡、鸭、鹅）霍乱、鸡白痢等细菌性病症有较好效果，而且对传染性腔上囊病、新城疫等病毒性疫病也有独特疗效；还有治疗母牛子宫内膜炎的中药复方制剂“清宫液”，经临床数万余例的治疗试验表明，有效率94.5%，治愈率87.84%，治后三个情期受胎率为84.97%。还有很多，在此不一一赘述。

12　绵羊循经信息的研究（略）

13　脾虚模型胃电图的研究（略）

中兽医国际培训班活动有关照片

2005 International Training Workshop on Traditional Chinese Veterinary Medicine and Techniques

2006 International Training Workshop on Traditional Chinese Veterinary Medicine and Techniques

2007 International Training Workshop on Traditional Chinese Veterinary Medicine and Techniques

2008 International Training Workshop on Traditional Chinese Veterinary Medicine and Techniques

2009 International Training Workshop on Traditional Chinese Veterinary Medicine and Techniques

2011 International Training Workshop on Traditional Chinese Veterinary Medicine and Techniques

Teaching

Teaching

Seminar

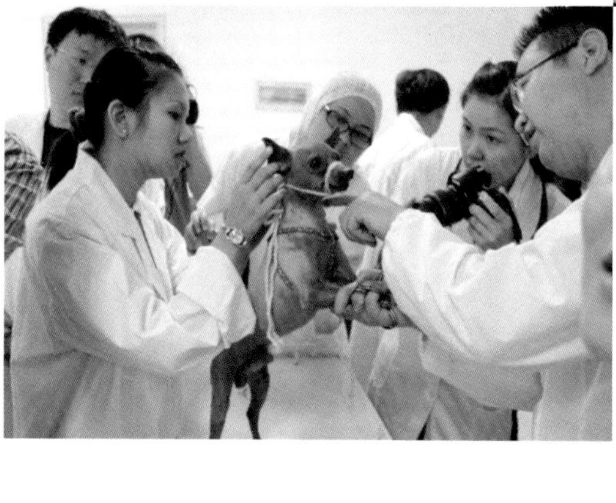

Acupuncture practice for dog

The lesson for extracting Chinese tradtional medicine

Acupuncture practice for donkey

Acupuncture practice for sheep

Graduation

To award certificates to the students

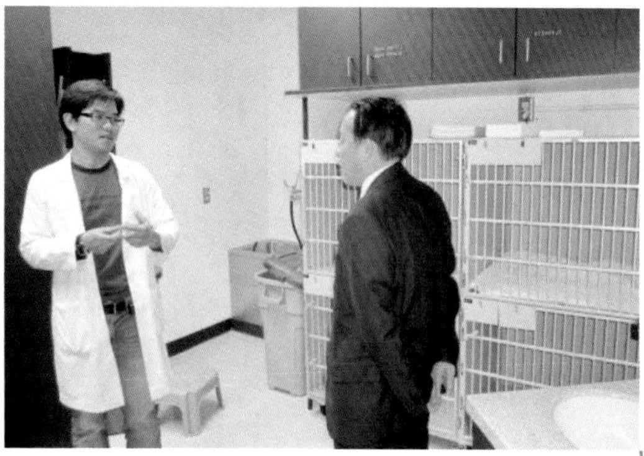

The student of International training study in pet clinic,Florida,USA

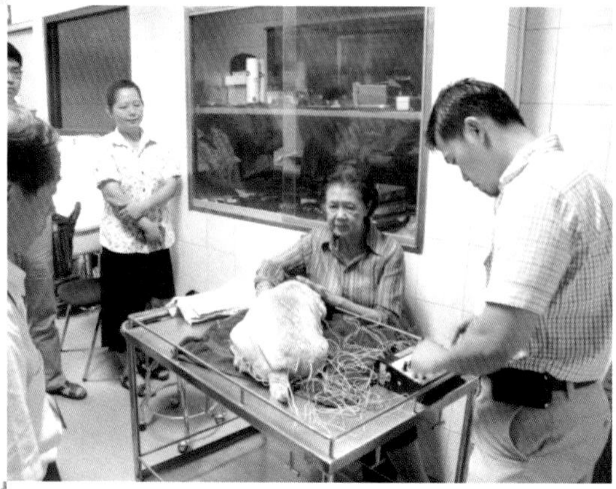

The student of International training work in horse clinic,Thai

Visit to an animal hospital

Studying Chinese Traditional Culture

Visiting students university in Thai

中兽医国际培训班教师简介

Yang Zhiqiang, professor ,Ph.D. tutor.,Lanzhou institute of Husbandry and Pharmaceutical Sciences, CAAS. He was honored as an advanced scientific worker of Gansu province of China, an advanced scientific worker of CAAS, As the chairman, 8 researching projects were awarded, 100 papers and 13 books were published. His research fields are on the toxicology, pharmacology of the traditional Chinese veterinary medicine and application of the traditional Chinese veterinary medicine in Diary cow disease,and also he is the famous national dairy cow disease scientists in China.

Zheng Jifang, professor of Lanzhou nstitute of Husbandry and Pharmaceutical Sciences, CAAS. He has been engaging in the research of Chinese traditional veterinary medicine and animal acupuncture. He has taken charge more than 20 the national or National Natural Fund Items of China successively,and is the editor in chief of 4 books and editor of 3 books. He lectured TCVM or animal acupuncture for the visitors from USA, France, England, Spain, Japan, India, Philippines, Indonesia, Korea, Thailand, Nepal, Turkey and Burma etc.

Wang Xuezhi, Ph.D.Master tutor. An associate professor of Lanzhou institute of Husbandry and Pharmaceutical Sciences, CAAS. He has been engaging in the research and management of the traditional Chinese veterinary medicine. He published 4 books (as co–author) and 30 papers.He has taken charge and taken part in more than 10 funds in Chinese traditional veterinary medicine field,andhe has been engaging in animal science and mangaement in International Training Workshop on Traditional Chinese Veterinary Medicine and Techniques for 6 times.

Luo Chaoying, professor of Lanzhou nstitute of Husbandry and Pharmaceutical Sciences, CAAS. He graduated from Veterinary Medicine specialty, Northwest College of Agriculture in 1982, and has been engaging in the research of integrate Traditional Chinese with Western Veterinary Medicine & Clinical Veterinary Medicine. He has taken charge or taken part in more than 20 the national or agricultural ministry items of China successively, and is the subeditor of 4 books and editor of 3 books. He lectured animal acupuncture for the visitors from USA, France, England, Spain, Japan, India, Philippines, Indonesia, Korea, Thailand, Nepal, Turkey and Burma etc.

Zhang Jiyu,Ph.D., professor, Deputy director of Lanzhou institute of Husbandry and Pharmaceutical Sciences, Chinese Academy of Agricultural Sciences. He was awarded of Ph.D. from Changchun University of Agriculture and Animal Sciences, China in 2002. He has taken charge and taken part in more than 20 funds in Chinese traditional veterinary medicine field, and also he is the editor of 3 books and has published more than 40 papersof scientific research.

Yang Ruile, Professor of Shanghai Veterinary Research Institute, CAAS. He graduated from Chinese traditional veterinary medicine specialty, College of Animal Sciences and Veterinary Medicine, Chinese University of Agriculture in 1990, and has been engaging in the research of and management on Chinese traditional veterinary medicine and animal acupuncture. He is the subeditor of 4 books.

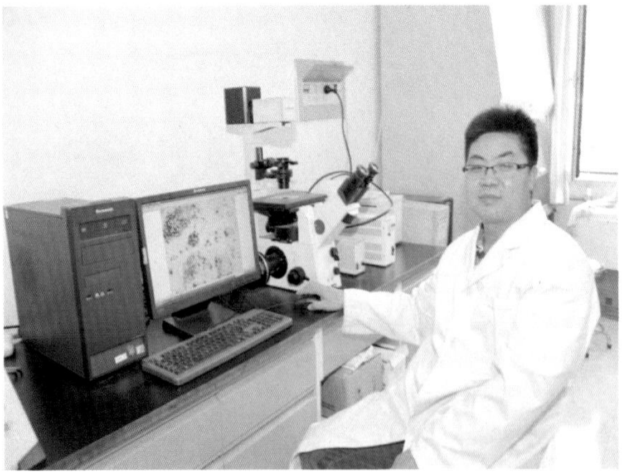

Li Jianxi, Master tutor. Professor of Lanzhou institute of Husbandry and Pharmaceutical Sciences, CAAS. He was born in 1971 and graduated from the Department of Veterinary Medicine, Gansu Agricultural University in 1995, and got Ph.D. from the Graduated School of Chinese academy of agricultural sciences in 2006. He has been engaging in the research of toxicology, pharmacology and fermentation of the traditional Chinese veterinary medicine. He is the leader of the research workgroup of the traditional Chinese veterinary medicine in CAAS. He published 80 papers and got 6 academic awards.

Li Zirui, PH.D., Associate professor, Director of the secretariat of CAAS.he has been engaging in the research of and management on Chinese traditional veterinary medicine and animal acupuncture and comprehensive affairs. And he is the coauthor of 7 books and has published more than 20 papers of scientific research.

Xie Jiasheng, associate professor of Lanzhou nstitute of Husbandry and Pharmaceutical Sciences, CAAS. He has been engaging in the research of Clinical Medicine of Traditional Chinese Veterinary Medicine, fodder additive of TCVM, etc. He has taken part in more than 10 the national or agricultural ministry items of China successively, and is the coauthor of 3 books and has published more than 50 papers of scientific research. He lectured clinical medicine of Traditional Chinese Veterinary Medicine, for the visitors from Spain, Japan, India, Philippines, Indonesia, Korea, Thailand, Nepal, Turkey and Burma etc.

Luo Yongjiang, Associate professor of Lanzhou Institute of Husbandry and Pharmaceutical Sciences, CAAS. He has been engaging in the research of Chinese traditional Veterinary Pharmaceutics. He has taken charge or taken part in more than 10 national or gansu province items about Chinese traditional veterinary medicine, and won 8 kinds of technological development prizes. He is the co-author of 7 books and has published more than 50 papers of scientific research.

Dong Pengcheng, he graduated from Department of Lan Zhou Medical College in 1999, and has been engaging in pathogenic microbe and management in technology management office. and he has been engaging in animal science and mangaement in International Training Workshop on Traditional Chinese Veterinary Medicine and Techniques for 6 times.

Zhou Lei, He graduated from Animal science and technology college, Hei Long-jiang August First Agriculture and Reclaim University in 2006,and he has been engaging in animal science and mangaement in International Training Workshop on Traditional Chinese Veterinary Medicine and Techniques for 6 times.

Shi Yin, She graduated from Lan Zhou University in 2008,and she has been engaging in animal science and mangaement in International Training Workshop on Traditional Chinese Veterinary Medicine and Techniques for 4 times.

Li Jin-yu, associate professor of Lanzhou Institute of Animal and Veterinary Pharmaceutics Sciences, Chinese Academy of Agricultural Sciences. He was born in 1973 and graduated from Gansu College of Traditional Chinese Medicine in 1997, and has been engaging in the research of veterinary medicine and veterinary acupuncture of Traditional Chinese Veterinary He has supported and participated in more than 10 countries and the Ministry of agriculture project such as the popularization and application of Traditional Chinese medicine preparation 'Jin Shi Weng Shao San' to anti avian infectious diseases etc.

Xin Ruihua, associate professor of Lanzhou Institute of Husbandry and Pharmaceutical Sciences, Chinese Academy of Agricultural Sciences. She was born in 1981 and graduated from pharmacology specialty, Lanhzou university in 1989, and has been engaging in the research of pharmacology of Traditional Chinese Veterinary Medicine, pharmaceutics technology of TCVM, etc. She assisted to lectured veterinary medicine of Traditional Chinese Veterinary Medicine, for the visitors from Spain, Japan, India, Philippines, Indonesia, Korea, Thailand, Nepal, Turkey and Burma etc.

Wang Guibo, associate professor of Lanzhou Institute of Husbandry and Pharmaceutical Sciences, Chinese Academy of Agricultural Sciences. He graduated from Clinical Veterinary Medicine specialty, Central China Agricultural University in 2009, and has been engaging in the research of Clinical Acupuncture of Traditional Chinese Veterinary, Clinical Medicine of TCVM, etc. He is the coauthor of 3 books and has published more than 20 papers of scientific research. He assisted to lecture clinical medicine of Traditional Chinese Veterinary Medicine, for the visitors from Spain, Japan, India, Philippines, Indonesia, Korea, Thailand, Nepal, Turkey and Burma etc.

Zhang Jingyan, research assistant of Lanzhou Institute of Husbandry and Pharmaceutical Sciences, Chinese Academy of Agricultural Sciences. She graduated from the Department of Veterinary, Gansu Agricultural University in 1989, and has been engaging in the research of new drugs development of Traditional Chinese Veterinary Medicine with some microorganism from rumen of dairy cows, and quality standard of TCVM. She has taken part in 3 the national or agricultural ministry items of China successively, such as "Study and application on some key technologies in traditional Chinese veterinary medicine manufacture, etc., and has published 6 papers of scientific research as the first author, and has received 1 invention patent in China.

Zeng Yufeng, PH.D, associate researcher of Lanzhou Institute of Husbandry and Pharmaceutical Sciences, Chinese Academy of Agricultural Sciences. He has worked in herbivore research in the areas of yak genetic diversity and sheep breeding.He has taken part in more than 10 the national or agricultural ministry items of China successively, He has been the coauthor of 6 books and has published more than 50 papers of scientific research and obtained 5 different kinds of the scientific technology advance of Gansu province and Lanzhou city. and he has been engaging in animal science and mangaement in International Training Workshop on Traditional Chinese Veterinary Medicine and Techniques for 3 times.

Qin Zhe, research assistant of Lanzhou Institute of Husbandry and Pharmaceutical Sciences, Chinese Academy of Agricultural Sciences. She got a doctorate on clinical veterinary from Gansu agricultural University in 2012, and has been engaging in the research of new drugs development of Traditional Chinese Veterinary Medicine with some microorganism from rumen of dairy cows. She has taken part in more than 5 the national or agricultural ministry items of China successively, and has published more than 5 papers of scientific research.

Wang Lei, research probationer of Lanzhou Institute of Husbandry and Pharmaceutical Sciences, Chinese Academy of Agricultural Sciences. She was born in 1985 and has been engaging in the research of Traditional Chinese Veterinary Medicine for dairy cows from 2013. She has taken part in 3 national or agricultural ministry items of China successively, such as "Study and application on some key technologies in traditional Chinese veterinary medicine manufacture" . She has published 3 papers of scientific research.

Kong Xiaojun, research research assistant of Lanzhou Institute of Husbandry and Pharmaceutical Science, Chinese Academy of Agricultural Sciences.He was occupied in the research of Clinical Medicine of Traditional Chinese Veterinary Medicine since he graduated from Chinese Academy of Agricultural Sciences in 2013, and has taken part in "the rescue and marshal of the chinese veterinarian medicine resources" , and is the co-auther of two books, and one of the scientific research he participated in has won provincial-level incentives.

Liu Lijuan, She graduated from Animal science and technology college, Shi Hezi University in 2014, and she has been engaging in the reaearch on animal genetics, breeding and reproduction and management in technology management office.